Communications in Computer and Information Science 985

Commenced Publication in 2007
Founding and Former Series Editors:
Phoebe Chen, Alfredo Cuzzocrea, Xiaoyong Du, Orhun Kara, Ting Liu,
Krishna M. Sivalingam, Dominik Ślęzak, Takashi Washio, and Xiaokang Yang

Editorial Board Members

More information about this series at http://www.springer.com/series/7899

Arun K. Somani · Seeram Ramakrishna ·
Anil Chaudhary · Chothmal Choudhary ·
Basant Agarwal (Eds.)

Emerging Technologies in Computer Engineering

Microservices in Big Data Analytics

Second International Conference, ICETCE 2019
Jaipur, India, February 1–2, 2019
Revised Selected Papers

 Springer

Editors
Arun K. Somani
College of Engineering
Iowa State University
Ames, IA, USA

Anil Chaudhary
Swami Keshvanand Institute of Technology
Management and Gramothan
Jaipur, India

Basant Agarwal
Swami Keshvanand Institute of Technology
Management and Gramothan
Jaipur, India

Seeram Ramakrishna
National University of Singapore
Singapore, Singapore

Chothmal Choudhary
Swami Keshvanand Institute of Technology
Management and Gramothan
Jaipur, India

ISSN 1865-0929   ISSN 1865-0937 (electronic)
Communications in Computer and Information Science
ISBN 978-981-13-8299-4   ISBN 978-981-13-8300-7 (eBook)
https://doi.org/10.1007/978-981-13-8300-7

This Springer imprint is published by the registered company Springer Nature Singapore Pte Ltd.
The registered company address is: 152 Beach Road, #21-01/04 Gateway East, Singapore 189721, Singapore

Preface

The ICETCE conference aims to showcase advanced technologies, techniques, innovations, and equipment in computer engineering. It provides a platform for researchers, scholars, experts, technicians, government officials, and industry personnel from all over the world to discuss and share their valuable ideas and experiences.

The Second International Conference on Emerging Technologies in Computer Engineering: Microservices in Big Data Analytics (ICETCE 2019) was held at Swami Keshvanand Institute of Technology, Management, and Gramothan (SKIT), Jaipur, Rajasthan, India during February 1–2, 2019. The main keynote addresses were given by Prof. (Dr.) Arun K. Somani, Associate Dean for Research, College of Engineering, Iowa State University, Ames, USA, and Prof. Seeram Ramakrishna, Vice President, Research Strategy, NUS, Singapore.

We received 253 submissions from all over the world including the USA, Singapore, France, Tunisia, to name a few. The papers were carefully reviewed by at least three reviewers from the Technical Program Committee as well as external reviewers. After the rigorous review process, 28 regular papers and one short paper were accepted for presentation at the conference and for publication in the conference proceedings.

We wish to thank all the reviewers for all their efforts. We are also thankful to the management of Swami Keshvanand Institute of Technology, Management and Gramothan (SKIT), Jaipur, Rajasthan, India, for providing the best infrastructure and logistics required to organize a successful conference. We also wish to express our thanks to Prof. Arun Somani and Prof. Seeram Ramakrishna for accepting our invitation to give keynote addresses. We are also very thankful for Springer for supporting ICETCE 2019, and Mrs. Suvira Srivastav and Nidhi Chandhoke for the continuous support and help.

We hope that these proceedings are very useful for all researchers working in the relevant areas.

April 2019

Arun K. Somani
Seeram Ramakrishna
Anil Chaudhary
Chothmal Choudhary
Basant Agarwal

Organization

General Chair

Seeram Ramakrishna   National University of Singapore (NUS), Singapore

Technical Program Chairs

Arun K. Somani	Iowa State University, Ames, USA
Virendra Singh	Indian Institute of Technology (IIT) Bombay, India
Basant Agarwal	Swami Keshvanand Institute of Technology Management and Gramothan, Jaipur, India

Organizing Chairs

Anil Chaudhary	SKIT Jaipur, India
Chothmal Choudhary	SKIT Jaipur, India

International Advisory Committee

Alexander Chernikov	Bauman Moscow State Technical University, Moscow, Russia
Ghasi Ram Verma	University of Rhode Island Kingston, USA
Raman M. Unnikrishnan	California State University, USA
Mauro Conti	University of Padua, Italy
Albert Dipanda	University of Bourgogne, France
Kokou Yetongnon	University of Bourgogne, France
Xiao-Zhi Gao	Lappeenranta University of Technology, Finland
Kamal Nayan Agrawal	Howard University, Washington, USA
Naveen Sharma	Rochester Institute of Technology, NY, USA
Sindhu Ghanta	ParallelM, Sunnyvale, CA, USA
Dinesh K. Sharma	Fayetteville State University, Fayetteville, USA

Technical Program Committee

Sandeep Sancheti	SRM University, Chennai, India
G. R. Sinha	International Institute of Information Technology (IIIT), Bangalore and IEEE Executive Council Member, MP Subsection, India
R. K. Joshi	IIT Bombay, India
Mitesh Khapara	IIT Madras, India
R. S. Shekhawat	Manipal University, India
Y. K. Vijay	VGU, Jaipur, India

Mahesh Chandra Govil	NIT Sikkim, India
Manoj Singh Gaur	Indian Institute of Technology, Jammu, India
Virendra Singh	IIT Bombay, India
Vijay Laxmi	MNIT, Jaipur, India
Neeta Nain	MNIT, Jaipur, India
Namita Mittal	MNIT, Jaipur, India
Emmanuel Shubhakar Pilli	MNIT, Jaipur, India
Mushtaq Ahmed	MNIT, Jaipur, India
Yogesh Meena	MNIT, Jaipur, India
S. C. Jain	RTU, Kota, India
Dayanand Kumar	Micron Technology, Taiwan
Mohammad Shariq	Jazan University, Saudi Arabia
Mohammd Salim	MNIT Jaipur, India
Ghanshyam Singh	MNIT Jaipur, India
Vipin Pal	NIT Meghalaya, India
Yogita	NIT Meghalaya, India
Sanjeev Kumar Metya	NIT Arunachal Pradesh, India
Abhimanyu Singh Garhwal	Massey University, Auckland, New Zealand
Mani Madhukar	University Relations, IBM India Pvt. Ltd.
Manish Pokharel	Kathmandu University, Nepal
Nistha Keshwani	Central University of Rajasthan, India
Shashnak Gupta	Department of Computer Science, BITS Pilani, India
Rajiv Ratn	IIIT Delhi, India
Vinod P. Nair	University of Padua, Italy
Adrian Will	National Technological University, Tucumán, Argentina
Chhagan Lal	University of Padua, Italy
Nisheeth Joshi	Banasthali Vidyapith, India
Maadvi Sinha	Birla Institute of Technology, India
Vaibhav Katewa	University of California, Riverside California, USA
Sugam Sharma	Iowa State University, USA
Xiao-Zhi Gao	Lappeenranta University of Technology, Finland
Deepak Garg	Bennett University, Greater Noida, India
Pranav Dass	Galgotias University, Uttar Pradesh, India
Linesh Raja	Amity University Rajasthan, India
Vijander Singh	Amity University Rajasthan, India
Piyush Maheshwari	Amity University, Dubai, UAE
Tapas Badal	Bennet University, India
Janos Arpad Kosa	Kecskemet College, Hungary
Dongxiao He	Tianjin University, China
Thoudam Doren Singh	Indian Institute of Information Technology Manipur, India
Dharm Singh Jat	Namibia University of Science and Technology, Namibia
Vishal Goyal	Punjabi University, India
Dushyant Singh	MNNIT Allahabad, India

Amit Kumar Gupta	DRDO, Hyderabad
Pallavi Kaliyar	University of Padua, Italy
Sumit Srivastava	Manipal University, India
Ripudaman (Director)	Natural Group, India
Sumit Srivastava	Pratham Software Pvt. Ltd., Jaipur, India
Nitin Purohit	Wollo University, Dessie, Ethiopia
Ankush Vasthistha	NUS, Singapore
Dinesh Goyal	Poornima University, India
Reena Dadhich	University of Kota, Rajasthan, India
Arvind K. Sharma	University of Kota, Rajasthan, India
Ramesh C. Poonia	AIIT, Amity University, Rajasthan, India
Vijendra Singh	North Cap University, India
Brijesh Kumbhani	IIT Ropar, India
Shashi Kant Sharma	IIIT Ranchi, India
Nikhil Deep Gupta	NIT Nagpur, India
Rimpy Bishnoi	LNMIIT, Jaipur, India
Ashok Chauhan	CEERI Pilani, India
Nidhi Chaturvedi	CEERI Pilani, India
Kunwar Pal	NIT Sikkim, India
Sandeep Saini	Myanmar Institute of Technology, Myanmar, India
C. Periasamy	MNIT Jaipur, India
Harish Indouria	Skillrock Technologies, India
Ashish Sharma	IIIT Kota, India
Rajat Goel	SKIT Jaipur, India
Niketa Sharma	SKIT Jaipur, India
Yogendra Gupta	SKIT Jaipur, India
Maninder Singh Nehra	Govt. Engineering College Bikaner, India
Sunita Gupta	SKIT Jaipur, India
Rakesh K. Bhujade	Technocrats Group of Institution, Bhopal MP, India
Mahesh Pareek	ONGC, New Delhi Area, India
Kamaljeet Kaur	Campus Connect and Agile, ETA, Chandigarh DC, Infosys Ltd.
Rupesh Jain	Wipro Technologies Ltd., India
Gaurav Singhal	Bennett University, Greater Noida, India
Lokesh Sharma	Galgotia University, Greater Noida, India
Ankit Vidhyarthi	Bennett University, Greater Noida, India
Vikas Tripathi	Graphic Era University, Dehradun
Prakash Choudhary	NIT Manipur, Manipur, India
Smita Naval	NIT Warangal, India
Subhash Panwar	Government Engineering College, Bikaner, India
Pankaj Jain	K.N. Modi University, Newai, Tonk, India
Manoj Bohra	Manipal University, India
Rajbir Kaur	The LNM Institute of Information Technology, India
Poonam Gera	The LNM Institute of Information Technology, India
Madan Mohan Agarwal	Birla Institute of Technology, India
Ajay Khunteta	Poornima College of Engineering, Jaipur, India

Subhash Gupta	Birla Institute of Technology, India
Baldev Singh	Vivekanand Institute of Technology, India
Kavita Choudhary	J.K. Laxmipat University, India
Rahul Dubey	Madhav Institute of Technology and Science, Gwalior, India
Sonal Jain	JK Lakshmipat University, India
Shrawan Ram	MBM Engineering College, Jodhpur, India
Sonali Vyas	AIIT, Amity University, Rajasthan, India

Contents

Innovative Mindset

Seeram Ramakrishna$^{(\boxtimes)}$

National University of Singapore, Singapore, Singapore
seeram@nus.edu.sg

Abstract. Human mindset is central to innovations. Deeper understanding of human mind gives us better insights to harness its innovation potential and perhaps, inculcate innovative mindsets. Mind supervenes on the brain organ. Mind is formless and non-physical, whereas the brain has the physical structure. Information exchange between the mind and body sensory organs and systems is coordinated via nervous system which comprises of brain, spinal cord and peripheral nerves. A simplistic metaphor to illustrate this is a computer. The mind is akin to software while the physical body is akin to biological hardware with brain as its biological central processing unit, CPU. Our thoughts, feelings, inspirations, sensory inputs and environment, experiences, culture, and education elicit the mind and often re-set mind. In other words, such external as well as internal inputs influence all things an individual can do with mind. Goal of this brief article is to capture the latest scientific advances in understanding of the human mind, and thus bringing us closer to gain a better sense of an innovative mindset.

1 Introduction

It is now a well-accepted global wisdom that scientific advances led innovations are backbone of new economic growth, new job creation and solutions to the sustainability challenges of Earth ecosystem. Governments, businesses and universities are pursing policies and programs to promote innovation (Ramakrishna and Ng 2011). Ardent critics observe that innovation is a culture that requires appropriate ecosystem to flourish. In this pursuit the human mind is often overlooked, which actually makes innovation possible. Deeper understanding of human mind gives us better insights to harness its innovation potential.

Humans are tiny in size and life span when compared to the size scale and life span of universe. Yet, it is human nature to grasp its own relationship with the vast universe, curious about existence, imagine future and create it. It is not possible for the humans to do this without mind. Human mind and physical body symbiotically co-exist and work in tandem (Fig. 1). A simplistic metaphor to illustrate this is a computer (Ramakrishna 2018). The mind is akin to software while the physical body is akin to biological hardware with brain as its biological central processing unit, CPU. Our thoughts, feelings, sensory inputs and environment, experiences, culture, and education elicit the mind and often re-set mind. In other words such external as well as internal inputs influence all things an individual can do with mind. Oxford dictionary lists over one hundred different mind functions or attributes. Human body and brain undergo

© Springer Nature Singapore Pte Ltd. 2019
A. K. Somani et al. (Eds.): ICETCE 2019, CCIS 985, pp. 1–6, 2019.
https://doi.org/10.1007/978-981-13-8300-7_1

several changes as we age from birth to death, with corresponding changes in mental functions. In addition to aging, the brain changes are influenced by the life style, nutrition, stress, physical disabilities, and diseases. Diverse mental functions include emotions, beliefs, thinking, memory, perceptions, judgement, reasoning, ideation, imagination, creativity, motivation and will. Performance of mind is a composite of diverse mental functions. Hence the most advanced artificial intelligence is nowhere close to the ingenious biological human mind which understands the experiences and grasps their meaning. In recent years, finding underlying biological mechanisms of each specific mental function is the focus of scores of scientists, supported by well-funded brain research programs. The biological mechanisms supporting our wandering minds are not yet well understood but research has begun to yield insights and demystifying mind like never before.

Mind supervenes on the brain organ. Mind is formless and non-physical, whereas the brain has the physical structure. Information exchange between the mind and external sensory organs eyes, nose, ears, tongue and skin, is coordinated via nervous system which comprises of brain, spinal cord and peripheral nerves. Neurons are the basic cells of nervous system. Human brain consists of more than 100 billion neurons which form over a trillion synapses or connections between neurons. When stimulated neurons transmit signals and information by means of changes in electrical and chemicals charges. Various mind functions emerging from conscious and unconscious mental processes are mediated by the functioning nervous system tethered with more than 100 biochemical messengers. For example, Serotonin regulates sleep and dreaming, appetite, mood, pain, aggression, anxiety and sexual behavior. Dopamine is associated with sensations of pleasure, feel good and reward. Cortisol is associated with boosting immune function, memory capacity, and tolerance to pain. Adrenalin to respond quickly to danger and stress. Glutamate in learning and memory. Imbalance of chemicals is related to health issues such as schizophrenia, Parkinson's, depression, Alzheimer's, anxiety, and obsessive compulsive disorder. Brain communications which underpin various mental functions are directed by synchronized electrical pulses emanating from masses of neurons in different structures of the brain. Electrical impulses transmitted across brain tissues aka brainwaves are categorized based on their frequencies or cycles per second, HZ. Beta waves in the range 12–38 Hz are characteristics of a strongly engaged mind. 8–12 Hz alpha waves are associated with calm and reflecting state of mind. Daydreaming state involves 3–8 Hz theta waves. Asleep state involves 0.5–3 Hz delta waves. Active dreaming is associated with altering of brain states from delta to theta. It is characterized by rapid eye movement, REM sleep. Brainwaves are picked up by electroencephalogram, EEG sensors. By tapping brainwaves it is now possible to digitally reconstruct the object perceived by the brain. Recently, functional magnetic resonance imaging, fMRI is able to pick-up frequencies less than 0.5 Hz, called infraslow waves during non-REM sleep or quiet sleep. They are thought to be the basic cortical rhythms that underlie higher brain functions. Endogenous electric fields, even though weak, facilitate subtle brain communications. It seems the brain is never inactive. The fastest of brain waves, 38–42 Hz Gamma waves are discovered reportedly in the states of compassion, altruism, and spiritual expanded consciousness.

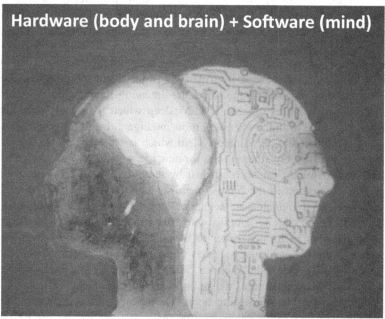

Fig. 1. (A) Professor Seeram Ramakrishna with Sophia robot. He ponders whether future robots will be innovative as well. (B) Physical body is the hardware with brain as CPU, and Mind is the software. Inner and outer data inputs such as thoughts, feelings, sensory inputs and environment, experiences, culture, and education, elicit the Mind and re-set the mind.

Scientists anticipate that all mind functions as derivatives of neurophysical and biochemical changes of brain and body systems, and they can be induced to a certain extent. For example, awareness of sensation or consciousness is related to the activation of certain brain structures by receiving information from the sensory receptors via mechanisms of action potentials of neurons followed by release of neurotransmitter chemicals at synapses of interneurons.

Human brain is highly complex with hundreds of billions of neurons communicating and interacting with other body systems. Therefore massive efforts are being made to map which tissues of brain, networks of neurons, and biomolecules involved in each specific mind function and state. They are correlating mental functions with specific brain neuronal circuits. Advanced magnetic resonance imaging, MRI scanners (Nowogrodzki 2018) are now able to reveal the structures smaller than 1 mm without opening the living brain. For example these studies revealed six layers of the cerebral cortex, which is responsible for cognition. They are able to study how the information is traveling among these fine layers and unique specialty of each layer. They are correlating an activity in an area to other activities in other areas of brain. In other words brain connectivity, correlating a particular mental function with others such as attention, memory and mood. Researchers found columnar organizations of the brain carry out computations and respond preferentially to a particular stimuli. They stimulate the activity of specific neurons with electrical, magnetic, optical, chemical, sound and physical means. For example, people are taught neurofeedback techniques to modulate brain wave activity. In other words, conscious thinking to change neuronal networks. Scientists recently developed brain memory implants. Quantum theories are adopted to explain mind.

Technologists mapped brain waves via artificial intelligence, AI and digitally recreated the dream experienced by a person. Brain scientists are investigating the science behind dreams and constraints on the dreaming imagination. Most dreaming takes place during rapid eye moment, REM sleep which is crucial for health. REM sleep is the play time of brain when it is most intelligent, insightful, creative and free. REM sleep is regulated by the limbic system which is a deep brain region. Dreaming imagination might be a breeding ground for new ideas as the brain experiments. Most people can improve their dream recall just by reminding themselves before bed that they want to. According to the Finnish neuroscientist Revonsuo (2000), most of the emotions experienced in dreams are negative. Recently neurologist Arnulf et al. (2017) compared students' dreaming patterns with their grades. Students who dreamt more often about the test, performed better in real life. Dreams are not governed by the normal laws of physics, and many aspects of dreaming yet remain mysterious.

William Miller (Miller and C'de Baca 2001) a psychology and psychiatry professor and co-author of book Quantum Change attribute Saha! moments or leaps of thought by the conscious mind to the resting brain. The resting brain circuitry is turned on, paradoxically, when we stop focusing on a problem after vexing about it. Resting brain burns 20 times the metabolic resources of conscious brain. In other words the resting brain employs the most creative mechanics and is a place to park the problem for the best solutions. Perhaps in the future science will enable us to separate the real happiness from fake happiness.

Memory is a recollection of information and experiences (Humphries 2016). Memory is central to who a person is. Human memory is approximated to be 2.5 petabytes. One petabyte is a million gigabytes. Memories are grouped into (a) sensory memory, (b) short-term memory and (c) long-term memory. As the name suggests the sensory memory holds sensory information less than a second, and it is an automatic response. The short-term memory or working memory with recall period in the range of several seconds to a minute. The long-term memory could potentially be as long as the whole life span. Nobel Laureate Susumu Tonegawa has been investigating how we form, store and recall memories. Proteins produced by a specific gene are linked to memory and cognitive processes. Hippocampus, a region of the brain critical for remembering experiences, is only a temporary repository. A memory that we keep for years is held in the neocortex. His team found brain circuitry that helps to form episodic memories. It feeds information about an experience to the hippocampus through distinct brain cells. A single memory is formed from a pattern of connections between a unique subset of brain cells. In animal models his team is able to artificially reactivate happy memories by stimulating specific brain cells. Several researchers are finding biological mechanisms of questions such as role of emotion on memory? Role of thinking on memory? Role of attention on memory? Role of stress on memory? Role of biological hardware on memory? Role of nutrients on memory? Effect of social media on mind? Influence of multitasking on memory?

Lipton (2015), a developmental biologist and author of book The Biology of Belief promotes the hypothesis that a cell's behavior is controlled by own genetic code as much as the environment. Perception proteins acts as the switches that integrate the function of cells with its environment. Although the perception proteins are manufactured through genetic mechanisms, the process is regulated by the environmental signals. The unconscious beliefs rooted in our minds early in life contribute to the health and fate of cells later on in life. He advocates the need for liberation from the ingrained stressful biology of belief so as to unlock the human capacity to be vital. As per The MIT Encyclopedia of the Cognitive Sciences, intelligence is the subject's abilities to adapt to, change and select the environment (Keil and Wilson 2001). Therefore through external interventions it is possible to alter the biologically set innate predispositions of mind.

Researchers are developing wrist bands, wearables and mobile phones to forecast positive and negative moods (Kaplan 2018). They measure variations in the electrical conductance of a person's skin, sleep, stress, activity and social interactions to predict moods. Models and AI algorithms are trained on carefully collected vast amounts of quality data sets to forecast happiness, calmness and state of mental health. Further work is needed as the actions to improve mood are different for different people.

Mindset of Steve Jobs, an awe inspiring innovator behind the world's first one trillion dollar company Apple, is a subject of intense interest to many. Common characteristics observed among innovators include curiosity, imaginative, passion, memory, alertness, attitude, energetic and joy. There are many underlying factors to these multi-faceted characteristic domains. A Stanford University professor Carol Dweck suggests that innovator's mindset is oriented towards creating value.

Nations and communities need people with innovative mindset. Innovative mindset can be inculcated while leveraging innate biological assets of people. Innovative mindset is a behavior that flourishes in conducive ecosystems and cultures. Biological studies of innovative mindset themselves are a goldmine for future innovations. They will in turnchange the face of future innovations. Evolving innovative mindsets and ecosystems will make the future innovations possible, which are not yet imagined.

2 Conclusions

We are far from precisely understanding the underlying mechanisms of an innovative mindset. But rapid progress is being made in recent years supported by extensive brain scientific research, which yields new insights and demystifies human mind like never before. By emulating brain and mind, the computer science and engineering and information sciences and technologies are being advanced. Artificial intelligence, AI enabled services, robots, and devices emerged in recent years are examples of upcoming future. Therefore, mind is the goldmine for innovations. Mind inspired technologies is a new frontier, and will change the way we live in the future (Ramakrishna et al. 2018). We are moving towards a better scientific understanding of human mind in general and innovative mindset in particular.

References

Nowogrodzki, A.: The strongest scanners. Nature **563**, 24–26 (2018)

Revonsuo, A.: The reinterpretation of dreams: an evolutionary hypothesis of the function of dreaming. Behav. Brain Sci. **2000**(23), 793–1121 (2000)

Lipton, B.H.: The Biology of Belief, Matter & Miracles, 464 p. (2015)

Humphries, C.: Tracing a memory. MIT Technol. Rev. (2016)

Keil, F.C., Wilson, R.A. (eds.): MIT Encyclopedia of the Cognitive Sciences, Intelligence, 1096 p. MIT Press (2001). ISBN 9780262731447

Arnulf, I., et al.: What does the sleeping brain say? Syntax and semantics of sleep talking in healthy subjects and in Parasomnia patients. Sleep **40**(11) (2017). https://doi.org/10.1093/sleep/zsx159

Kaplan, M.: Happy with a 20% chance of sadness. Nature **563**, 20–22 (2018)

Ramakrishna, S.: Power of the Human Mind, Tabla, p. 15. SPH Publishers, Singapore (2018)

Ramakrishna, S., Ng, D.: The Changing Face of Innovation, 250 p. World Scientific Publishers (2011). ISBN-13 978-9814291583

Ramakrishna, S., He, L., Wu, W.: Editorial overview: neural interfaces and neuro-prosthetics. Curr. Opin. Biomed. Eng. **6**, vi–viii (2018). https://doi.org/10.1016/j.cobme.2018.06.002

Miller, W.R., C'de Baca, J.: Quantum Change, 212 p. Guildford Press, 2 May 2001. ISBN 9781572305052

Design of IoT Blockchain Based Smart Agriculture for Enlightening Safety and Security

M. Shyamala Devi$^{(\boxtimes)}$ ⓘ, R. Suguna ⓘ, Aparna Shashikant Joshi ⓘ, and Rupali Amit Bagate ⓘ

Department of CSE,
Vel Tech Rangarajan Dr. Sagunthala R&D Institute of Science
and Technology, Avadi, Chennai, TamilNadu, India
shyamalapmr@gmail.com

Abstract. The Internet of Things is evolving as a complete matured technology to be used in all the Smart applications and it establishes itself in the future generations of internet. As like Internet of Things, Blockchain is the blooming technology in which each node involved in the blockchain contains the distributed ledger which enhances the security and data transparency. Illegal users are not able to perform any fault transaction in the blockchain network due to its ability of performing smart contract and consensus. The Internet of Things can be merged with the blockchain to improve the performance of the application in real time. However, managing the devices connected to the sensors in IoT environment and mining the block chain remains the technical challenge forever. With this background, we make an attempt to survey the core details of blockchain technology and its features. In this paper, we have proposed design architecture by merging IoT and BlockChain for Smart Agriculture and ended up with some new architectural framework.

Keywords: Blockchain · IoT · Mining · Distributed ledger · Sensors

1 Introduction

Blockchain consists of any number of nodes, which individually has distributed ledgers that allow multiple nodes to access and update a single edition of a ledger along with shared control maintenance. The nodes in the Blockchain contain a distributed ledger that can record transactions between the nodes in a secure and permanent way. By sharing the databases between the nodes, blockchain [1] technology avoid the existence of third party intermediate that were previously required to act as trusted agents to authenticate, trace, store and synchronize the transactions. By progressing the technology from centralized systems to a decentralized and then to distributed network system, blockchain successfully release data from the ledger [2] that was previously kept in secured way as shown below (see Figs. 1, 2). Blockchain technology can be used in business networks. A business network portrays any group of association or individuals that connect with a desire to transfer or share the assets. Those assets can be tangible, such as food, raw materials, equipment's or manufactured goods, or digital,

© Springer Nature Singapore Pte Ltd. 2019
A. K. Somani et al. (Eds.): ICETCE 2019, CCIS 985, pp. 7–19, 2019.
https://doi.org/10.1007/978-981-13-8300-7_2

such as music or data. These assets can be transferred between members, by tracking the items using a common, shared distributed ledger [3] that is distributed across the business network, assets can be transferred between members, with each member having a record of the transaction and access to the latest version of the ledger. Blockchain creates belief across a business network by the combination of a distributed ledger, smart contracts [4], and consensus [5]. The distributed ledger contains the current state of assets and the history of all transactions. Transactions can only be added to the distributed ledger. Once the transaction is added to the distributed ledger [6], it cannot be removed from it. All the transaction in the distributed ledger is encrypted so that no unauthorized users can be tampered with it. The blockchain also alters the distributed ledger and it is immutable, allowing the distributed ledger to be the source of exactness, proof within the network.

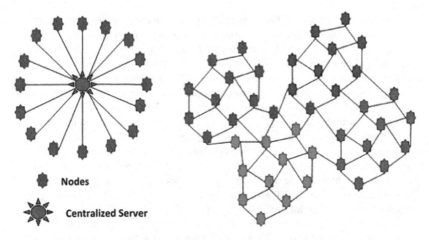

Fig. 1. (a) Centralized system architecture (b) Distributed system architecture

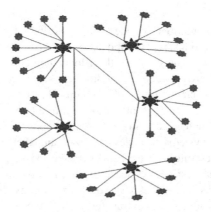

Fig. 2. DeCentralized system architecture

2 Preliminaries

2.1 Blockchain

A blockchain is a data structure that is used to create a digital ledger of transactions and is shared with the distributed network of nodes [7]. Each node user on the network performs manipulation on the distributed ledger in a secure way without the need for a central authority using cryptographic techniques [8]. The Components of the block chain are Previous Block Header, Timestamp, Nonce and Merkle Root Hash.

2.2 Elements of Blockchain

The Merkle root [8] is created in the way that is shown in the tree formation. The hash of each transaction is formed. This process is further continued until the final transactions. As new transactions are added, the Merkle hash root varies with the new hash value and is highly secure. The key elements of blockchain is shown (see Fig. 3).

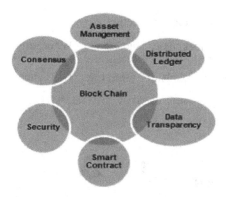

Fig. 3. Blockchain elements

2.3 Blockchain Architecture

The blockchain architecture (see Fig. 4) allows a distributed network of nodes to reach consensus without the need for this central authentication.

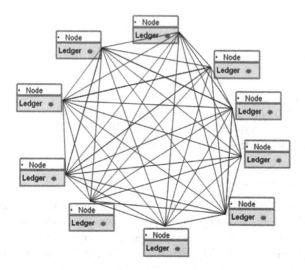

Fig. 4. Blockchain architecture

The categories of Blockchain architecture are,

1. Public Permissionless Blockchains (see Fig. 5) - Where any users can participate in the blockchain network
2. Private Permissioned Blockchains (see Fig. 5) - Where participants must be authorized.

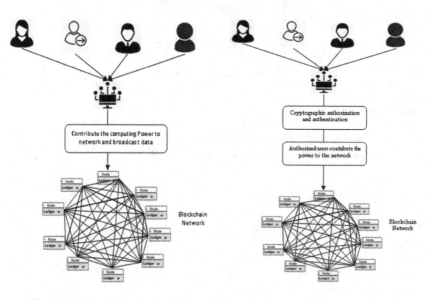

Fig. 5. (Left) public permissionless (Right) private permissioned blockchains

2.4 Blockchain Process

The steps that are process in the blockchain network is shown (see Fig. 6).

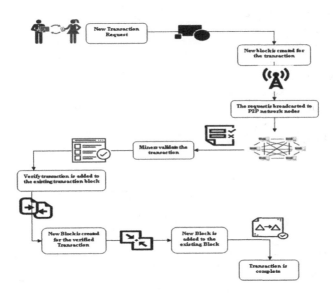

Fig. 6. Blockchain process

2.5 Internet of Things

Internet-connected devices need to be designed to ensure that the data provided by them is sufficiently reliable for its intended use, such as big data analytics [9]. The Internet of Things make the devices to be controlled by the end devices like computer or mobile. Infinite number of devices can be connected in the IoT network. The information about the things are captured by the sensors and they are accessed by the remote devices that are connected with the devices through internet gateways. The end devices involved in the IoT network without the human intervention. The scalability [8] is very high in the internet of Things. But the only disadvantage is that, it is subjective to high risk of loss of accessing and data from the IoT devices. The sensors and actuators are connected to the devices, which you want to control and monitor.

3 Proposed Design of IoT Blockchain Based Smart Agriculture

In this paper, we attempt to design an architecture for IoT based blockchain for implementing the Smart Agriculture. In Smart Home, the nodes involved in the blockchain receives the information from the sensors that are connected to the things involves in the Smart Agriculture monitoring process [10, 11]. The nodes involved in the IoT Blockchain based Smart Agriculture are as follows and is shown (see Fig. 7). The node components are IoT Blockchain based Smart Agriculture are as follows,

- Temperature Sensor Node
- Pressure Sensor Node
- Illuminance Control Node
- Wind speed, Air Control Node
- CO2, Pressure Control Node
- Pollution Control Node
- Moisture Water Control Node
- Smoke, Fire Control Node
- PH Control Node

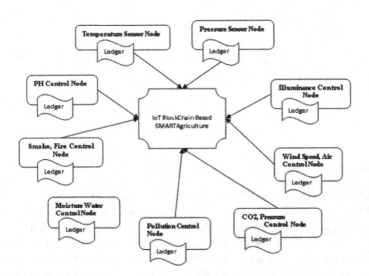

Fig. 7. Nodes in the IoT blockchain based smart agriculture

3.1 Proposed Architecture of Proposed IoT Blockchain Based Smart

The overall architecture is designed for IoT Blockchain based Smart Agriculture and is shown (see Fig. 8).

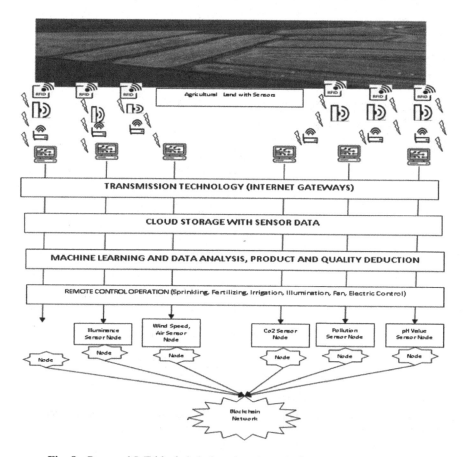

Fig. 8. Proposed IoT blockchain based smart agriculture system architecture

Here we have described the process of Smoke fire control node.

3.2 Smoke Fire Control Node

Each node in the network acts as a miner. Each node maintain the local copy of the blockchain with all the approved transactions. The transactions that are involved in each node are accessing, storing and monitoring the sensor data. Update of the information is involved along with the storage process itself. Now let us take a single node namely Smoke Fire Control Node. The Operations of the Smoke Fire Control Node are as follows and the use case diagram of Access the Smoke Fire Details Transaction in IoT Blockchain Based Smart Agriculture is shown (see Fig. 9).

1. Smoke Fire Control Transaction
2. Store Smoke Fire Details Transaction
3. Access the Smoke Fire details Transaction
4. Monitor the Smoke Fire Status Transaction

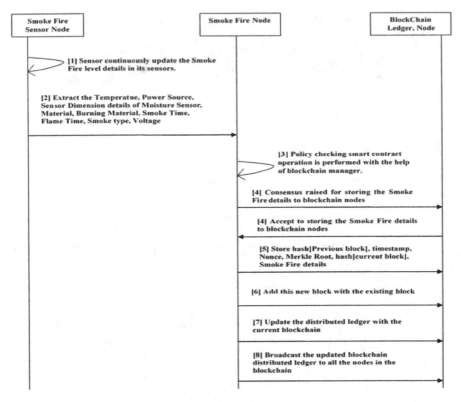

Fig. 9. Proposed use case diagram of access smoke fire details transcation in IoT blockchain based smart agriculture

Algorithm 1: Smoke Fire Control Process

1. Start the process of Remote Switch on of Smoke Fire Detector Sensor that is installed on the agricultural Farm in the places near water motor, land, Power Supply Transformer, Grass Dense Area, Trees.
2. Smoke Fire Detector Sensor continuously update the moisture level details in its sensors.
3. Extract the Smoke chemical, Fire Temperature, Sensor Dimension, Temperature, Power Source details of Smoke Fire Detector Sensor.
4. Receive the Smoke Fire details from the Smoke Fire Detector Sensor through internet gateways.
5. Smoke Fire details data is stored in the cloud storage for data analytics.
6. Smoke Fire processed data is retrieved by the Smoke Fire Control Node.

Algorithm 2: Store Smoke Fire Details Transaction

1. Get the Smoke Fire details of the installed places in the agriculture from the Sensor node
2. Policy checking smart contract operation is performed in to store in the blockchain
3. Consensus raised for storing the Smoke Fire details from other blockchain nodes
4. Allow the Smoke Fire details to be updated in the blockchain
5. Store hash[Previous block], timestamp, Nonce, Merkle Root, hash[current block], Smoke Fire details.
6. Add this new block with the existing block.
7. Update the distributed ledger with the current blockchain.
8. Broadcast the updated blockchain distributed ledger to all nodes in the blockchain

Algorithm 3: Monitor Smoke Fire Status Transaction

1. Track the frequent change of Smoke Fire details of devices from the Sensor node
2. Get the Smoke Fire details subsequently within certain time interval based on the smart contract policy.
 3. Policy checking smart contract operation is performed to monitor the Smoke Fire details
4. If the required Smoke Fire details is above the idle level, then automate the alarm to on and automate the water pump
5. If the required time is expired, receive the Smoke Fire details from sensor node.
6. Consensus raised for storing the Smoke Fire details from other blockchain nodes
7. Allow the Smoke Fire details to be updated in the blockchain
8. Store hash[Previous block], timestamp, Nonce, Merkle Root, hash[current block], Smoke Fire details.
9. Add this new block with the existing block.
10. Update the distributed ledger with the current blockchain.
11. Broadcast the updated blockchain distributed ledger to all nodes in blockchain

Algorithm 4: Access Smoke Fire Details Transaction

1. Get the Smoke Fire details of the agricultural land from the Sensor node
2. Extract the following details (Battery Power, Sensor Dimensions, Sensor Material (Copper, MS Body, Plastic), Burning Material, Smoke Visible Time, Flame Visible Time, Smoke Type(Spero, Boss, Conspec), Voltage level)
3. Policy checking smart contract operation is performed to store in the blockchain
4. Consensus raised for storing the Smoke Fire details from other blockchain nodes
5. Allow the Smoke Fire details to be updated in the blockchain
6. Store hash[Previous block], timestamp, Nonce, Merkle Root, hash[current block], Smoke Fire details.
7. Add this new block with the existing block.
8. Update the distributed ledger with the current blockchain.
9. Broadcast the updated blockchain distributed ledger to all nodes in blockchain.

4 Implementation of Proposed Work

The Proof of Concept for the proposed system is implemented with Ethereum Private Blockchain network under a genesis block in this paper. There are two types of accounts in the Ethereum. They are externally owned account and Contract account that is controlled by the Contract code. Here all the nine nodes in this system are the externally owned account and their entire smart contract are deployed under the Contract account Monitoring Agriculture is shown (see Fig. 10). The code is executed when the Contract account receives a message.

IoT devices are implemented through LibCoAP Library which is the C implementation of CoAP. The LibCoAP code was modified to automatically generate a public/private key per device which identifies the IoT devices uniquely. To test the performance of this system, the benchmark tool CoAPBench is used in this paper. It sends the confirmable requests and waits for the response before issuing the next request. The Management hub is the java script interface that makes the IoT device to connect with the block chain network. It receives the request and fetches the information from the block chain through RPC and return the response to the IoT devices. The throughput of the management hub is evaluated that affects the latency of the operations of the blockchain and is shown (see Figs. 11, 12).

Fig. 10. Proposed use case diagram of access smoke fire details transcation in IoT blockchain based smart agriculture

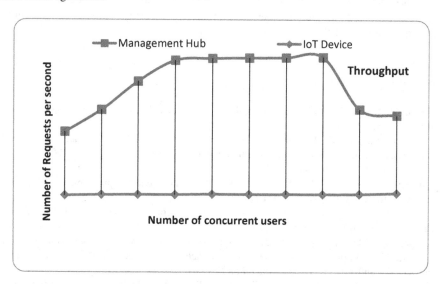

Fig. 11. Throughput of management hub transcation of proposed IoT blockchain based smart agriculture

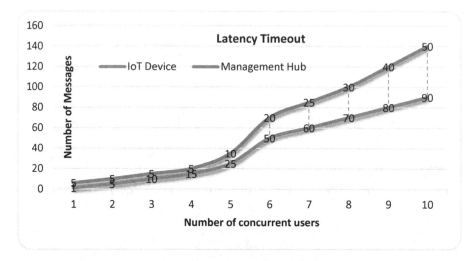

Fig. 12. Timeout latency of management hub of proposed IoT blockchain based smart agriculture

5   Conclusion

In this paper, we make an attempt to survey the core details of blockchain technology and its features. In this paper, we have proposed the design architecture namely IoT Blockchain Based Smart Agriculture and ended up with some new architectural framework which enhances the performance of security and data transparency. We have clearly shown the nodes involved in blockchain along with the architecture design. Further we try to enhance the same by applying various consensus algorithms for the same to predict the performance parameters.

References

1. Novo, O.: Blockchain meets IoT: an architecture for scalable access management in IoT. IEEE Internet. Things J. **5**(2), 1184–1195 (2018)
2. Peters, G.W., Panayi, E., Chapelle, A.: Trends in crypto-currencies and blockchain technologies: a monetary theory and regulation perspective. SSRN Electron. J. (2015). http://dx.doi.org/10.2139/ssrn.2646618
3. Lamport, L., Shostak, R., Pease, M.: The Byzantine generals problem. ACM Trans. Program. Lang. Syst. **4**(3), 382–401 (1982)
4. Nguyen, G.-T., Kim, K.: A survey about consensus algorithms used in blockchain. J. Inf. Process. Syst. **14**(1), 101–128 (2018)
5. Kosba, A., Miller, A., Shi, E., Wen, Z., Papamanthou, C.: Hawk: the blockchain model of cryptography and privacy-preserving smart contracts. In: Proceedings of IEEE Symposium on Security and Privacy (SP), pp. 839–858 (2016)
6. Zheng, Z., Xie, S., Dai, H., Chen, X., Wang, H.: Blockchain challenges and opportunities: a survey. Int. J. Web Grid Serv. **14**(4), 352–375 (2018)

7. Habib, K., Torjusen, A., Leister, W.: Security analysis of a patient monitoring system for the Internet of Things in eHealth. In: The Seventh International Conference on eHealth, Telemedicine, and Social Medicine, pp. 73–78 (2015)

8. Zheng, Z., Xie, S., Dai, H., Chen, X., Wang, H.: An overview of blockchain technology: architecture, consensus, and future trends. In: IEEE International Congress on Big Data (BigData Congress), pp. 557–564 (2017)

9. Garg, R., Mittal, M., Son, L.H.: Reliability and energy efficient workflow scheduling in cloud environment. In: Garg, R., Mittal, M., Son, L.H. (eds.) Cluster Computing. Springer, New York (2019). https://doi.org/10.1007/s10586-019-02911-7

10. Bhatnagar, V., Poonia, R.C.: Sustainable development in agriculture: past and present scenario of Indian agriculture. In: Smart Farming Technologies for Sustainable Agricultural Development, pp. 40–66 (2019). https://doi.org/10.4018/978-1-5225-5909-2.ch003

11. Kumar, S., Sharma, B., Sharma, V.K., Poonia, R.C.: Automated soil prediction using bag-of-features and chaotic spider monkey optimization algorithm. Evol. Intel. (2018). https://doi.org/10.1007/s12065-018-0186-9

An Efficient and Adaptive Method for Collision Probability of Ships, Icebergs Using CNN and DBSCAN Clustering Algorithm

Syed Zishan Ali[(✉)], Monica Makhija, Daljeet Choudhary,
and Hitesh Singh

Bhilai Institute of Technology Raipur, Raipur, Chhattisgarh, India
zishan786s@gmail.com, monicamakhija20@gmail.com,
daljeet2510@gmail.com, hiteshnick4l@gmail.com

Abstract. Collision between ships and icebergs is a major problem in glacial area, where large to small icebergs becomes a threat to cargo ships, tankers, fishing ships etc. In this paper, we have devised a new approach for the detection of icebergs and movement of ships to predict their probability of collision. In this proposed work, an adaptive method is used to detect the presence of icebergs and the velocity of ships, followed by integrating the obtained data and applying the Bayesian algorithm we have successfully computed the collision probability. This work exhibits effective results against reduced visibility due to fog. Besides, we have acquired all the foreground authentic data from valid resources. So, the results will help in marking the safe and unsafe zones in the form of clusters by using DBSCAN algorithm.

Keywords: Ships · Icebergs · Convolution neural network ·
Collision probability · Cluster

1 Introduction

Icebergs are created from the separation of large glaciers which are found in cold regions of the globe. Due to global warming, there is a tremendous breakdown in the glaciers [1] which results in the formation of large to small icebergs. These icebergs possess threat to sea traffic and may cause accidents such as collision with moving ships, which results in loss of man and material [9]. Numerous methods are used to navigate ships in large water bodies. In this paper we have contributed an adaptive method which will detect the presence of icebergs and predict the collision probability between ships and icebergs. Also we have included a module which indicates whether a region is safe or not for the sea traffic.

Previous works have been done on detecting the icebergs and proposed a computer based method to identify and track icebergs at higher resolution satellite generated synthetic-aperture radar (SAR) images [4]. This iceberg detection uses the approach of edge detection method. These detected results are retained to use this data for tracking the outcomes at different times. While performing tracking; objects which were formerly identified are matched with images acquired at different times and locations. This

© Springer Nature Singapore Pte Ltd. 2019
A. K. Somani et al. (Eds.): ICETCE 2019, CCIS 985, pp. 20–33, 2019.
https://doi.org/10.1007/978-981-13-8300-7_3

was done by ranking them in terms of their shape resemblance, later an iceberg database is generated having the same attributes and parameters which were used previously during their identification. The method presented here can be used to analyse iceberg distribution and their movements. The application involves the estimation of iceberg fluxing through choke points at coastal areas, studies of iceberg calving [3] and studies of iceberg abundance and distribution.

An effective work has been focused on clustering of moving objects which catches interesting pattern changes while moving and identifying moving micro cluster technique. Clustering over moving objects unveils interesting and effective applications like weather forecasting, cyclone clustering, and traffic jam tracking etc. Objects that are grouped altogether while being in motion are considered for working over this algorithm [2]. In this paper, we have done analysis over moving objects which provides us some interesting patterns. The idea of clustering was proposed to visualize some regularity of moving objects and many efficient algorithms were applied to keep them geographically small. Whereas, combining the concepts of collisions over moving micro clusters provides us with superficial results.

In this work, a new structure of detection, prediction and cluster formation is proposed in which CNN technique is utilized along with Bayesian algorithm. CNN is basically used to train datasets to predict whether the targeted image contains iceberg or not. Thereafter, using those resulting attributes as inputs for Bayesian algorithm we computed the collision probability for ships and icebergs, and lastly depending upon the computed results we can indicate them as a safe flag or an unsafe flag.

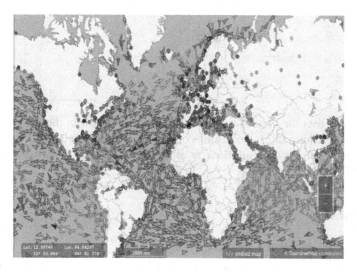

Fig. 1. Ship traffic

Figure 1 shows [5] the ships traffic across the world, where different categories of ships like cruiser, cargo, fishing etc. are moving in their respective areas. The distance between these ships is measure through latitudes and longitudes, where each degree is approximately 69 miles apart. The traffic situation becomes alarming in harsh weather conditions like in winters.

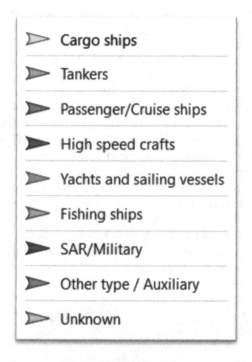

Fig. 2. Categories of ships

Figure 2 [5] lists the various categories of ships as shown in the image Fig. 1. It contains normal fishing ships, cruise ships or may include crude oil ships. The purpose of this Figure is to show the information of all the moving ships in the ocean or in glacier-filled areas.

2   Algorithms

2.1   Image Classification Using CNN

Convolution neural network model works precisely on the application where image classification is to be used. The process is simple. By passing various test images on which the AI can train itself to the prediction accuracy which depends exponentially on number of iterations in training the model as mentioned in below Eq. (1). Trained model (i.e. JSON file) is then inflicted towards the program to counter the individual images supplied. The output gives the accuracy or précised chances in percentage for whether there's an iceberg or not (Eq. 1). The various stages involved in this part are:

$$E(w) = \frac{1}{K \times N_L} \sum_{k=1}^{K} \sum_{n=1}^{N_L} \left(y_n^k - d_n^k \right)^2 \qquad (1)$$

(a) Convolution layer (some changes in finalizing): The first layer involves taking input as 2D image which is performed using python functions. These are fed with adjustable entries called weights. A mask of appropriate unit is added which is termed as convolution mask. The product is a matrix which is then applied with bias and sigmoid functions [6].

(b) Sub sampling layer: The number of planes is same as convolution layer. The layer divides input into non- overlapping blocks of 2 × 2 pixels. For each block, sum of each pixel is calculated and multiplied with weights followed by adding with the bias. Activation function uses these results to produce same block size of 2 × 2. With every sub sampling layer the size of input is reduced by half. Each of these layers is connected to the next layer in the succeeding dimension.

(c) Last convolution layer: Each plane maps exactly on one unique feature. The convolution mask in the first layer is again used as they hold identical size to input data mapping, producing exactly one scalar output.

(d) Output layer-: The layer uses sigmoid functions and neurons or radial-based functions to produce network output. In this work, the outputs indicate the category of the image.

2.2 Bayesian Framework Inference to Predict Probability of Collision

Bayes theorem or Bayes rule is generally used to calculate conditional probability. The theorem states that if any condition to satisfy the event had already occurred then finding the probability of occurrences of that event is calculated by the formula given below [7].

$$P(A|B) = \frac{P(B|A) * P(A)}{P(B)}$$

This work uses Bayes theorem to find the collision probability of ships in iceberg prone areas, provided that the event has already fulfilled the condition of iceberg presence and threshold velocity. The derived formula is given below with P1 which indicates the presence of iceberg and P (occurrence) indicates the condition for collision which is satisfied.

$$P(velocity\ exceed | occurence) = \frac{P1 * P(velocity\ exceed)}{P(occurence)}$$

2.3   DBSCAN Clustering

We have used clustering for pointing out the safe zones in the form of clusters. The first stage being formation of clusters with counter value or conditional value, where the attributes are supplied to Bayesian theorem. Each attribute holds a separate clustering function using DBSCAN to form clusters. The names are C1, C2, C3 for respectively speed of ship, location of ship and accuracy of presence of iceberg supplied from CNN image classifier [12].

In the second stage, we introduced the system to integrate all independent clusters formed into a super cluster of which the condition function being the independent clusters itself, which means it would treat clusters as objects first and the objects within those as sub-objects. The method implemented here successfully interacts with all the variables fed and clusters the safe zones for the ships. The safe zone is defined as the minimum probability of the ship having collision or having a safe route across the frame in consideration. For implementing DBSCAN algorithm throughout the steps involved, it includes marking a centre of the sphere (imaginary) and grouping is done for the objects within the radius specified as the conditional function for algorithm. The number of objects in the spheres formed is called as density numbers which is the root of the algorithm. The selection of the appropriate cluster or say sphere (as considered in this section) is as per the highest density number in that proximity [8, 12].

3   Methodology

The methodology of this research is based on taking different attributes like velocity and coordinates of ships, in order to calculate the iceberg occurrence probability and its collision with moving ships. Once the detection and probability computation is successfully completed, the clusters are formed which represent two types of flags i.e. safe and unsafe flags. The step by step procedure of this work is mentioned below (Fig. 3):

Step1. Input the Satellite image of glacial area [11].
Step2. Using Trained CNN Model, it will predict the probability P1 of the presence of icebergs in specified image.
Step3. Extract the data of ships like coordinates and velocity [5].
Step4. Set Threshold value of ships [10] and icebergs for P1 probability, coordinates and velocity.
Step5. Using Bayesian algorithm, probability of collision P2 is predicted.
Step6. Applying DBSCAN algorithm on P2 to form clusters of different probabilities.
Step7. Marking the clusters in two categories i.e., Safe and Unsafe.

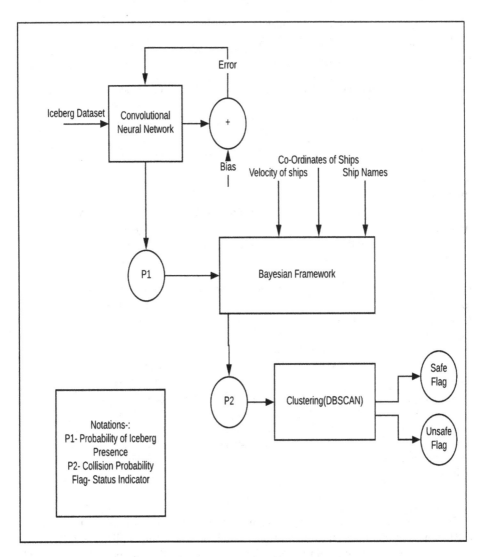

Fig. 3. Proposed architecture

Table 1. Calculation of collision probability

Ship name and type	Velocity of ships (KN)/percentage exceeding the threshold velocity (7 KN)	P1 (from CNN Trained prediction model in percentage)	Coordinates of ships (Latitude, Longitude)	Probability of collision by statistics
Nuka arctica (Container ship)	14/100	84.76333333333	66.26686, −55.32715	Y
Masik sioraq (Hopperdredger)	0	80.32222222222	63.89873 −53.21777	N
Am quebec (Bulk carrier)	11.9/70	60.13423333333	65.21989 −53.21777	N
Aqviq (Fishing vessel)	6.1/12	82.76666666666	65.69448 −62.00684	Y
Norse spirit (Crude oil tanker)	0.3/95.7	76.12121212121	66.16051 −60.95125	Y
Federal weser (Bulk carrier)	0	67.51345600000	65.69448 −62.27061	N
Beaumont hamel (Passenger ship)	8.3/18.5	70.98411111111	65.47651 −62.35840	Y
Msxt capella (Bulk carrier)	11.5/64.3	81.43654222222	64.34562 −61.42331	Y
Minerva virgo (Chemical/oil products tanker)	13.4/91.4	84.87777777777	66.45387 −55.45764	Y
Azamara pursuit (Passenger cruise ship)	18.7/167.1	55.14563333333	64.59145 −52.45138	N

The above Table 1 consists of ship names, their velocity, coordinates and the value of P1 (which is obtained after training the model) as well as the collision probability. The speed probability threshold is 7KN [10]. This threshold value is a standard speed which indicates that the speed of the ships should not decrease below 7KN otherwise it wouldn't be considered moving towards the icebergs.

$$\text{Collision chances} = \frac{|x - 7| * 100}{7}$$

Where x is the velocity of ship.

Table 2. Collision probability

Ship name and type	P2 (collision probability in %age)
Nuka arctica (Container ship)	140
Masik sioraq (Hopper dredger)	0
Am quebec (Bulk carrier)	70
Aqviq (Fishing vessel)	16.6
Norse spirit (Crude oil tanker)	121.6
Federal weser (Bulk carrier)	0
Beaumont hamel (Passenger ship)	22
Msxt capella (Bulk carrier)	86
Minerva virgo (Chemical/oil products tanker)	129
Azamara pursuit (Passenger cruise ship)	153

As per the calculation mentioned in section Bayesian theorem the collision occurrence is related to probability of collision statistics (Table 2).

- If speed is 0 KN then, no collision.
- If the iceberg's presence is above 65% then the probability of collision is 'Yes', otherwise 'No'.

Fig. 4. Description of ships

Fig. 5. Description of ships

| MINERVA VIRGO | | | AZAMARA PURSUIT | | |
| Chemical/Oil Products Tanker | | | Passenger (Cruise) Ship | | |

AIS DATA			AIS DATA		
Course	Speed	Current draught	Course	Speed	Current draught
186.0°	13.4 kn	10 m	275.0°	18.7 kn	5.9 m
GT	Built	IMO number	GT	Built	IMO number
28960	2006	9307827	30277	2001	9210220
DWT	Size	MMSI	DWT	Size	MMSI
50921	182 x 32 m	240730000	2700	181 x 26 m	248762000
	(i)			(j)	

Fig. 6. Description of ships

4   Experimental Results

The above given figures are extracted from source [5], which shows the necessary information about ships as well as vital information regarding its statistics (Figs. 4, 5 and 6).

Graphs and Work Snippets
(See Figs. 7, 8, 9 and 10)

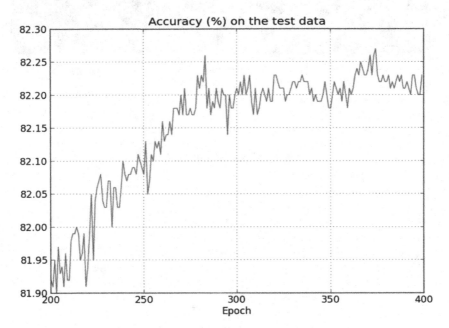

Fig. 7. Accuracy vs. Epoch

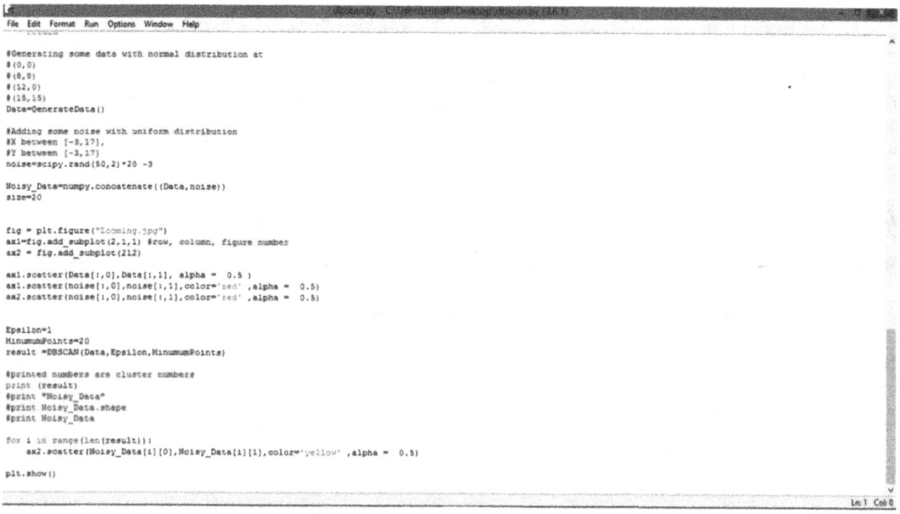

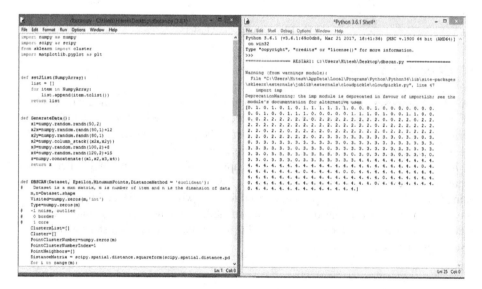

Fig. 8. Code snippets

Fig. 9. Python code for data cluster

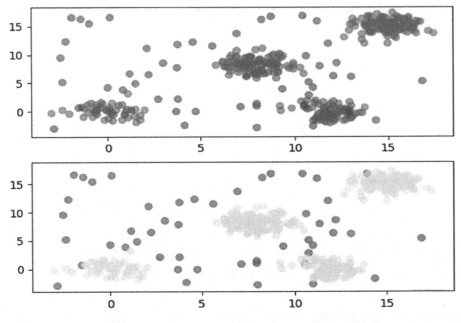

Fig. 10. Actual data clusters

5   Conclusion

This paper proposes an adaptive method in which we have trained our model to identify the iceberg, capture the ship velocity and location of the ship. By using these vital attributes the results predicts the probability of iceberg and ship collision which fulfills the need of necessary information to avoid disasters. These disasters cause a huge loss to human and other resources. The work appears to be efficient in predicting an accurate measure of the probabilities resulted. The prediction of iceberg in a still image is inefficient to the real world problems, therefore the work includes additional layer of Bayesian interface to define the collision probability for the ship in actual scenario. After the prediction we have used clustering technique to create clusters which represents safe and unsafe flags of ships.

References

1. Khan, A.A.: Why would sea-level rise for global warming and polar ice-melt, China University of Geosciences (Beijing and Peking University), Elsevier B.V. http://doi.org/10.1016/j.jsf.2018.01.008
2. Li, Y., Han, J., Yang. J.: Clustering moving objects, Department of Computer Science University of Illinois UrbanaChampaign. ACM (2004). 1-58113-888-1/04/0008
3. Wesche, C., Dierking, W.: From ice shelves to icebergs: classification of calving fronts, iceberg monitoring and drift simulation. In: 2014 IEEE Geoscience and Remote Sensing Symposium (2014). https://doi.org/10.1109/igarss.2014.6946410

4. Tiago, A.M., Silva, G.R.B.: Computer-based identification and tracking of Antarctic icebergs in SAR images, Department Street of Geography, Sheffield S10 2TN UK. Elsevier http://doi.org/10.1016/j.rse.2004.10.002
5. http://www.vesselfinder.com
6. Phung, S.L., Bouzerdoum, A.: Matlab library for convolutional neural networks (2009)
7. Zhang, M.-L., Pena, J.M., Robles, V.: Feature selection for multi-label naive Bayes classification. Inf. Sci. **179**(19), 3218–3229 (2009)
8. Soman, K.P., Diwakar, S., Ajay, V.: Insight into data mining. PHI Publication (2009)
9. www.nsidc.org
10. http://www.ccg-gcc.gc.ca/Icebreaking/Ice-Navigation-Canadian-Waters/Navigation-in-ice-cov ered-waters
11. https://maps.google.com
12. Mittal, M., Goyal, L.M., Hemanth, D.J., Sethi, J.K.: Clustering approaches for high-dimensional databases: a review. WIREs Data Min. Knowl. Discov. (2019). John Wiley & Sons. https://doi.org/10.1002/widm.1300

Robust Moving Object Detection and Tracking Framework Using Linear Phase FIR Filter

Tanmay Saxena$^{(\boxtimes)}$, Vikas Tripathi, Apoorv Chandola, and Sarthak Garg

Graphic Era (Deemed to be University), Dehradun, India
tanmaysaxena2904@gmail.com

Abstract. Moving Object detection is a technique in computer vision in which multiple consecutive frames from a video are compared by applying various detection techniques to determine movement of an object. The most challenging task in motion detection is object tracking. In this paper we have proposed an effective framework for tracking down the moving object in which least square linear-phase Finite Impulse Response Filter is used for smoothing of image while optical flow estimation is used to calculate the motion between two images, median filter removes noise from a frame and motion vector estimation is used to find the variation in pixel movement using successive frames.

Keywords: Object detection · Object tracking · Optical flow ·
Least square Linear-Phase FIR Filter · Median filter · Motion vector estimation

1 Introduction

Nowadays, many researchers are actively involved in the development of computer vision systems in which real time detection and tracking of moving objects can be used for tasks, such as counting people, tracking trajectory of a moving object [1]. Moving object detection has many phases and one of them is to extract the movable object because objects are present in the frames of a video [2]. There are three vital phases first, analyzation of video-identification of the moving object second, frame by frame tracking of the object and last, behavior tracking of object [3].

Image Processing is an approach which is similar to detection of objects that deals with objects cite (like cars, humans, etc.). In the broad region of computer vision, detection of object serves a lot of applications including image retrieval and video surveillance [2]. Object detection are of two types: Static object detection-In this type of object detection the background remains static and objects move and Moving object detection-In this type of object detection there is a dynamic background [1]. Object detection algorithms focuses on finding objects of some importance in a scene. There are many difficulties which arise in making these algorithms, some of these difficulties are changing orientation of an object with time and detection of objects which are moving relatively faster than their frame rate [2]. To get rid of these difficulties, tracking of target object in continuous video frames can be done which estimates the trajectory of the moving object and the target object is given as output [1]. There are many algorithms that can be used for object detection and tracking, every algorithm has

© Springer Nature Singapore Pte Ltd. 2019
A. K. Somani et al. (Eds.): ICETCE 2019, CCIS 985, pp. 34–44, 2019.
https://doi.org/10.1007/978-981-13-8300-7_4

its own merits and demerits. It depends on the user, which algorithm works best according to their needs [4].

At the primary stage, motion detection of the moving object is done. Motion is basically the action of changing location or position, explained in terms of velocity, direction, displacement, acceleration and time. Motion detection is a process to detect any change in object's position relative to its environment or vice-versa. In video surveillance, tracking is done by detection of motion of object and by capturing interesting regions in which the object that needs to be tracked is present. In [5], motion detection is categorized into two types, first one is based on pixel detection (based on binary difference) like temporal difference model and the second one depends on region detection (based on special point detection) like background modelling. A special point detection like background modelling is included in motion detection of region.

Further this paper is divided into the following sections, Sect. 2 gives an overview of all the different framework which are already being used for object detection, Sect. 3 comprises of our proposed methodology which uses LSLP FIR Filter, Sect. 4 consists of the result which we obtained from our framework and in Sect. 5 we concluded that our framework gave better result than the existing frameworks and where it has future scope of improvement.

2 Literature Review

Many algorithms for tracking down an object have been proposed by a number of researchers in the previous years but they all have shortcomings of their own. Tracking an object can be done by using background cues [6], with the help of Kalman filter and optical flow [3], by the use of particle filter [2], by using Lucas-Kanade optical flow [7]. Every method has its own shortcomings and the most difficult task is to develop a tracker which shows accurate results. In [6], the tracking of object is done using background cues. Traditionally background cues were one of the most important features which were used for tracking down the object. Like most of the algorithms, it also has some drawbacks, such as it is delicate to sudden light changes, camera movement and occlusion.

Fig. 1. Represents a frame with its threshold. (from left to right)

Kalman filter and optical flow are used in algorithm [3] in which optical flow is used for the detection of the moving object and Kalman Filter for tracking that object.

Optical flow works well with the changing velocity but not with the change in intensity of environment. On the other hand, Kalman Filter works well with changing intensity but not with changing velocity, whereas in algorithm [2], detection and tracking is done by using particle filter and optical flow. A large computation time is required for this algorithm. In tracking of moving object environment and low resolution input, the particle filter is considered as an inefficient tool. Further in algorithm [7], tracking of the object is done using optical flow estimation and division of objects is done using blob analysis. In this algorithm median filter is used to remove noises and for evaluating undesirable objects using threshold (as seen in Fig. 1). For tracking objects, this algorithm is not that much efficient and it is also sensitive to camera movement. But in algorithm [8], optical flow and motion vector estimation are used to identify and to track the object respectively but it is not very sensitive to camera movements. In algorithm [9], detection and tracking is done using Farneback technique for the estimation of Dense Flow and the amplitude of the motion is normalized in range from 0 to 255 which is assigned to each pixel of optical flow map. But in the case of algorithm [10], optical flow is not able to track the faster and large moving objects.

Our proposed framework is the advanced version of framework [8] in which detection and tracking of objects is done using Least Square Linear Phase (LSLP) FIR (Finite Impulse Response) Filter along with optical flow, median filter and motion vector estimation for better results. The Linear-Phase FIR Filter along with the proposed framework provides better accuracy than framework [8].

3   Methodology

Several methods are available to identify and to track the moving object by analyzing every frame of the video. There are three vital phases in the analyzation of a video firstly, identification of the moving object is then secondly, tracking of the object using successive frames is done and lastly, analyzation of the behavior of object which is being tracked [3] is done. This paper introduces an enhanced framework for tracking down objects which gives better results than framework [8].

For robust visual surveillance system, the dynamic environmental conditions like object shadows and lightning changes etc. needs to be handled well. An even more complex problem with shadow emerges if the shadow of an object is cast on another object, like two humans walking parallel to each other. The first step in this procedure is to consider these dynamic environmental conditions and filter out the undesired area. This helps in performing better object detection which further increases the tracking accuracy.

A new tracking framework is proposed which uses Least Square Linear Phase (LSLP) FIR Filter for smoothing of image along with optical flow estimation, which allows motion estimation as either instantaneous image velocities or discrete image displacements followed by median filter which is used to remove noise from binary images and smoothens out small speckles of noise. Median filter is a non-linear filtering technique, it has wide applications in digital image processing because, it preserves edges as well while removing noise, it also has applications in signal processing. Median filter considers each pixel of the image by looking at its neighboring pixels to decide whether or not that pixel is a representative of its surroundings. Tracking of the moving object is done using motion vector estimation (see Fig. 2) that gives out vector direction in which object is moving, which gives more precise results than the results of algorithm [8].

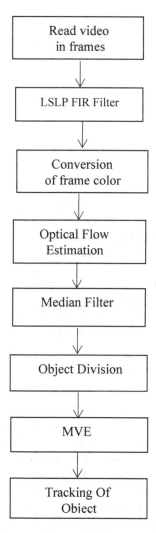

Fig. 2. Proposed framework for object detection

3.1 LSLP FIR Filter

As soon as the frame is read, the FIR filter is applied to it. Advantages of the filter: absolute linear phase, stable, linear design methods, efficient performance, suitable for multi-rate applications. Disadvantage of this filter are: slow computation speed, filtering requires higher coefficients compared to IIR (Infinite impulse response) filter. Default values of the designed filter is indicated in the table (see Table 1):

- Filter structure - Direct Form FIR Filter
- Filter Length - 11
- Stable and linear phase (Type 1)

Table 1. Default criteria and values of filter

Criteria	Values
Multipliers	11
Adders	10
States	10
Multiplication per input sample	11
Addition per input sample	10

FDA tools of MATLAB are used in making this filter. The used filter comes under the category of filter designs, 'firls' is the function used for designing a linear-phase FIR filter in MATLAB. The type 1 filter is used here because based on symmetry of the filter there are many inherent limitations on linear phase filter's frequency response but this type of the filter doesn't have these limitations. Amplitude, frequency and weight vectors are determined by hit and trial method and those values are chosen while designing the filter which will result in a better filtered video. In designing of a filter there are few factors that needs to be taken in account like at the primary stage initialization of frequency, amplitude and weight vectors needs to be done then after this calculations for coefficients of filter and at last the creation of filtered object.

3.2 Object Detection via Optical Flow

Motion can be described by optical flow and it is used to get measurable values of images by analyzing the motion. There are two categories of motion: object is moving while background is static and object is moving with a dynamic background [1]. Optical flow is a technique which uses two concerned frames of a video at different intervals of time to get the motion. To calculate the field vector for optical flow of a moving object, the technique (optical flow) separates the foreground objects from the background. Optical flow estimation has different methods like Lucas-Kanade, Black-Jepson, Buxton-Buxton and Horn-Schunck. In terms of accuracy and efficiency, Lucas-Kanade is the best method for evaluating optical flow estimation. Comparison has been done with all the different techniques to find the most reliable one and Lucas-Kanade [11] is found to be the most efficient one. Among all the computational techniques of optical flow the implementation of Lucas-Kanade in real world is much faster. This method has a bunch of advantages like little processing cost and great tracking ability.

This framework basically evaluates the movement between the frames taken at time s and $s + \partial s$ for every pixel location [11]. Pixel at position (a, q, z) with intensity T(a, q, z) will move between two frames by ∂p, ∂q and ∂r. As shown in Eq. (1):

$$T(a, q, z) = T(a + \partial a, q + \partial q, z + \partial z) \tag{1}$$

Considering the movement to be minimum, the image constraints at T(a, q, z) can be derived with the help of Taylor series as shown in Eq. (2):

$$T(a + \partial a, +q + \partial q, z + \partial z) = T(a, q, z) + \frac{\partial T}{\partial a}\partial a + \frac{\partial T}{\partial q}\partial q + \frac{\partial T}{\partial z}\partial z + HOT \tag{2}$$

Here HOT stands for higher order terms which have minimum values and can be eliminated. Using Eqs. (1) and (2), Eqs. (3) and (4) can be deduced:

$$\frac{\partial T}{\partial a}\partial a + \frac{\partial T}{\partial q}\partial q + \frac{\partial T}{\partial z}\partial z = 0 \tag{3}$$

$$\frac{\partial T}{\partial a}\frac{\partial a}{\partial z} + \frac{\partial T}{\partial q}\frac{\partial q}{\partial z} + \frac{\partial T}{\partial z}\frac{\partial z}{\partial z} = 0 \tag{4}$$

Equation (5) is deduced from Eqs. (3) and (4):

$$\frac{\partial T}{\partial z}U_a + \frac{\partial T}{\partial z}U_q + \frac{\partial T}{\partial z}U_z = 0 \tag{5}$$

Where Ua and Uq are a and q constituent of velocity or optical flow of T(a, q, z) and $\frac{\partial T}{\partial a}$, $\frac{\partial T}{\partial q}$ and $\frac{\partial T}{\partial z}$ are the derivatives of image at (a, q, z) in the desired directions. Optical flow constraints on the constituent is represented by Eqs. (5), optical flow constraint equation is given as:

$$T_a U_a + T_q U_q = -T_z \tag{6}$$

Our main objective is to calculate Ua and Uq but we have only one equation to calculate two unknowns thus, the unknowns cannot be calculated using a single equation. So other constraints are taken into account to give different set of equations which will be used for finding the optical flow.

In the definition of this algorithm it is stated that motion vectors never change for a given area it only gets shifted from one to other position. Let us consider the flow (Ua, Uq) to be constant for a small time period of 1 * 1 with 1 > 1 i.e. concentrated at (a, q) and numbering of pixels is done as 1, 2, 3 ... upto n and the derived equations are shown in (7):

$$\begin{aligned} T_{a1}U_a + T_{q1}U_q &= -T_{z1} \\ T_{a2}U_a + T_{q2}U_q &= -T_{z2} \\ &\vdots \\ T_{an}U_a + T_{qn}U_q &= -T_{zn} \end{aligned} \tag{7}$$

The system becomes over-determined because of more than three unfamiliar equations in (7). Hence:

$$\begin{pmatrix} T_{a1} & T_{q1} \\ T_{a2} & T_{q2} \end{pmatrix} \begin{pmatrix} U_a \\ U_q \end{pmatrix} = \begin{pmatrix} -T_{z1} \\ -T_{z2} \\ -T_{z3} \end{pmatrix} \tag{8}$$

OR

$$X \vec{u} = -y \tag{9}$$

Least square method is used to solve the above system of equations:

$$X^T X \vec{u} = X^T(-y) \tag{10}$$

$$\vec{u} = \left[X^T X \right] X^T(-y) \tag{11}$$

OR

$$\begin{bmatrix} U_x \\ U_y \end{bmatrix} = \begin{bmatrix} \Sigma P_{xi}^2 & \Sigma P_{xi} P_{yi} \\ \Sigma P_{xi} P_{yi} & \Sigma P_{yi}^2 \end{bmatrix}^{-1} \begin{bmatrix} \Sigma P_{xi} P_{di} \\ \Sigma P_{yi} P_{di} \end{bmatrix} \tag{12}$$

Now the sums from i = 1 to n will be taken into account for performing the calculation to find out motion vector (12) and there is a limit condition which is given below as shown in Eq. (13):

$$X^T X = \begin{bmatrix} \Sigma P_{xi}^2 & \Sigma P_{xi} P_{yi} \\ \Sigma P_{xi} P_{yi} & \Sigma P_{yi}^2 \end{bmatrix} \tag{13}$$

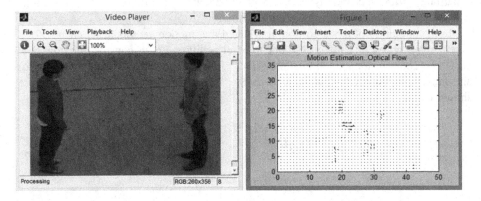

Fig. 3. Blue marks in Fig. 1 represents direction and magnitude of motion for the given input video using optical flow estimation (Color figure online)

Using Eq. (13) image derivatives are going to be determined for optical flow in the three directions: u-direction, v-direction and direction of time (see in Fig. 3). The main benefit of this method is that it is robust and it does accurate detection in the presence of noise [11].

3.3   Object Segmentation

Object segmentation means moving objects gets segmented i.e. locating objects and its boundaries from the background. Captured frames contains some noise for which the median filter is used, mostly these noises are speckle noise from the captured frames. After filtering, the useful objects are passed to optical flow vectors for thresholding operation and the unwanted objects gets removed. Some holes are created during the filtering process and for filling them, morphological close operation is used. After this, blob analysis is performed on the frame in which bounding boxes are drawn around the moving objects [8].

3.4   Object Tracking Using Motion Vector Estimation

There is no use in applying different algorithms if its motion estimation is not properly handled. Image is processed in small blocks called macro blocks. Macro blocks of the associated frame gets operated by the motion vector estimation technique [8] in which every frame gets divided in macro blocks and the selected macro blocks of associated frame gets compared with the target frame [8]. The two selected frames and the macro blocks are taken in account to calculate the mean absolute difference (MAD) and the one which has minimum absolute difference is considered to be the best match. The following expression is used in the calculation of MAD [8]:

$$\text{MAD} = \frac{1}{Z} \sum_{x=0}^{Z-1} \sum_{y=0}^{Z-1} T(a+m+x, b+n+y) - R(a+m+b+n) \qquad (14)$$

Where, Z - Size of the macro block,
m and n - indices for pixels in the macro block,
x and y- horizontal and vertical displacements,
T (a + m + x, b + n + y) - pixels in macro block of Target frame,
R (a + m, b + n) - pixels in macro block of associated frame.

If we take reference frame 'm' then the movement of the object is in frame 'm + 1'. By this technique we can track moving objects more efficiently and it makes it more robust. This technique is applied on low resolution videos to check accuracy and it provides a better result.

4   Experimental Results

This segment comprises of test results in protest identification over the following dataset. For the assessment of our framework the standard arrangements of UT-Interaction dataset [12] is utilized.

Comparison of results is done between the proposed framework and the framework described in [8]. LSLP FIR filter is applied on the captured frame to remove noises like undesirable frequencies followed by optical flow estimation (Lucas and Kanade) then median filter and then motion vector estimation is applied to get the results.

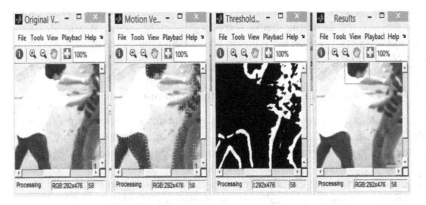

Fig. 4.   Result of proposed framework

Figure 4 consists of four windows in which original video is shown in the first window while calculated motion vectors is shown in the second window and the third window shows the segmented object due to thresholding and last window shows the tracked object.

One execution measure in this dataset of protest, following is precision of the algorithm. Following is the articulation which is utilized to ascertain the exactness of the question recognition and following:

$$\text{Accuracy} = \left(\frac{\text{Number of moving objects detected by framework}}{\text{Total number of moving objects present in the frame}} \right) * 100 \quad (15)$$

By using the above formula the accuracy is evaluated for both the frameworks (proposed framework and framework described in [8]) on different input video samples of UT interaction dataset and the table below shows the precision of both the frameworks (see Table 2 and Fig. 5).

Table 2. Precision of both frameworks in percentage

Videos (UT-interaction dataset)	Framework described in [8]	Our proposed framework
Test 1	87.03%	96.77%
Test 2	93.93%	97.22%
Test 3	90.41%	96.55%
Test 4	92.00%	97.36%
Test 5	91.89%	97.77%
Test 6	88.80%	94.11%
Test 7	91.66%	96.55%
Test 8	93.75%	96.96%
Test 9	91.37%	96.92%
Test 10	84.61%	90.00%

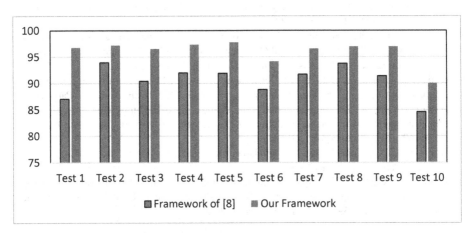

Fig. 5. Comparison of results of both frameworks

5 Conclusion

In this paper, we have proposed a robust framework for detection and tracking of moving objects. Our framework uses LSLP FIR filter along with optical flow, motion vector estimation and median filter for the best results. This framework provides an effective accuracy of 96.012%. This framework can be used in vehicle detection, human detection, and in Unmanned Aerial Vehicle (UAV) for precise tracking results. This framework can be further enhanced in the future, as it suffers in tracking still objects and in tracking moving objects that becomes still for a short period of time.

References

1. Verma, R.: A review of object detection and tracking methods. Int. J. Adv. Eng. Res. Dev. **4** (10), 569–578 (2017)
2. Abdelali, H.A., Essannouni, F., Aboutajdine, D.: Object tracking in video via particle filter. Int. J. Intell. Eng. Inf. **4**(3–4), 340–353 (2016)
3. Shantaiya, S., Verma, K., Mehta, K.: Multiple object tracking using Kalman filter and optical flow. Eur. J. Adv. Eng. Technol. **2**(2), 34–39 (2015)
4. Vekariya, D., Shah, H.R., Sodha, N.: Implementation of object tracking using camera. Int. J. Adv. Eng. Res. Dev. **2**(6), 225–237 (2015)
5. Charadva, M.J., Sejpal, R.V., Sarwade, N.P.: A study of motion detection method for smart home system. Int. J. Innov. Res. Adv. Eng. (IJIRAE) **1**(5), 148–151 (2014)
6. Li, A., Yan, S.: Object tracking with only background cues. IEEE Trans. Circ. Syst. Video Technol. **24**(11), 1911–1919 (2014)
7. Murugan, A.S., Devi, K.S., Sivaranjani, A., Srinivasan, P.: A study on various methods used for video summarization and moving object detection for video surveillance applications. Multimed. Tools Appl. **77**(18), 23273–23290 (2018)
8. Kale, K., Pawar, S., Dhulekar, P.: Moving object tracking using optical flow and motion vector estimation. In: 4th International Conference on Reliability, Infocom Technologies and Optimization (ICRITO) (Trends and Future Directions), pp. 1–6. IEEE, Noida (2015)
9. Walker, J., Gupta, A., Hebert, M.: Dense optical flow prediction from a static image. In: International Conference on Computer Vision, pp. 2443–2451. IEEE, Araucano Park (2015)
10. Wang, Z., Yang, X.: Moving target detection and tracking based on pyramid Lucas-Kanade optical flow. In: 3rd International Conference on Image, Vision and Computing (ICIVC), pp. 66–69. IEEE, Chongqing (2018)
11. Thota, S.D., Vemulapalli, K.S., Chintalapati, K., Gudipudi, P.S.S.: Int. J. Eng. Trends Technol. **4**(10), 4507–4511 (2013)
12. Ryoo, M.S., Aggarwal, J.K.: Spatio-temporal relationship match: video structure comparison for recognition of complex human activities. In: IEEE International Conference on Computer Vision (ICCV), pp. 1593–1600. IEEE, Kyoto (2009)

Advance Persistent Threat Detection Using Long Short Term Memory (LSTM) Neural Networks

P. V. Sai Charan$^{(\boxtimes)}$, T. Gireesh Kumar, and P. Mohan Anand

TIFAC-CORE in Cyber Security, Amrita school of Engineering,
Amrita Vishwa Vidyapeetham, Amrita University, Coimbatore, India
pvsaicharan2011@gmail.com, gireeshkumart@gmail.com,
mohananand1997@gmail.com

Abstract. Advance Persistent Threat (APT) is a malware attack on sensitive corporate, banking networks and stays there for a long time undetected. In real time corporate networks, identifying the presence of intruder is a big challenging task to security experts. Recent APT attacks like Carbanak and The Big Bang ringing alarms globally. New methods for data exfiltration and evolving malware techniques are two main reasons for rapid and robust APT evolution. In this paper, we propose a method for APT detection System for real time corporate and banking organizations by using Long Short Term Memory (LSTM) Neural networks in order to analyze huge amount of SIEM (Security Information and Event Management) system event logs.

Keywords: LSTM · APT · Hadoop · Splunk · Hive

1 Introduction

APT is a combination of several sophisticated attacks which are composed by a professional attacker on a specific sensitive organizations. Usually, Security specialists consider this APT as undetected shady RAT (Remote Access Trojan) operations. According to McAffe survey on APT, 83% of these APT attacks are due to out going sessions through TCP (Transmission Control Protocol) port numbers 80 and 443 [1,2]. Some of organizations counter this thing by web proxies which can inspect HTTP traffic i.e port number 80 but 443 is mostly untouched by most of organizations. Added to this, using reverse shell instead of bind shell for injecting commands into the victims machines is very difficult to identify at both network and host level in real time. There are broadly 6 phases in which APT will infect and spread in a particular targeted network [3].

1.1 Phase1 - Reconnaissance

In this particular phase he tries to gather information about targeted organization by using some social engineering techniques. The key objective of attacker

© Springer Nature Singapore Pte Ltd. 2019
A. K. Somani et al. (Eds.): ICETCE 2019, CCIS 985, pp. 45–54, 2019.
https://doi.org/10.1007/978-981-13-8300-7_5

in this phase is to infiltrate into organizations network by exploiting some vulnerability.

1.2   Phase 2 – Gaining Access

In this particular phase, payload injection occurs into network by directly exploiting remote access backdoor or making the user to click on malicious link in the form of a phising/spam mail etc. According to IBM researchers, attackers are using deceptive and highly targeted attack tools embedded with Artificial Intelligence which reveals its identity to a specific targeted victim [4]. That means the malicious payload of malware will be hidden in a normal day to day applications to avoid detection by most antivirus and malware scanners. Usually, attackers using video conferencing software, file sharing software until it reaches specific victims, who are identified via indicators such as geo location, voice or facial recognition and other system-level features.

1.3   Phase 3 - Lateral Movement

In this phase, attacker tries to move from initial target to specific targeted part of network by exploiting compromised privileges and as well as configuration weakness in that particular network.

1.4   Phase 4 – Gathering Information

In this phase, attacker captures and observes different workflow patterns in that organization and tries to capture them in the form of logging keystrokes, capturing the screen recordings of user workflows of specific intended targets.

1.5   Phase 5 – Data Exfiltration

In this particular phase, attacker ties to push gathered information to external Control and Command Server. In phase also attackers gained a lead than security experts in order to hide exporting sensitive contents to C2C servers bypassing robust firewalls and Intrusion Detection Systems at Host and Network levels as well. For example, attackers bypassed Google, Adobe firewalls in operation Aurora by sending traffic over TCP with a custom encrypted protocol instead of suing SSL [5].

1.6   Phase 6 - Cleaning

In this particular phase, the attacker tries to clean his operations to cover his identity in order to escape from any further legal actions from targeted organization end.

Rapid growth in the design of sophisticated malware tools not only causing a great matter of threat for global IT industry but also defense and banking industry as well. Recently, the word APT has become the common tool for cyber warfare between countries. Carbank attack in 2015 has literally infected thousands of victims all over the world. This Carbanak APT totally effected almost equal to $1 Billion though various operations like generating bogus accounts and using fake services to collect the money, transferring money to cybercriminals using the SWIFT (Society for Worldwide Interbank Financial Telecommunication) network etc. Kaspersky dig into very same issue and published a detailed report which gives detailed analysis about attack pattern [6]. But recently, The Big Bang APT in 2018 came up with much more robust and sophisticated way which mainly focused on targeting the Palestinian Authority [7]. This APT malware contains a number of modules that perform certain functionalities such as taking screenshots, obtaining a list of files, retrieving system information, restarting the system and self-deletion. The malware will fetch additional modules from the Command and Control server if it finds something of interest. Because of merge of cybercrime and APT, emerging malware techniques, hybrid methods of data exfiltration like pass by hash, fragmentation of bigger APT groups, embedding Artificial Intelligence in APT framework to reach specific intended target, all these kind of scenarios making APT attacks great matter of concern globally. In order to deal with target specified next generation intelligent APT attack we propose a novel method for APT detection mechanism by using Long Short Term Memory (LSTM) Neural network.

This paper is organized as follows. Section 1 details about APT overview and Phases of APT, followed by the challenges that corporate network face with APT. Section 2 explains Related works and problems involved in current methodologies. Problem statement and Proposed system for APT detection using LSTM Neural networks explained under Sect. 3. Implementation and results are detailed in Sect. 4. Conclusion and future work is discussed under Sect. 5.

2 Related Works

Research work on various APT detection methods have been evolving rapidly from last 5 years. Many hybrid techniques have been developed by integrating both network level detection and behavior based abnormality identification techniques. Fabio Pierazzi has proposed a novel method for APT detection analysis of high volumes of network traffic from different network probes [8]. By assigning scores for different traffic flows, they identified the APT behavior at data exfiltration phase. Similarly, Guodongzhao proposed a method in which a dedicated system in network will detect the APT malware based on malicious DNS [9]. It uses a combination of anomaly based detection and signature based detection to find the malicious APT C2C domains. But in the real environments time it's very difficult to rely on DNS based APT detection methods because latest APT malwares are not using malicious flux service or DGA (Domain Generation Algorithm) domains. On the other hand, there is plenty of research work going on to detect APT kind of persistent hidden malware by observing abnormal user behavior by auditing large amount of semi structure server log files in real time corporate networks [10, 11].

Recently, Artur Rot has proposed a multi-layered approach to detect APT in which the seven-layer model based on OSI creates an environment where each layer can't defend an APT attack on its own, but their combination gives us fruitful results [12]. Although many layers exist in this model, Sandboxing layer plays a key role in detection of APT. In one way sandboxing seems efficient in APT detection, its have its own disadvantages as well. Sandboxing technique is still vulnerable to the zero-day vulnerabilities and the high probability for the inclusion of new evasion measures. Along with this new generation APT are Anti-Sandbox resistant in nature. So, traditional sandboxing techniques may not work in future in the case of APT [13]. Added to this, there are many big players in market like QRadar IBM, Q1 Labs Qradar, and NetIQ Security Manager SIEM tools which can able to deal with different varieties of log files in real time APT detection [14,15]. On the other side of coin Roman Jasek has proposed a novel method for detection of APT by using Honeypot [16,17]. In this method, Honeypot agents installed to trap the malicious users and also these technique will be very efficient to observe the attack patterns of malicious users especially in long run which is exactly suitable for APT attack scenarios.

3   Proposed Work

Although many traditional and hybrid methods are available in detection of APT malware, still number of APT attacks increasing rapidly at global level. Attackers made APT sandboxing resistant so that traditional sand boxing techniques may not work with robust new generation APT detection. Next generation AI based APT techniques are very hard to identify by any of existing APT detection methods. In this paper, we propose a novel method for APT detection in real time by using Long Short Term Neural Networks which is a varient of Recurrent Neural network (RNN) with added storage capabilities.

In this paper, we have collected huge system event log files from APT infected machines by installing splunk forwarder at host level [18]. In the splunk forwarder we specify the IP address for splunk server machine so that all those logs generated at the client side will be pushed to server machine. Usually, in real time these log files are unstructured and very huge in its size. In the initial stage we have loaded this log file to HDFS for further processing. We have used open source big data tool named HIVE to extract event codes, corresponding message along with the time stamp [19,20].

3.1   Extracted Event Code Structure from HIVE Tool

"4/10/18 6:22:23:00 PM, 7036, The WinHTTP Web Proxy Auto-Discovery Service service entered the stopped state"

Sample extracted event code structure is given below. After processing at HIVE tool, the output log file consists of time stamp followed by event code generated at that particular time and the system event message corresponding to that particular event code [21] (Table 1).

Table 1. Observed APT pattern from event code sequence

S.no	Event code	Event code meaning
1	4648	This event is generated when a process attempts to logon
2	4624	This event is generated when a logon session is created
3	4672	Special privileges assigned to new logon
		SecurityPrivilege
		TakeOwnershipPrivilege
		LoadDriverPrivilege
		BackupPrivilege
		RestorePrivilege
		DebugPrivilege
		SystemEnvironmentPrivilege
		ImpersonatePrivilege
4	7036	This event indicates that the firewall has been moved to stopped state
5	1014	This event Name resolution for some external C&C server
6	7036	This event indicates WinHTTP Web Proxy Auto-Discovery Service service entered the stopped state
7	7036	This event indicates the Application Experience service entered the stopped state

Once log processing completed at HIVE level, a series of dependent event at the Reducer level. We Have identified a event pattern (series of events) which exactly replicate APT behavior in the real time. The series of Event codes which indicates that behavior of APT are actually Interdependent on each other. That means if we need to detect APT and raise a alarm based on a particular event code then we need to remember all the previous six event codes. In order to address this problem we have used LSTM Neural Networks to efficiently train these interdependent event pattern in our case [22].

3.2 Long Short Term Memory (LSTM) Neural Network in APT Detection

LSTM is a special kind of Recurrent Neural Networks (RNN) capable of learning long term dependencies proposed by Hochreiter and Schmidhuber to overcome vanishing gradient problem in standard RNN [23]. LSTM consists of one input layer, one output layer, and one recurrent hidden layer. The hidden layer contains a memory cell with self-connections in order to memorize the temporal state. The two gates named input gate and output gate is used for regulating the information flow through the cell. Constant Error Carousel (CEC) will be considered as the core of memory cell which is a recurrently self-connected linear unit, and the cell state usually represented by the activation of the CEC. Multi-

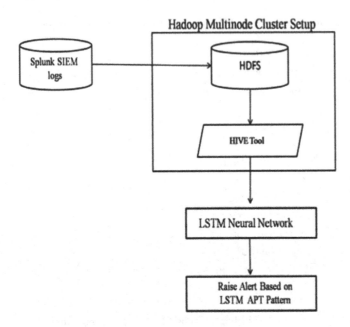

Fig. 1. Proposed system archirtecture

plicative gates can learn to open and close because of this CEC, which indirectly helps LSTM NN to solve the vanishing gradient problem (Fig. 1).

The input is denoted as $x = (x_1, x_2, \ldots, x_T)$, and the output denoted as $y = (y_1, y_2, \ldots, y_T)$ and T will be the time taken for identifying exact APT event code pattern. In our APT detection case, x can be considered as series of event code data that we extracted fom the Splunk machine and y is the estimated time to detect exact APT event code pattern. The objective of LSTM NN is to detect exact APT event code pattern based on previous event codes without specifying information about how many steps should be traced back. In real time implementation, we will get the exact APT detection time by iteratively calculating the following the equations at the background level.

$$f_t = \sigma(W_{fx}x_t + W_{fm}m_{t-1} + W_{fc}C_{t-1} + b_f) \tag{1}$$

$$C_t = f_t \bullet C_{t-1} + I_t \bullet g(W_{cx}X_t + W_{cm}m_{t-1} + b) \tag{2}$$

$$O_t = \sigma(W_{ox}x_t + W_{om}mt - 1 + W_{oc}C_t + b_0) \tag{3}$$

$$m_t = O_t \bullet h(C_t) \tag{4}$$

$$y_t = W_{ym}m_t + b_y \tag{5}$$

$$\sigma(x) = \frac{1}{1 + e^{-x}} \tag{6}$$

Each block of LSTM contains an input gate, an output gate, and a forget gate. i_t, o_t, f_t are outputs of those three gates are respectively. c_t and m_t represents the

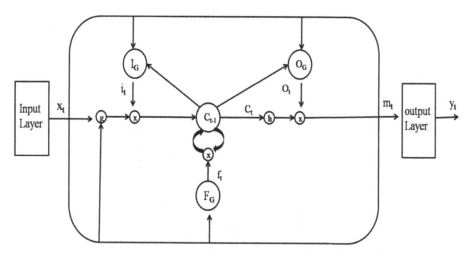

Fig. 2. LSTM cell archirtecture [23]

activation vectors and memory block for each cell. To build connection between the cell block, input and output layers bias vectors b and weight matrices W will be used normally.

• indicates the scalar product of two vectors, $\sigma(x)$ indicates logistic sigmoid function is denoted and centered logistic sigmoid function with range $[-2, 2]$ is denoted with g(x), centered logistic sigmoid function with range $[-1, 1]$ denoted with h(x) in Fig. 2. For detailed execution steps of LSTM cell architecture please refer Xiaolei work [24]. In our case, we need to remember all previous 6 states information while predicting the APT from a huge amount event codes which are continously generating in the real time. We have trained LSTM neural network to predict APT event code pattern. At the output cell all the errors are truncated, where these errors can flow back which makes error tends decay exponentially. In this way LSTM network in our case deals with long term dependencies of system event codes in order to detect exact APT pattern.

4 Implementation and Results

The processed output from the HIVE engine will be divided as 40% for training remaining 60% for the testing LSTM neural network. In our case, we have used open source machine learning library named Tensorflow which is developed by Google for performing dataflow programming of LSTM neural networks to predict the APT pattern from System Event code logs [25]. The output from the HIVE phase is segregated as two individual datasets i.e sample-1 and sample-2 respectivly. Sample-1 dataset is splitted as 60:40 ratio for training and testing phases of LSTM netowrk. Similarly, sample-2 dataset is splitted as 70:30 ratio for training and testing phases in our case. We have trained LSTM with these two different datasets and results are tabulated in Table 2.

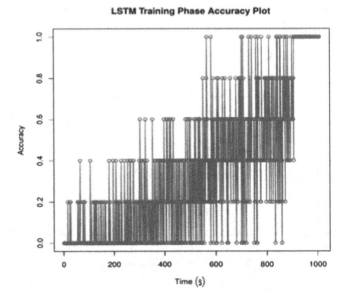

Fig. 3. LSTM network training phase accuracy plot for sample-1 dataset

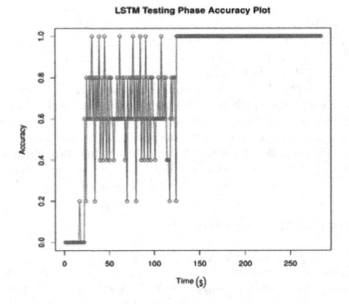

Fig. 4. LSTM network testing phase accuracy plot for sample-1 dataset

In both training and testing phases for LSTM neural network, we can clearly understand the way in which LSTM neural network is identifying the APT event code pattern by observing the accuracy value pattern in the Figs. 3 and 4.

Table 2. Implementation results for different data samples

Sample name	Phase (training/testing)	Percentage of data considered	Time taken for APT detection (sec)
Sample-1	Training	60	900 (~15 Min)
	Testing	40	129 (~2.2 Min)
Sample-2	Training	70	1028 (~17.13 Min)
	Testing	30	117 (~1.95 Min)

Initially Neural network accuracy is 0 is both the cases but eventually LSTM neurons will start learning and identifying the patterns so that we can see a gradual improvement in accuracy in both of the cases. In training phase, LSTM neural network nearly took 15 min and 17.13 min to identify the pattern for sample-1 as well as sample-2 data sets respectivly. But in the testing phase, for both data sets we can observe a drastic improvement i.e approximately 2 min and 1.95 min to identifying the APT event code pattern which will be highly suitable for the real time APT detection.

5 Conclusion and Future Work

In this paper, we propose a technique to detect APT in the real time with the help of LSTM neural network by analyzing large amount of SIEM system event log files which are collected from APT infected machine. From the Fig. 4. We can clearly identify that LSTM can able to detect the APT pattern with in a span of minutes which we can further optimize using high end processors at the hardware level. As an additional enrichment to this system, we can further train different event code patterns to this LSTM network so that we can build much robust system against next generation malware attacks. Although, many Intrusion detection systems and sandboxing techniques are available in market which operates at host and network layer, they are unable to cope up with the next generation APT techniques which are inbuilt AI components. In order to deal with evolving malware techniques, we can integrate our proposed model as a AI component level for traditional APT detection methods to build next generation intelligent and robust malware detection systems.

References

1. Kaspersky Lab: The Great Bank Robbery: The Carbanak APT (Detailed Investigation Report) (2015). https://securelist.com/the-great-bank-robbery-the-carbanak-apt/6873/
2. McAfee Labs Threats Report, June 2018. https://www.mcafee.com/enterprise/en-us/assets/reports/rp-quarterly-threats-jun-2018.pdf
3. Messaoud, B.I.D., et al.: Advanced persistent threat: new analysis driven by life cycle phases and their challenges. In: International Conference on Advanced Communication Systems and Information Security (ACOSIS). IEEE (2016)

4. DeepLocker: How AI Can Power a Stealthy New Breed of Malware (2018). https://securityintelligence.com/deeplocker-how-ai-can-power-a-stealthy-new-bred-of-malware/

5. Kharitonov, D., Ibatullin, O.: Extended security risks in IP networks. arXiv preprint arXiv:1309.5997 (2013)

6. Kaspersky Security Bulletin (2015). https://securelist.com/kaspersky-security-bulletin-2015-overall-statistics-for-2015/73038/

7. The Big Bang APT (2018). https://research.checkpoint.com/apt-attack-middle-east-big-bang/

8. Marchetti, M., et al.: Analysis of high volumes of network traffic for advanced persistent threat detection. Comput. Netw. **109**, 127–141 (2016)

9. Zhao, G., et al.: Detecting APT malware infections based on malicious DNS and traffic analysis. IEEE Access **3**, 1132–1142 (2015)

10. Kayacik, H.G., et al.: Detecting Anomalous Hypertext Transfer Protocol (HTTP) Events from Semi-Structured Data. U.S. Patent Application No. 15/420,560

11. Sai Charan, P.V.: Abnormal user pattern detection using semi-structured server log file analysis. In: Satapathy, S.C., Bhateja, V., Das, S. (eds.) Smart Intelligent Computing and Applications. SIST, vol. 104, pp. 97–105. Springer, Singapore (2019). https://doi.org/10.1007/978-981-13-1921-1_10

12. Rot, A., Olszewski, B.: Advanced persistent threats attacks in cyberspace. Threats, vulnerabilities, methods of protection. In: 2017 Federated Conference on Computer Science and Information Systems, vol. 13 (2017)

13. Brickell, E.F., et al.: Method of improving computer security through sandboxing. U.S. Patent No. 7,908,653, 15 March 2011

14. IBM QRadar (The Intelligent SIEM). https://www.ibm.com/security/security-intelligence/qradar

15. NetIQ. https://www.netiq.com/de-de/

16. Jasek, R., Kolarik, M., Vymola, T.: APT detection system using honeypots. In: Proceedings of the 13th International Conference on Applied Informatics and Communications (AIC 2013), WSEAS Press (2013)

17. Ali, P.D., Gireesh Kumar, T.: Malware capturing and detection in dionaea honeypot. In: 2017 Innovations in Power and Advanced Computing Technologies (i-PACT). IEEE (2017)

18. Anastasov, I.: DancoDavcev.: SIEM implementation for global and distributed environments. In: 2014 World Congress on Computer Applications and Information Systems (WCCAIS). IEEE (2014)

19. Apache-Hadoop. http://Hadoop.apache.org

20. Apache-Hive. https://hive.apache.org/

21. Armour, D.J., Kalki, J.: Determining computer system usage from logged events. U.S. Patent No. 8,185,353, 22 May 2012

22. Hochreiter, S., Schmidhuber, J.: Long short-term memory. Neural Comput. **9**(8), 1735–1780 (1997)

23. Hochreiter, S., Schmidhuber, J.: LSTM can solve hard long time lag problems. In: Advances in Neural Information Processing Systems (1997)

24. Ma, X., et al.: Long short-term memory neural network for traffic speed prediction using remote microwave sensor data. Transp. Res. Part C: Emerg. Technol. **54**, 187–197 (2015)

25. Tensorflow. https://www.tensorflow.org/

Detection and Analysis of Life Style based Diseases in Early Phase of Life: A Survey

Pankaj Ramakant Kunekar, Mukesh Gupta, and Basant Agarwal[✉]

Department of Computer Science and Engineering,
Swami Keshvanand Institute of Technology Management & Gramothan,
Jaipur, Rajasthan, India
kunekarpankaj30@gmail.com,
{mukeshgupta,basant}@skit.ac.in

Abstract. In India there is big transition in life style due to industrialization and western influence. Life style diseases are on surging rate with it affect across all age borders. According to a recent health survey almost 60% of all death reported in India are due to life style and non-communicable diseases (NCD) with life style contributing the major part in it. Early screening and predictive analysis is way forward to put a break on surging life style diseases. In this work a survey on scalable technologies assisting for early screening and predictive analysis for life style diseases is done. Each of technologies is analyzed in perceptive of multiple parameters like effectiveness, cost, convenience, adaptability rate etc. and open areas identified for further research.

Keywords: Disease prediction · Machine learning · IOT · Big data

1 Introduction

Over last two decade Life style diseases have been on a surging rate and have replaced traditional health risk factors in India. Traditional risk factors like unsafe water, sanitation and child malnutrition are now controlled due to proactive measures from Governments and NGO's. Health loss due to metabolic risk factors has doubled. Among the metabolic risk factors high blood pressure, blood sugar, cholesterol, poor diet and alcohol use are the major contributors. According to Public Health Foundation of India (PHFI) report "Metabolic risk factors like high blood pressure, blood sugar and cholesterol, along with unhealthy diets and smoking are responsible for about 5.2 million premature deaths in India every year".

Some of most common types of life style diseases in India are Obesity, type 2 diabetes, blood pressure, cancer, Arteriosclerosis, stroke and Cirrhosis. Unhealthy living habits, sedentary lifestyle, stress and pollution are the main reasons for rapid surge in Life style diseases. Early screening and proactive predictive analysis of life style diseases at early onset and timely precautions are the best way to protect against life style diseases. Monitoring at health centers and regular body checkup test are not effective as it is time consuming and people lose interest in waiting for checkups. Motivated by fact of surging life style diseases, we attempt to analyze current diagnosis method for earlier detection of disease as to reduce the health impacts (Fig. 1).

© Springer Nature Singapore Pte Ltd. 2019
A. K. Somani et al. (Eds.): ICETCE 2019, CCIS 985, pp. 55–69, 2019.
https://doi.org/10.1007/978-981-13-8300-7_6

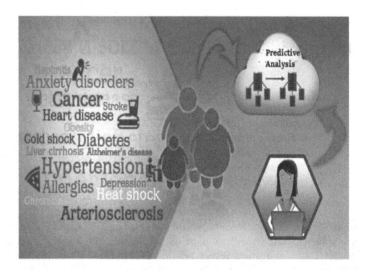

Fig. 1. Technology assisted screening [1]

Technologies like non intrusive monitoring with sensors, IOT (Internet of Things) and predictive analysis aids for effective screening. With these technologies, personalized health care service is at the disposal of users. The technologies collect data from users in a non intrusive manner and send to server where predictive analysis on data yields valuable insights into the health of the users and helps in preventive care at earlier onset [30, 33].

In this work the current state of methods for technology assisted screening and health care management for diseases in general is analyzed in detail and the research gaps for better adaptation of these technologies for Life Style diseases in Indian conditions are identified. The importance of this work is that the research gaps identified in this work will motivate for effective solution design for proactive and timely monitoring of life style diseases at best cost and adaptiveness needed for Indian conditions.

2   Survey

The existing solutions for technology assisted health care management are discussed in detail in this section. The study is conducted in following categories

A. Predictive Analysis
B. Feature Analysis
C. Data collection Analysis
D. Big Data Analysis

A. Predictive Analysis

In [2] author proposed disease risk prediction model using convolutional neural network (CNN) which works based on structured and unstructured data. The advantage of this approach is that it handles both unstructured and structured data. The features dimension is high in this approach and CNN complexity is high, it could have been reduced using feature dimension reduction techniques [32].

In [3] authors analyzed the different deep learning models applied in health care informatics. The advantage in deep learning models is that high level features and semantic interpretation is learnt automatically. Six different architectures of deep learning models are analyzed and their pros and cons identified. The analyzed architectures were

1. Deep Neural Network
2. Deep Autoencoder
3. Deep Belief Network
4. Deep Boltzmann Machine
5. Recurrent Neural Network
6. Convolutional Neural Network

The common issue in deep learning models is:

Large volume of training data set is needed and errors in one layer get cascaded to multiple layer. The initial set of configuration hyperparameters that control the architecture of deep learning models, especially the size, the number of filters, depth etc. is still a blind exploration process [31]. Due to error cascading, even some wrong input is enough to affect the accuracy of Deep learning models, so proper outlier detection in the input data sets provided to Deep learning models is a must. In [4] author deceived the deep learning models by applying evolutionary algorithms on input and made the deep learning model to classify erroneously.

In [5] author modeled the health of population as a dynamic system and predicted the time evolution of the new diagnosis of cancer and chronic disease. In this model population is segmented into three classes

N – The number of person those without any of the seven chronic conditions and occupy least severe health state
D - The number of patients with one or more of the listed chronic diseases
C- The number of cancer patients

Over the period of time people pass from one state to another say some people develop chronic conditions and others develop cancer. The model calculates and forecast the count of population after a set period of time. The model was evaluated against US health statistics and deviation of model from actual observation was less than 3%. Even though this model is done for limited diseases, this kind of models must be developed for Indian conditions, so that outbreak and surge rate can be acutely forecasted and precautionary mechanisms can be developed. The concept in this model can be developed for individuals based on his life style transitions for disease forecast in advance.

In [6] authors correlated Mindful Attention Awareness Scale (MAAS) score for prediction of Type 2 Diabetes. The score used parameters like age, sex, race/ethnicity, family history of diabetes, and childhood socioeconomic status. Based on these parameters score is calculated reflecting glycemic control and self-care. An increased value of MAAS score is an indication of high risk for Type 2 Diabetes. Even though

MAAS score does not fully indicate the care and control for Type 2 Diabetes, a health score like this would be very important for preventive health care. As of now there is no health score like this for life style diseases and most suited for preventive health care.

In [9] authors proposed a modeling technique to extract medical concept relations from unstructured documents called Code2Vec. The model applies the concept of Distributional Hypothesis. It declares words appearing in similar contexts tend to have similar semantic meaning. The diagnosis and procedures are replaced with International medical codes in the unstructured document. A co-occurrence matrix is created with vertices corresponding to diagnosis and edges corresponding to sum of co-occurrences scores modeled in a configurable window. From the co-occurrence matrix cost function minimization is done according to Pennington [10].

$$J = \sum_{i,j=1}^{V} f(X_{ij}) \left(w_i^T w_j + b_i + b_j - \log(X_{ij}) \right)^2$$

Where V is the number of distinct codes, co-occurrence between medical code (between i and j) is given as X_{ij}. $f(X_{ij})$ is the weight function on co-occurrence score. w_i, w_j are embedding vectors associated with i^{th} and j^{th} medicinal codes, b_i, b_j are the bias attached towards the medicinal codes (for i and j).

Using this method the associations between different diagnoses can be learnt and prediction of new disease based on past medical history can be done. Even though the accuracy is less in this way of prediction, it can be used to reduce the search space of screening process.

In [13] authors applied a hybrid system for diagnosis of Type 2 Diabetes. The novelty in this approach is that after building a Support Vector Machine (SVM) classifiers using labeled training set, rules are extracted from the SVM using SQRex-SVM and Eclectic Rule Extraction. The first approach extracts rules using modified sequential covering algorithm and ordered search of most discriminatory features. The second approach extracts using C5 decision trees. A sample of decision rules learnt for Type 2 Diabetes is given below

If FBS > 106.2 then diabetic

If waist circumference > 91 and BPDIAS > 90 then diabetic

These kinds of rules makes screen process easier for non expert and especially in India where non expert staffs are employed for preliminary screen, this approach would be very useful. The accuracy of SVM is low compared to Deep learning, so if Deep learning results can be used to build the decision rules, the accuracy of classification will be higher.

In [14] author proposed a novel approach of personalized classifier model for each patient instead of one classifier for all adopted in previous works. Each of biomarkers for disease classification are analyzed against local weighting rule and based on it classifier model is tailored for each patients requirement. The flow of process in this approach is shown below (Fig. 2):

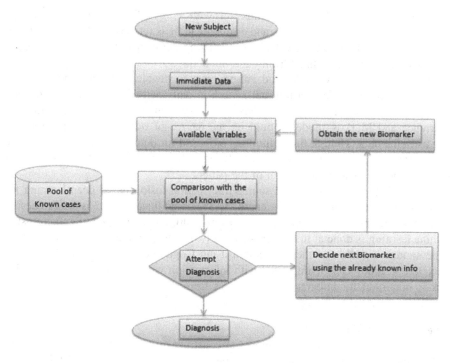

Fig. 2. Flowchart for [14]

The salient feature in this approach is instead of choosing all features for a patient; the subset of features which works well for classification of diseases for that patient is experimented. The author proposed the solution for Alzheimer disease but the take away is can it be applied for life style diseases too. Even if the classifier is not personalized for each user, clustering of user can be done and personalized classifier can made for each cluster. It suits for Indian conditions due to its demographics variants.

In [16] authors proposed a solution for hyper-parameter selection problem in Deep Learning models. To reduce the search space progressive sampling, filtering and fine tuning is done. Quick trails with limited samples of data is done to filter unworthy combinations and iterated over promising ones to fine tune the parameters. The approach has high time complexity and the one of demerit is that it does not address the data sampling. When a limited data set is used for trails, the sample quality is important and the paper failed to address the sample quality.

In [18] authors proposed a state transition based health model learnt from medical history using end to end deep dynamic neural networks. The future risk is predicted based on Historical admissions, attention and illness dynamics as below in [18]

$$P(y|u_{1..n}) = P\big(nnet_y(pool\{C - LSTM(u_{1:n})\})\big)$$

Where $u_{1..n}$ is the historical admission with $C - LSTM$ for modeling illness dynamics. The approach is complex and error cascading is uncontrolled. But the idea of modeling the health as state machine and predicting the next state based on history is very important for personalized health care modeling.

In [21] used Markov model and entropy rate to develop dementia management system based on the detecting unusual pattern in daily activities of dementia patient. Entropy rate was used for on collected daily activity to quantify the degree of random occurrence and Markov model is built on the degree of occurrence to find unusual patterns in the data. Though this solution is proposed for dementia, the concept can be applied for post disease management of life style diseases by building same kind of models on activities and medication of the patient.

In [22] authors proposed multi source information fusion based deep learning for health risk prediction. Situation awareness (SAW) based fusion model is used to design a three layer framework. Data acquisition from multiple sources and alignment using machine learning is done at first layer. At level two feature extractions is done for each of the sources signals and then integrated using Cognitive Neural Network. Disease prediction accuracy can be improved with information with multiple sources and this work is important for integrating the features from different sources for designing a deep learning model with high accuracy.

B. Feature Analysis

In [2] author used latent factor model to construct missing data, so as to make the data complete for deep learning. The patient's features are represented in a two dimensional matrix $R_{m \times n}$ where number of patients is m and the number of features is n. The matrix is approximated as below with assumption of k latent factors

$$R_{mXn} \cong P_{mXk} \, Q_{nXk}^{T}$$

Each element value in the matrix was written as

$$r_{uv} = p_u^T q_v$$

Where p_u^T indicates patients preference to potential factors and q_v is the attribute feature vector.

Using stochastic gradient descent method missing value is fitted as

$$\min \left(\sum_{u,v} \left(r_{uv} - p_u^T q_v \right)^2 + \lambda_1 p_u^2 + \lambda_2 q_v^2 \right)$$

Patient features have some degree of sparseness and this work attempted to solve it as a error minimization optimization. But the approach does not work well in case of low volume of dataset.

In [7] authors used laboratory results over years 2005 to 2013 in southeastern Pennsylvania, of 4.1 million individuals to identity the attributes describing the health status of individual. Totally 42,000 variables were identified. The features were collected from Laboratory tests, ICD9 (International Classification of Diseases) history, NDC (National Drug Code) medication history and Health care utilization records. From this initial set of attributes L1 regularization method is done to filter irrelevant attributes and detect most relevant parameters for Type 2 diabetes prediction. This kind of exhaustive datasets is not available for Indian conditions but the methods applied to derive the models (San-Antonio Model) can be used. Due to demographics varsity the models derived from the data set cannot be applied for Indian conditions.

In [8] authors applied concepts from Ayurveda medical system for modeling the life style diseases. Ayurveda relates the tridosha parameters in body to probability of metabolic and chronic diseases. Tridosha parameters is determined at the time of birth and the parameters vary based on diet consumed by body and lifestyle. Tridosha proportion variations causes diseases. Therefore, assessment of Tridosha analysis will help in diagnosis and screening of life style diseases.

In [11] authors proposed a solution to prevent medication errors using a unsupervised learning method DDC-Outlier (Density-Distance-Centrality). The method is able to detect overdose and under dose from electronic prescriptions.DDC models the relation between the dose and frequency as a graph $G = (V, E)$ where V is unique dose/frequency and E is the edges connect the pair of vertex. From the matrix, Weighted Page rank is used to centrality score of each vertex. Considering the mean centrality score as threshold, the dose corresponding to node whose centrality scores less than threshold is detected as outlier. The centrality score is calculated as

$$WPR(u) = \sum_{v \varepsilon B_u} W(v, u) \frac{WPR(v)}{N_v}$$

Where B_u are the neighbor nodes of u with links to it and N_v is the number of neighborhoods.$W(v, u)$ is the weight of out link from v to u. This method of detecting the outlier is effective and can be used as preprocessing step to filter outlier in data who deviation from average behavior is suspicious.

In [15] author trained the image pixels as whole in Cognitive Neural Network for skin disease classification. Features extraction methods don't work well due micro scale variations in skin lesions, so entire image is used for training and classification. A salient observation in this approach is that for case of fine grained variability, it is better to train as whole without extracting the features.

In [17] authors proposed a multiple kernel learning (MKL) based approach for feature sub set selection. The relevancy between the features and class labels is measured by using the optimization function of MKL. For each feature in the feature set J value is calculated as

$$J = \underset{d_m}{\underbrace{min}} \, \underset{\alpha}{\underbrace{max}} \, W(\alpha, d_m) \; such \; that \; \sum_{m=1}^{M} d_m = 1$$

The features are ranked based on J by sorting in ascending order. Top n features selected from the sorted list. The time complexity is high due to use multiple SVM models for calculating the W. Statistical correlation based feature selection methods are comparatively faster.

In [19] authors proposed a general purpose patient representation from electronic health records using unsupervised deep learning. Preprocessing is done on electronic health records to identify clinically relevant phenotypes and grouped into patient vectors. The patient vector is passed to deep feature learning model to extract high level general descriptors. These high level general descriptors are called as deep patient, the classifier can be trained using the deep patient. The advantage in this approach is the classification becomes faster and deep features representation is very secure and can be shared without leaking any privacy.

In [20] author proposed a similar approach like [19] for deep representation of features. The solution applied principal component analysis (PCA) on the patient data for dimension reduction and gets a reduced representation. From this representation, Auto encoder neural network is applied to learn the feature set. Auto encoder is a feed forward neural network to learn effective encoding of input data. It is comparatively faster than [19] due to use of PCA for dimensionality reduction.

C. Data Collection Analysis

Currently portable health monitoring devices fixed in watches, bracelets, and belts collect human physiological signals. These data collection methods have following problems:

1. Shorter Life span typically 2 to 3 months
2. Higher False judgments
3. Limited acquisition
4. Noise interference and erroneous

The sensors attached to various consumables are

1. Smart Glasses
2. Smart Watch
3. Smart Bracelet
4. Smart Finger
5. Smart Ring
6. Smart Belt
7. Smart Pants
8. Smart Socks
9. Smart Shoes
10. Bluetooth Key tracker
11. SGPS/GPRS Body Control
12. Smart Shirt

In [12] author introduced the concept of wearable clothing. The micro sensors part of cloth fabric collects human physiological signals. Wearable clothing reduce the problems in traditional wearable device for data acquisition. Dry textile electrodes are integrated into clothing to monitor the signal in non pervasive manner. The advantage in this wearable clothing is that it is soft, breathable, and washable. It is durable against prolonged use compared to traditional electrode. This method of data collection is best as the errors are limited and life span of device is higher.

D. Big Data Analysis

In [23] authors applied Apache Spark Big data platform for health status prediction. Apache Spark is an open source big data platform. It is faster than well-known Hadoop platform. To enhance the performance of Hadoop, Apache Spark implements in-memory processing for maintaining all data in memory instead of disk memory. Apache Spark also uses Resilient Data set for efficient parallel computations. In additions Apache Spark has a streaming extension to process live data streams. The comparison of different open source big data platforms is given in (Table 1). It can be found that Apache Spark is a better platform for Big data analytics for health care management.

Table 1. Comparison of open source big data platforms

No	Features	Hadoop	Spark	Flink	Storm	Samza
1	Batch processing	Supports	Supports	Supports	Don't support	Don't support
2	Stream processing	Don't support	Supports	Supports	Supports	Supports
3	Scalability	Scalable but not as that of spark	Highest scalable, Largest known cluster size is 8000	Scalable but not as that of Spark	Scalable but not as that of Spark	Scalable but not as that of Spark
4	Supporting languages	Java, C, C++, RubyGroovy, Perl and Python	Java, Scala, Python and R	Java, Scala, Python and R	Java	Java, Scala
5	Latency	Higher latency than spark and flink	Lower latency than Hadoop	Low latency similar to Spark	Lowest latency among all platforms	Low latency
6	Cost	Can run on less expensive hardware's	Lot of RAM needed for in-memory computation, it increases the hardware cost	Lot of RAM in needed and this increases the hardware cost	Lot of RAM in needed and this increases the hardware cost	Lot of RAM in needed and this increases the hardware cost
7	Security	Due to Kerberos authentication, security is difficult to manage	Simple authentication supported by shared secret	Uses Hadoop/Kerberos Authentication and difficult to manage	User Kerberos based authentication Difficult to manage	Apache Hadoop Yarn is used for authentication which is comparatively easier to manage than Kerberos

3 Issues

The summary of survey is presented in Table 2. The open issues in the existing solutions for technology assisted health informatics is listed below

1. Deep Learning models can be deceived.
2. Best Models for missing data Imputation
3. Lack of Personalization in Forecast Models
4. Lack of Health score models
5. No works on Features from Indian Traditional medicine systems for Deep Learning

Table 2. Summary of feature and predictive analysis

Solution	Methodology	Issues	Take away
Feature analysis			
[2]	Latent factor model to find missing data	Does not work for large dataset	Use of gradient descent method to reduce error in missing data estimation
[7]	Fusion of data from multiple sources	The data is not specific to Indian conditions	Using ICD and NDC code for efficient representation
[8]	Ayurveda medical system for modeling diseases	Data set is limited	Use of tridosha parameters
[11]	Detect outliers in data	Works only on statistical data and can be deceived	Can be used to preprocess data set and remove outliers
[15]	Image features for disease classification	Complexity of classifier is high	Fine grained variability can be accounted
[17]	Feature subset selection using multiple kernel learning	Time complexity is high	Statistical correlation methods are comparatively faster and can be used
[19]	Deep learning based feature extraction	No issue	The advantage in this approach is the classification becomes faster and deep features representation is very secure and can be shared without leaking any privacy
Predictive analysis			
[2]	CNN based disease risk prediction	Features dimension & CNN complexity is high	Works for unstructured data.
[3]	Six different architectures of deep learning models are analyzed and their pros and cons identified	Deep learning models can be deceived	Data preprocessing must be done to remove outliers
[5]	Population disease model is proposed to model the evolution of disease count over time	Works only for two disease	Can be adapted for life style diseases

(continued)

Table 2. (*continued*)

Solution	Methodology	Issues	Take away
Feature analysis			
[6]	MAAS score for Type 2 Diabetes.	Scoring misses many features of diseases and not accurate	Can be adapted for life style diseases
[9]	Modeling technique to extract medical concept relations from unstructured documents	Data missing over time intervals is not accommodated	Can be used for predictive rule learning from life style diseases
[13]	Hybrid system for diagnosis of Type 2 Diabetes	The accuracy is lower than deep learning models	Decision rules are provided as output from SVM model and these rules can be used for easy screening process
[14]	Personalized classifier model for each patient	Works only for Alzheimer disease.	Idea of personalized classifier model can be adapted for life style diseases
[16]	Solution for hyper-parameter selection problem in deep learning models.	Sample quality is not considered	The deep learning configuration parameters can be identified using this work
[18]	State transition based health model	Complex approach and error cascading is uncontrolled	Idea of continuous state model for each patient must be adapted for life style diseases.
[21]	Markov model and entropy rate based dementia management	The approach fails if the sensor data fails. There is no recovery strategy proposed	Can be adapted for post disease management of life style diseases
[22]	Multi source information fusion based deep learning	Error cascading is not considered at fusion level	Idea of use multiple source of information to build disease prediction will have high accuracy.

4 Discussion on Open Issues

Issue 1: Deep Learning models are the latest classification methods that can be applied in health care informatics. Due to its approaches in learning the features automatically and incorporation of semantic information, it is best for health care informatics domain. But the problem in this deep learning model is that it can be deceived easily either purposely or without by contradictory training data or by mild changes in data with use of evolutionary algorithms. In Indian conditions this is quite common as most of the data collected through survey from patients will have contradictions and the model learnt from this data may not be accurate. Thus it requires efficient outlier detection models to process the datasets.

Issue 2: Missing data is a general characteristic of health care dataset. Various statistical based methods are proposed for missing data imputation like Expectation Maximization. But for health care domain, demographics association modeling has to be applied in addition to statistics for missing data imputation. Without it the data replaced will not be accurate enough and when this data is used for deep learning the error cascaded in multiple layers are cascaded and it affects the classification accuracy.

Issue 3: Personalized models are needed better prediction in case of preventive health care and post disease care management for life style diseases. The personalization can be in form of specialized classifiers or state based health transition models customized for each patient or cluster of patients with properties such as same demographics, same race etc.

Issue 4: Like MAAS Score, there is lack of scores for life style diseases. The score can be calculated based on food, exercise, clinical and other parameters. The score is a valuable indicator in preventive health care for patients can take precautionary measures. Typical BMI score is not a good indicator as it can be seen that for stress induced life style diseases, the BMI values are still satisfactory.

Issue 5: The current features used for disease classification are only clinical features laid out by modern Allopathic medical system. Traditional Indian medical systems have a proven track record of over 5000 years and have identified parameters for general wellness. There is no preventive health care technique using these parameters separately or in fusion with clinical parameters for prediction of diseases. The parameters identified using traditional systems can be fused with clinical parameters and deep learning models can be constructed using the approach similar to [22] for increased accuracy of screening process during prevention and analyzing the curing abilities during after disease management.

The summary of existing works for diagnosis of life style diseases is below (Tables 3 and 4).

Table 3. Diagnosis of life style diseases

Work	Disease	Methodology	Performance
[24]	Diagnosis of type 2 diabetes	Random Forest and Support Vector Machine	Accuracy = 94.2%
[25]	Fasting plasma glucose status	Linear Regression and Naïve Baiyes Classifier	Accuracy = 74%
[26]	Hypoglycemic episodes for type 1 diabetes children	Neural Network	Accuracy = 78%
[27]	Peripheral vascular occlusion	Support Vector Machine	Accuracy = 100%
[28]	Healthy control + cardiomyopathy + myocardial infarction	Least Square Support Vector Machine	Accuracy = 90.34%
[29]	Heart Beat	Support Vector Machine	Accuracy = 98.3%
[30]	Obesity	Random forest and Neural Networks	Accuracy = 87%

Table 4. Features used for diagnosis

Work	Disease	Features
[24]	Diagnosis of type 2 diabetes	China health and nutrition survey data
[25]	Fasting plasma glucose status	Data from 4870 subjects (2955 females and 1915 males) aged 31–90 years from the Korean Health and Genome Epidemiology study database (KHGES). The KHGES routinely measures weight, height, and circumferences of regional sites of the body by trained observers according to standardized protocols
[26]	Hypoglycemic episodes for type 1 diabetes children	ECG signals are used
[27]	Peripheral vascular occlusion	Photoplethysmography (PPG) pulses of feat is used
[28]	Healthy control + cardiomyopathy + myocardial infarction	ECG signals are used
[29]	Heart Beat	ECG signals are used
[30]	Obesity	Age, Gender, Height, Weight, BMI

5 Conclusion

The paper summarizes the current works in technology enabled health care management. The survey focused on core technologies in categories of Predictive Analysis, Feature Analysis, Data collection Analysis and Big Data Analysis. More importance was given to Predictive Analysis and Feature Analysis; as improvement in these areas can contribute to a bigger gain in health care informatics. Data collection Analysis is focused mainly on identification of sensors technology to collect real time signals useful for prediction and management of life style diseases. Big Data analysis is focused on identification of latest data analytics platform for increased scalability of processing large volumes of data. The paper identified issues in Predictive Analysis and Feature Analysis and proposed prospective solutions for solving them in brief. As a future work, the identified prospective solutions will be detailed and experimented for their effectiveness in solving the issues.

References

1. Indian Council of Medical Research (ICMR): India State-Level Disease Burden Study report (2017)
2. Chen, M.: Disease prediction by machine learning over big data from healthcare communities. IEEE Access **5**, 8869–8879 (2016)
3. Ravi, D., Wong, C.: Deep learning for health informatics. IEEE J. Biomed. Health Inf. **21**(1), 4–21 (2017)

4. Nguyen, A, Yosinski J, Clune J.: Deep neural networks are easily fooled: high confidence predictions for unrecognizable images. In: Computer Vision and Pattern Recognition (CVPR 2015). IEEE (2015)
5. Kuriyan, J., Cobb, N.: Forecasts of cancer and chronic patients: big data metrics of population health, Cornell University library, pp. 1–26 (2013)
6. Wang, A.C.A.: Big data analytics as applied to diabetes management. Eur. J. Clin. Biomed. Sci. **2**(5), 29–38 (2016)
7. Razavian, N., Blecker, S., Schmidt, A.M.: Population-level prediction of type 2 diabetes from claims data and analysis of risk factors. Big Data **3**(4), 277–282 (2015)
8. Dey, S., Pahwa, P.: Prakriti and its associations with metabolism, chronic diseases, and genotypes: possibilities of new born screening and a lifetime of personalized prevention. J. Ayurveda Integr. Med. **5**(1), 15 (2014)
9. Christensen, T., Frandsen, A.: Machine learning methods for disease prediction with claims data. In: IEEE International Conference on Healthcare Informatics (2018)
10. Pennington, J., Socher, R., Manning, C,D.: Glove: global vectors for word representation. In: Empirical Methods in Natural Language Processing (EMNLP) (2014)
11. dos Santos, H.D., Ana Helena, D.P.S., Ulbrich, A.H.: DDC-outlier: preventing medication errors using unsupervised learning. J. Med. **14**(8), 874–881 (2018)
12. Ma, Y., Wang, Y., Yang, J.: Big health application system based on health Internet of Things and big data. Special Section on Healthcare Big Data, October 2016
13. Nahla, H., Barakat, M.N., Bradley, A.P.: Intelligible support vector machines for diagnosis of diabetes mellitus. IEEE Trans. Inf Technol. Biomed. **14**(4), 1114–1120 (2010)
14. Escudero, J., Ifeachor, E.: Machine learning-based method for personalized and cost-effective detection of Alzheimer's disease. IEEE Trans. Biomed. Eng. **60**(1), 164–168 (2013)
15. Esteva, A., Kuprel, B.: Dermatologist-level classification of skin cancer with deep neural networks. Letter Macmillan Publishers Limited, part of Springer Nature (2017)
16. Luo, G.: PredicT-ML: a tool for automating machine learning model building with big clinical data. Health Inf. Sci. Syst. **4**(1), 5 (2016)
17. Du, W.: A feature selection method based on multiple kernel learning with expression profiles of different types. Biodata Min. **10**(1), 4 (2017)
18. Pham, T., Tran, T., Phung, D., Venkatesh, S.: Predicting healthcare trajectories from medical records: a deep learning approach. J. Biomed. Inf. **69**, 218–229 (2017)
19. Miotto, R., Li, L.: Deep patient: an unsupervised representation to predict the future of patients from the electronic health records. Sci. Rep. **6**, 26094 (2016)
20. Zhang, D., Zou, L.: Integrating feature selection and feature extraction methods with deep learning to predict clinical outcome of breast cancer. IEEE, March 2018
21. Enshaeifar, S., Zoha, A., Markides, A.: Health management and pattern analysis of daily living activities of people with dementia using in-home sensors and machine learning techniques. PLoS ONE **13**(5), e0195605 (2018)
22. Zhong, H., Xiao, J.: Enhancing health risk prediction with deep learning on big data and revised fusion node paradigm. Sci. Prog. **2017**, 18 (2017)
23. Nair, L.R., Shetty, S.D.: Applying spark based machine learning model on streaming big data for health status prediction. Elsevier, March 2017
24. Han, L., Luo, S., Yu, J., Pan, L., Chen, S.: Rule extraction from support vector machines using ensemble learning approach: an application for diagnosis of diabetes. IEEE J. Biomed. Health Inform. **19**, 728–734 (2015)
25. Lee, B.J., Ku, B., Nam, J., Pham, D.D., Kim, J.Y.: Prediction of fasting plasma glucose status using anthropometric measures for diagnosing type 2 diabetes. IEEE J. Biomed. Health Inform. **18**, 555–561 (2014)

26. Ling, S.H., San, P.P., Nguyen, H.T.: Non-invasive hypoglycemia monitoring system using extreme learning machine for type 1 diabetes. ISA Trans. **64**, 440–446 (2016)
27. Li, C.M., et al.: Synchronizing chaotification with support vector machine and wolf pack search algorithm for estimation of peripheral vascular occlusion in diabetes mellitus. Biomed. Signal Process. Control **9**, 45–55 (2014)
28. Tripathy, R.K., Sharma, L.N., Dandapat, S.: A new way of quantifying diagnostic information from multilead electrocardiogram for cardiac disease classification. Healthcare Technol. Lett. **1**, 98–103 (2014)
29. Oster, J., Behar, J., Sayadi, O., Nemati, S., Johnson, A.E., Clifford, G.D.: Semisupervised ECG ventricular beat classification with novelty detection based on switching Kalman filters. IEEE Trans. Biomed. Eng. **62**(9), 2125–2134 (2015)
30. Montañez, C.A.C.: Machine learning approaches for the prediction of obesity using publicly available genetic profiles. In: International Joint Conference on Neural Networks (IJCNN) (2017)
31. Singh, R., Gahlot, A., Mittal, M.: IoT based intelligent robot for various disasters monitoring and prevention with visual data manipulating. Int. J. Tomogr. Simul. **32**(1), 89–99 (2019)
32. Jain, G., Sharma, M., Agarwal, B.: Spam detection in social media using convolutional and long short term memory neural network. Ann. Math. Artif. Intell. **85**(1), 21–44 (2019). https://doi.org/10.1007/s10472-018-9612-z
33. Bhatnagar, V., Poonia, R.C.: Design of prototype model for irrigation based decision support system. J. Inf. Optim. Sci. **39**(7), 1607–1612 (2018). https://doi.org/10.1080/02522667.2018.1507763

3D Trajectory Reconstruction Using Color-Based Optical Flow and Stereo Vision

Rachna Verma[1(✉)] and Arvind Kumar Verma[2]

[1] Department of CSE, Faculty of Engineering, J.N.V. University, Jodhpur, Rajasthan, India
rachnaverma@jnvu.edu.in
[2] Department of PI, Faculty of Engineering, J.N.V. University, Jodhpur, Rajasthan, India
akverma.pi@jnvu.edu.in

Abstract. Automatic trajectory estimation of a moving object in a video is one of the most active research areas of computer vision, which finds many practical applications, such as development of sport playing robots, predicting trajectory for avoiding obstacle collision, automatic navigation of driverless vehicles, monitoring target hitting, etc. However, most of the work reported in literature only considers monocular videos. Due to the availability of low price stereo cameras, many applications take their advantages by incorporating depth information. In this paper, the 3D trajectory of a primary-color (red or green or blue) object is estimated using color-based optical flow and stereo vision. The purpose of using stereo vision is to gain depth information for generating 3D trajectory. The system has been tested on many stereo videos and experimental results are quite accurate. Besides, the low computation time required for finding depth of the tracked path makes it suitable for real time applications.

Keywords: Object detection · Object tracking · Optical flow · 3D trajectory · Stereo vision

1 Introduction

Trajectory estimation has received a considerable attention by researchers due to many practical applications, such as predicting trajectory for avoiding obstacle collision, location estimation of a robot, tracing the trajectory of a moving vehicle for lane violation, etc. Trajectory generation is a two-step process: (1) detect the object in the initial frame of a video and (2) then locate the object in the subsequent frames.

It is established that the 3D trajectories of moving objects in a scene is impossible from monocular videos without prior assumptions about the scene, which restricts its practical utilities [1]. However, it is possible to generate 3D trajectories by combining stereo vision with any object tracking technique. In stereo vision, two views of the same scene are captured by a pair of cameras and the disparities of scene points in the captured images are used to calculate depths of scene points, using the principle of triangulation. The calculated depths are utilized to construct the 3D trajectories of the objects of interest.

© Springer Nature Singapore Pte Ltd. 2019
A. K. Somani et al. (Eds.): ICETCE 2019, CCIS 985, pp. 70–80, 2019.
https://doi.org/10.1007/978-981-13-8300-7_7

Optical flow [2] is a widely used technique to track moving objects in a static background with constant brightness in mono-color videos. However, it performs poorly in a dynamic background. This paper combines color feature with the optical flow method for tracking the object in a dynamic background. The work presented in this paper is limited to track only single primary (red or green or blue) color object, and it is an extension of the work reported in [3], which is limited to monocular videos.

In an effort to generate fast and accurate 3D trajectory, this paper combines the color-based optical flow and the stereo vision concept. The optical flow is used to locate the centroids of the object of interest in both frames of the stereo video, separately. Instead of using dense disparity map, it is proposed to use the disparity of the centroids of the object to generate 3D location of the object. This makes the whole process very efficient and 3D trajectory can be generated in real time.

The organization of the paper is as follows: Sect. 2 presents the related work. Section 3 discusses color-based optical flow for object detection and tracking. Section 4 describes stereo vision system and perspective projection equations for generating 3D trajectory. Section 5 presents the proposed method and experimental results, and finally, Sect. 6 concludes the paper.

2 Related Work

In a typical object tracking system, some discriminating features of the object, such as color, texture, motion, edges, etc., are used to extract the object of interest in a video frame and then the location of the object is found in the successive frames to track it. Object tracking has many practical applications and has been an active research area from the last two decades. Many methods have been reported for object tracking, such as background subtraction, Kalman filter, particle filter, optical flow, etc. Extensive reviews of the past research in object tracking are well presented in [4–6].

Optical flow is a technique to find high motion areas in two successive video frames. It finds moving objects from two successive video frames by observing the motion of intensities of pixels, under the assumption that the intensities of objects of interest are not changing from frame to frame. Based on the assumption of the conservation of brightness intensity, the two most widely used methods are: Lucas-Kanade and Horn-Schunck, refer [7] for more details about these methods. Choi et al. [8] used optical flow to find the feature points of the object for tracking contour of objects. Xiang et al. [9] computed optical flow from sample points of the templates for tracking and used quality of optical flow to decide whether to keep tracking the target or not in a multiple object tracking scenario. Beaupre et al. [10] used optical flow for object tracking and detecting occlusion of objects moving in opposite direction.

Most of the research for tracking is limited to 2D trajectory generation. For extending 2D trajectory to 3D trajectory generation, many researchers proposed to use stereo vision. Tsutsui et al. [11] used stereo vision and optical flow to create 3D trajectory. Their method uses multiple cameras for handling occlusion. Zhou et al. [12] used stereoscopy in combination with camshaft algorithm to reconstruct 3D trajectory. They used dynamic programming for correspondence matching and disparity calculation. Park et al. [13] used multiple perspective projections of a scene to reconstruct

3D trajectory of moving objects. They used coordinate independent basis vectors derived from stationary areas of the scene, which reduces computational complexity of trajectory reconstruction.

3 Color-Based Optical Flow

The limitation of the optical flow method is that it detects all moving objects in a static scene. In order to detect single primary-color moving object in a dynamic scene, the optical flow method is combined with a color-based detector [14], discussed next that efficiently tracks an object of primary-color shades.

3.1 Primary-Color Based Object Detection and Tracking

For tracking a primary-color object (color is widely used feature for tracking), the formula proposed in [14] is used, which handles dull and bright color objects of different shades of primary-color. The formula, called color-component-ratio (CCR), converts a RGB video frame to a CCR video frame by enhancing the primary-color of interest and suppressing other colors. For example, to extract red color shades in an image, Eq. (1) is used that suppresses colors other than red.

$$IR = \frac{R * R}{G * B} \tag{1}$$

where R = red value, G = green value, and B = blue value of the pixel intensity.

Following additional operations are performed to further enhance the red color:

(a) The pixels with R/G and R/B values greater than a specified threshold (T1) are considered for further processing.

(b) Further, the pixels with R values greater than a specified threshold (T2) are only considered.

Finally, for extracting a red color object, Eq. (2) is used to calculate CCR at each pixel location:

$$\text{CCR} = \left(\text{IR*}\left(\frac{R}{G} > T1\right) * \left(\frac{R}{B} > T1\right) * (R > T2) \right) > TH \tag{2}$$

T1, T2 and TH are the thresholds. TH is set to value equal to 2. Based on the various experimentation with different real videos, the values of the thresholds which are found to be most effective in discriminating the primary-color object from the background are T1 = 1.5 and T2 = 70 for bright color objects, whereas T1 = 1 and T2 = 50 are used for dull color objects. For example, consider a case of detecting a bright red color object. All the pixels of this bright red object will have higher value of red component and higher value of R/B and R/G ratio. Whereas, dull color objects of red color shades have lower value of red color component and lower value of R/B and R/G ratio. Similar is the case with bright green and dull green color objects and bright blue and dull blue color objects. The result of Eq. (2) when applied to various test scenarios for extracting a primary-color object is given in Fig. 1. The advantage of using this formula is that it facilitates automatic detection of primary-color objects in a static as well as in a dynamic background.

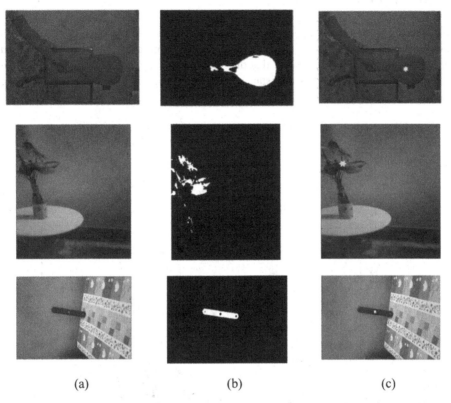

<table>
<tr><td>(a)</td><td>(b)</td><td>(c)</td></tr>
</table>

Fig. 1. (a) Original frame, (b) the CCR image, (c) the detected object in the frame with its centroid, marked as a yellow star. (Color figure online)

3.2 Optical Flow

Optical flow is the motion of intensity values of pixels, assuming that intensities of pixels of interest are constant, between two consecutive video frames [15]. The assumption of constant intensity is very impractical and limits the use of this method. Due to this assumption, the method cannot handle object tracking in a dynamic background. For more details and mathematical formulations of optical flow methods, readers can refer [2] and MATLAB online help.

The traditional optical flow is capable of tracking apparent intensity motion only in intensity images (gray video) and cannot handle colored images, as the intensity images generated by the widely used intensity conversion scheme has no discriminating power to highlight a particular color. To extend the capability of the optical flow method to handle RGB images and dynamic background, [3] combines the CCR formula, Eq. (2), and the optical flow method. The optical flow method used in this work is the Horn-Schunck optical flow method [7].

4 Stereo Vision

In a stereo vision system, a pair of two identical cameras are placed side by side with non-zero baseline distance, b. The images of a scene point, P, in the scene are captured by the two cameras and are projected at different locations, P_l and P_r respectively, in the two images, Fig. 2 [16]. P_l is at distance x_l from left camera axis and P_r is at distance x_r from right camera axis.

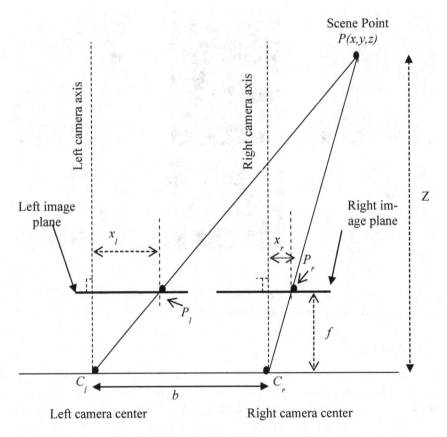

Fig. 2. Depth estimation from a stereo system

The displacement $x_l–x_r$ is called as disparity, d, which is inversely proportional to the depth of the scene point from the camera planes, Eq. (3) [17]:

$$z = \frac{bf}{d} \tag{3}$$

Where b = cameras baseline distance, f = focal length, d = disparity, z = depth of the scene point. The meaning of the various symbols used in Eq. (3) is shown in Fig. 2.

Once, the depth (z coordinate) of a scene point is available using stereo vision the remaining x and y coordinates of the scene point is obtained using Eq. (4) of perspective projection [17].

$$\begin{cases} x = \frac{x'z}{f} \\ y = \frac{y'z}{f} \end{cases} \tag{4}$$

Where (x', y') are x and y co-ordinates of the scene point in the projected image, f = focal length of the camera, and z = depth of the scene point, calculated using Eq. (3).

5 Proposed Method and Experimental Results

In this paper, it is proposed to combine the optical flow method and stereo vision concepts to generate 3D trajectory of a moving object in a scene. Figure 3 shows the flowchart of the proposed concept. The proposed scheme is a two-step process. In the first step, the optical flow is used to find the centroids of the object in the rectified stereo video frames, separately using the method described in Sect. 5.1. In the second step, the stereo vision concept is used to calculate the 3D location of the object centroid from the centroids calculated in the first step. This step is described in Sect. 5.2.

The experimental setup for capturing stereo videos consists of a stereo camera, a laptop, a primary colored object and a chair with known dimensions placed at a known distance from the cameras (for verification of the results). A normal household normal lighting condition is used during video capturing. For 3D trajectory generation, a stereo video of 150 frames is captured, during which a red colored bat is moved along the chair handles and the back support edges. The video frame resolution is kept at the camera default, which is equal to 640 × 480. The frame capture rate is set to 30. The ground truth data for the tracked object is measured by a measuring tape. OpenCV under Visual C++ environment is used for capturing videos. MATLAB is used as the programming environment.

5.1 2D Trajectory Generation

The steps used for 2D trajectory generation are described below, for which either left stereo video or right stereo video can be used.

Step 1: Read a video frame from the video.
Step 2: Convert the image frame into the CCR image frame using Eq. (2).
Step 3: Estimate the optical flow of the current CCR image frame in reference to the previous CCR image frame.

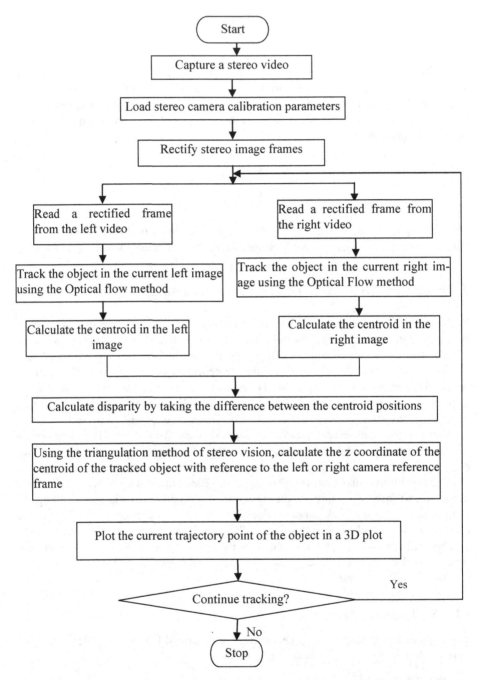

Fig. 3. Flow chart of 3D trajectory reconstruction using Optical flow method

Step 4: Locate the object of interest based on the optical flow with the specified threshold and find the centroid of the object based on the optical flow.

Step 5: Plot the centroid of the object in a 2D plot and repeat the above steps for the remaining frames.

Figure 4 (top row) shows a stereo frame (left and right videos) along with optical flow vectors as blue arrows on the detected object. Figure 4 (bottom row) shows the 2D trajectories of the centroids of the tracked object (in left and right videos) superimposed on the last frame of the stereo video.

Left video frame   Right video frame

Fig. 4. Top row shows optical flow vectors as blue arrows on the detected object. Bottom row shows the 2D trajectory (in red color) of the centroids of the tracked object superimposed on the last frame of the stereo video. (Color figure online)

5.2   3D Trajectory Generation

Before using the stereo camera for 3D trajectory generation, it is calibrated by capturing stereo images using a checker board image, which is placed at different locations and orientations. Using the stereo camera calibration toolbox of MATLAB, the calibration parameters of the stereo camera setup are computed, which are shown in Table 1. These parameters are further used for image rectification and 3D trajectory generation. For the detailed description of these parameters and camera calibration procedure, readers can refer MATLAB online help of the stereo camera calibration tool box.

The basic flow diagram of 3D trajectory generation is given in Fig. 3. The inputs to the prototype system are the camera parameters and a stereo video. The stereo video frames are then rectified using camera parameters. The advantage of image rectification is that the corresponding points lie on the same horizontal scanline. After this step, two independent object trackers are initialized. One tracker tracks the object in the left video and the other in the right video.

Table 1. Calibration parameters of the stereo camera

Parameter	Value
CameraParameters1 in pixels (Intrinsic parameters)	
Focal length	[559.19 609.81];
Principle point	[348.21 270.88];
CameraParameters2 in pixels (Intrinsic parameters)	
Focal length	[558.48 610.87];
Principle point	[342.56 250.37];
Inter-camera Geometry (Extrinsic parameters)	
Focal length	[558.48 610.87];
Principle point	[342.56 250.37];
Rotation of camera 2	$\begin{bmatrix} 0.9995 & 0.0087 & -0.0270 \\ -0.0084 & 0.9998 & 0.0137 \\ 0.0271 & -0.0135 & 0.9995 \end{bmatrix}$
Translation of Camera2	[− 154.5946 0.3739 0.6614]

In the next step, one rectified frame of the left video and one rectified frame of the right video are read. Once the rectified frames are available, the color-based optical flow trackers are used to track the locations of the object in both the image frames, and the centroid locations of the tracked object in both the images are calculated.

In the final step, calculate the z coordinate of the centroid of the tracked object with reference to the left or right camera reference frame using the triangulation method of stereo vision as given by Eq. (3). Finally, the 3D trajectory is plotted as a 3D plot as shown in Fig. 5. Figure 5 shows the computed z coordinates with reference to image coordinates, i.e., the z values are in mm, whereas x and y are image coordinates. Once, the z coordinates are available, for 3D reconstruction of trajectories Eq. (4), obtained using the perspective projection concept, is used. Figure 6 shows the 3D trajectory world coordinates (in mm). In this work, disparity is computed only for the tracked path points, which reduces computational effort, making is suitable for real time applications.

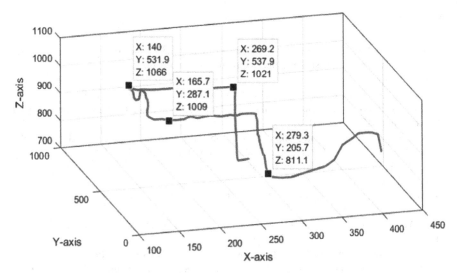

Fig. 5. 3D trajectory generated using color-based optical flow (Z-axis in mm and X and Y axes are in pixels). (Color figure online)

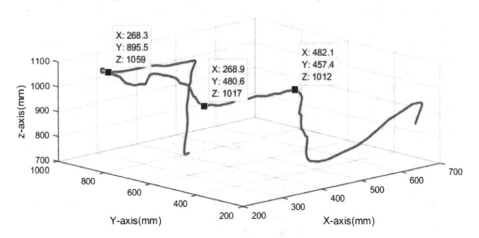

Fig. 6. 3D trajectory in world coordinates, X, Y and Z axes are in mm.

6   Conclusion

This paper presents object detection and tracking system for 3D trajectory generation by combining color-based optical flow with stereo vision. The use of color in the optical flow method makes the method capable to handle dynamic background. Further, for generating 3D trajectory stereo vision is combined with the tracking method as it is not possible to generate 3D trajectory without making some assumptions about the scene from monocular videos. Since depth is computed for the tracked path only, thus making the method suitable for applications that require 3D trajectory generation in real time.

References

1. Heath, K., Guibas, L.: Multi-person tracking from sparse 3D trajectories in a camera sensor network. In: Second ACM/IEEE International Conference on Distributed Smart Cameras, ICDSC (2008)
2. Kale, K., Pawar, S., Dhulekar, P.: Moving object tracking using optical flow and motion vector estimation. In: 4th International Conference on Reliability, Infocom Technologies and Optimization (ICRITO) (2015)
3. Verma, R.: Detecting and tracking a moving object in a dynamic background using color-based optical flow. Int. J. Adv. Res. Comput. Eng. Technol. **6**(11), 1758–1763 (2017)
4. Pan, Z., Liu, S., Fu, W.: A review of visual moving target tracking. Multimedia Tools Appl. **76**(16), 16989–17018 (2017)
5. Verma, R.: A review of object detection and tracking methods. Int. J. Adv. Eng. Res. Dev. **4**(10), 569–578 (2017)
6. Fiaz, M., Mahmoody, A., Jungz, S.K.: Tracking Noisy Targets: A Review of Recent Object Tracking Approaches", arXiv:1802.03098v2 [cs.CV] (2018)
7. Aslani, S., Mahdavi-Nasab, H.: Optical flow based moving object detection and tracking for traffic surveillance. Int. J. Electr. Comput. Energe. Electron. Commun. Eng. **7**(9), 789–793 (2013)
8. Choi, J.W., Whangbo, T.K., Kim, C.G.: A contour tracking method of large motion object using optical flow and active contour model. Multimedia Tools Appl. **74**(1), 199–210 (2015)
9. Xiang, Y., Alahi, A., Savarese, S.: Learning to track: online multi-object tracking by decision making. In: ICCV (2015)
10. Beaupre, D., Bilodeau, G., Saunier, N.: Improving Multiple Object Tracking With Optical Flow And Edge Preprocessing, arXiv:1801.09646v1 [cs.CV] (2018)
11. Tsutsui, H., Miura, J., Shirai, J.: Optical flow-based person tracking by multiple cameras. In: International Conference on Multisensor Fusion and Integration for Intelligent Systems, pp. 91–96 (2001)
12. Zhou, Z., Xu, M., Fu, W., Zhao, J.: Object tracking and positioning based on stereo vision. Appl. Mech. Mater. **303–306**, 313–317 (2013)
13. Park, H.S., Shiratori, T., Matthews, I., Sheikh, Y.: 3D trajectory reconstruction under perspective projection. Int. J. Comput. Vis. **115**(2), 115–135 (2015)
14. Verma, R.: An efficient color-based object detection and tracking in videos. Int. J. Comput. Eng. Appl. **XI**(XI), 172–178 (2017)
15. Aires, K.R.T., Santana, A.M., Medeiros, A.A.D.: Optical flow using color information: preliminary results. ACM, New York (2008). ISBN 978-1-59593-753-7
16. Jain, R., Kasturi, R., Schunck, B.G.: Machine Vision. McGraw Hill International Edition (1995)
17. Forsyth, D.A., Ponce, J.: Computer vision: A Modern Approach, 2nd edn. Pearson, London (2012)

Medical Image Analysis Using Deep Learning: A Systematic Literature Review

E. Sudheer Kumar[(⊠)] and C. Shoba Bindu

Department of Computer Science and Engineering, JNTUA, Ananthapuramu,
Andhra Pradesh, India
sudheerkumar.e@gmail.com, shobabindhu.cse@jntua.ac.in

Abstract. The field of big data analytics has started playing a vital role in the advancement of Medical Image Analysis (MIA) over the last decades very quickly. Healthcare is a major example of how the three Vs of data i.e., velocity, variety, and volume, are an important feature of the data it generates. In medical imaging (MI), the exact diagnosis of the disease and/or assessment of disease relies on both image collection and interpretation. Image interpretation by the human experts is a bit difficult with respect to its discrimination, the complication of the image, and also the prevalent variations exist across various analyzers. Recent improvements in Machine Learning (ML), specifically in Deep Learning (DL), help in identifying, classifying and measuring patterns in medical images. This paper is focused on the Systematic Literature Review (SLR) of various microservice events like image localization, segmentation, detection, and classification tasks.

Keywords: Big data analytics · Medical Image Analysis ·
Deep Learning · Localization · Segmentation · Detection ·
Classification

1 Introduction

Now a day's the big data analytics has been applied for supporting the progression of care delivery and disease exploration. An aspect of healthcare innovation that has recently gained importance is in addressing some of the growing pains in introducing concepts of big data analytics to MIA. An image is a spatial map of various physical properties of anatomy where the pixel intensity represents the worth of a physical property of the anatomy at that point. Imaging the anatomy is an approach to record spatial information, structure, and context information. In this situation, the anatomy could be basically anything. But the objective of imaging is simple and straightforward: convert some scene of the real world into some kind of array of pixels that represents that scene and it will be stored in a computer. For now, it is all about biomedical images, which are a subspace of images that pertain to some form of a biological specimen, which is generally some part of human or animal anatomy.

© Springer Nature Singapore Pte Ltd. 2019
A. K. Somani et al. (Eds.): ICETCE 2019, CCIS 985, pp. 81–97, 2019.
https://doi.org/10.1007/978-981-13-8300-7_8

The categorization of MI modalities has shown in Fig. 1. Such modalities have various purposes like to have an image inner side of the body without harming the body or to have image specimens that are too short to be viewed with the naked eye. The purpose of MI is to assist radiologists and doctors to diagnose and to provide treatment further efficiently. The mechanisms of the building block of our system, the cell, can now be viewed with the aid of the latest computing equipment meant for MI. But being capable of viewing these phenomena is not sufficient, and generating quantitative information through image analysis is very much needed for diagnosis. Voluminous data can be better handled by biomedical image analysis. Such analysis methods obtain quantitative measurements and inferences from images. So, it is feasible to find and monitor certain biological processes and extract information about them. And also, it comprises lot more challenges because images are diverse, complicated, and of uneven shapes.

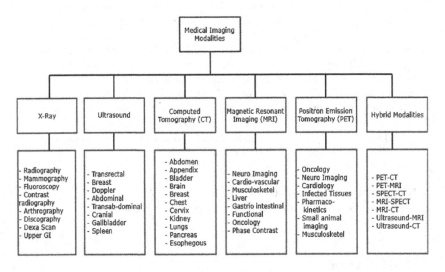

Fig. 1. Categorization of medical imaging modalities [1].

ML and Artificial Intelligence (AI) will facilitate doctors for investigating and predicting the uncertainty of diseases precisely and more quickly. These approaches strengthen the capabilities of doctors and researchers to perceive how to interpret the general differences that lead to the disease. Although automatic identification of diseases based on traditional practices in MI has shown momentous correctness for decades, new developments in ML Techniques has attracted many researchers to increase the effectiveness of MIA.

In case of anatomical structures performing localization and presentation tasks which are the key phases in the work flow process of radiologist on MI. The radiologist carry out both the tasks by finding definite anatomical signatures like image features which can differentiate from one structure to the other structures. Segmentation of organs and/or various substructures allows the measurable investigation of clinical factors related to size and contour of MI [2]. This is often first stage in computer aided diagnosis (CADx) pipeline process. The segmentation task is usually defined as classifying a set of pixels which compose of either the edge of an object or interior of the object as a next task in MIA. Detection consistently recognized as Computer Aided Detection (CADe), in which it is also an intense field of studying about a missing lesion on the scan which will have more importance for both patient and clinician in their work flow process. In the classification task there is a possibility of one or more images as input for analyzing with one diagnostic value as output i.e., whether the disease is present or not. Among all the above discussed tasks, DL methods have enlightened its performance from time to time.

The most strong models for image analysis till date are the Convolutional Neural Networks (CNN). A single CNN model can have many layers working on identifying edges and general features on inner layers and more in-depth features in deeper layers. An image is convolved with filters (some refer to it as kernels) and later pooling is applied, this process may go on for some layers and may eventually find a more recognizable feature [3]. CNN's first real-world application is handwritten recognition in LeNet (1998). MI is an unexplored area and there is a lot of researches to be conducted, hopefully, DL will have a great impact on MI as a whole.

Rest of the paper were discussed as follows. Section 2 illuminates about history of MIA, standards relevant to image formats, representations and about PACS system. Section 3 discusses research methodology, search criteria and search process for this review. Section 4 describes the influences of DL to certain tasks in MIA. Finally, Sect. 5 gives a detail discussion on attained results and research challenges in various application areas for further improvements.

2   Evolution of Medical Image Analysis and Its Standards

The AI standard in 1970s has driven towards the implementation of Rule-based and Expert systems. In the medicine domain MYCIN system [4] from Shortliffe was the first implemented a system which produces various regimes of antibiotic therapies to the patients. From the period 2015–2017 the more number of the algorithms are focused on unsupervised ML, then that is being researched towards supervised ML methods namely Convolutional Neural Networks (CNN) [5]. The first artificial neuron concept was described by McCulloch and Pitts [6] in the year 1943. Later it turned for implementation as perceptron from Rosenblatt [8] in the year of 1958. The purpose of the deep neural network (DNN) is to identify important low-level features (LLF) (i.e., lines or edges) automatically and then combine high-level features (i.e., shapes) in the layers [7].

CNNs can have it's boundaries with the concept of Neocognitron suggested from Fukushima [9] in the year of 1982, but later the author LeCun et al. [10] is the one who established CNNs with the support of error Backpropagation described from Rumelhart et al. [11] in order to implement automatic recognition of hand written digits effectively. This approach has become most popular after winning the 2012 Imagenet Large Scale Visual Recognition Challenge (ILSVRC) by Krizhevsky et al. [11] with very less error rate as 15% as distinguished with the error rate of second place as 26%. Krizhevsky et al. presented many promising concepts for CNN like: Rectified Linear Unit (RELU) function, data augmentation and dropout. Consequently, there is a intense upsurge in count of papers related to CNN architecture and its applications, with this perspective CNNs became principal architecture in MIA.

Development of image analytics and quantification methods is originated upon common standards associated with image formats, data representation, and capturing of meta-data required for downstream analysis. Digital Imaging and Communications in Medicine (DICOM) [12] is a widely used standard that helps to achieve for organizing, storing, printing, and transmitting MI data.

While High Level Seven (HL7) [13] is a more general standard used for interchange, incorporation, distribution, and recovery of electronic healthcare information. It defines standards not just for data but also for application interfaces that use electronic healthcare data. The Integrating the Healthcare Enterprise (IHE) [14] initiative drives the promotion and adoption of DICOM and HL7 standard for improved clinical care and better integration of the healthcare enterprise. MI data is generally gathered and handled using specialized systems known as Picture Archiving and Communications System (PACS) [15]. PACS systems house medical images from most imaging modalities and in addition it can also contain electronic reports and radiologist annotations in encapsulated form. Commercial PACS systems not only allows to perform search, query retrieve, display and visualize imaging data, but often also contain sophisticated post-processing and analysis tools for image data exploration, analysis, and interpretation.

3   Research Methodology

SLR is a process of identifying, interpreting and evaluating all possibly available research resources which are related to a specific research question, or area of topic, or of specific interest. SLR is also called as a secondary study. With this approach, it is possible to optimize the actual evidence of a technology, which helps to determine any research gaps that should be addressed [16,17].

3.1   Research Questions

The main aim of this SLR is to address this research question: How can we perform better identification and segmentation of organ or substructure and also how much accurate can we do classification task by using DL? This research question further divided in to four sub questions, they are:

RQ1: Which DL techniques are being used to perform the detection of Organ, Region and Landmark localization?

RQ2: What are the various DL methods considered to perform the segmentation of organs and other substructures?

RQ3: How to discover or localize anomalous/incredulous regions in structural images using DL techniques?

RQ4: What are the various DL approaches considered for Image/Exam classification?

3.2 Search Strategy

Since the index terms act as "keys" to isolate the scientific articles. Hence, it is necessary to consider appropriate keywords, which can legitimately observe related articles and refine the surplus material [18]. Therefore, the deliberated index terms are: "Medical Image Analysis", "Medical Imaging", "Localization", "Segmentation", "Classification", "Deep Learning", "Machine Learning", and "CNN". Based on our familiarity's with journals, we referred databases that traditionally publish articles on the subject. The following databases were selected: PubMed, ArXiv, IEEE Xplore Digital Library (IEEE), Web of Science, Scopus.

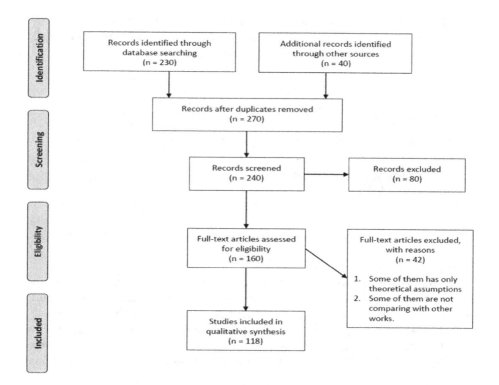

Fig. 2. PRISMA flow chart diagram for selection of research papers [22].

In order to address these RQs, a search string has specified with the help of PICO approach, which distributes the RQ into 4 sub parts: population, intervention, comparison and outcome [17]. The comparison phase was neglected because of the SLR is worried with identification. Remaining are expressed as follows:

1. **Population:** Refers to review research about MIA that helps to infer some information regarding diseases from various imaging modalities. Localization keywords are considered because it is a primary activity for segmentation task, and then segmenting the region of interest to classify the progression/grade of disease.
2. **Intervention:** ML or CNN techniques are more prominent nowadays for image analysis. The ML keyword was selected from the branch of ML Approaches and the CNN keyword are selected from the branch of DL approaches.
3. **Outcome:** Detecting whether the disease is present or not and identify the progression of disease, if present.

3.3   Search Results

The PRISMA (Preferred Reporting Items for Systematic Reviews and Meta-Analyses) Flow chart [19] defines the choice of articles as presented in Fig. 2. This investigation emerged on a overall of 270 articles.

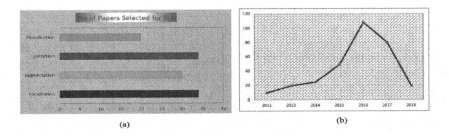

(a)   (b)

Fig. 3. (a) Total number of considered publications (b) Year wise publications

The resulting condition used in this work is: having MIA as a major description, not only the use of a software that performs Localization, Segmentation, Detection, and Classification; and the work was existing for analysis, i.e., accessing to the complete paper. Next to this phase, the selected number of articles were 118, which required an exhaustive assessment. The designated works were classified conferring to the key method used with respect to the segmentation procedure. Finally, respective tables were created in chronological order and summarized with the highlighted works. The below Fig. 3 shows the number of papers published per year from 2012–2018.

4   Deep Learning Applications in Medical Imaging

Traditional ML based approaches had shown encouraging results in various CV tasks, but still there is a little amount of human involvement is required for solving certain tasks. This can be effectively addressed with the help of DL based approaches these days and also DL outperformed in various image analysis tasks with high accuracy and efficiency. These achievements from DL has prompted majority of the researchers working towards DL based MIA for handling certain complicated tasks more efficiently within the stipulated time. In this section, we briefly discuss about the successful applications of DL for Localization, Segmentation, Detection and Classification tasks. As shown in the example Fig. 4 a CNN takes an input image of raw pixels, and transforms it via Convolutional Layers, Rectified Linear Unit (RELU) Layers and Pooling Layers. This feeds into a last Fully Connected Layer which refers class scores or probabilities, thus classifying the input to the class with the maximum probability.

pagination

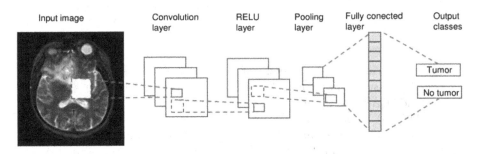

Fig. 4. Flow diagram of CNN model for Brain Tumor disease Classification Task [7].

4.1   Organ, Region and Landmark Localization

The localization task is used to specify a bounding box over a unique object inside the image by identifying specific image features which helps us to distinguish from one anatomical structure from other structures. To report this issue we have framed a research question (RQ1): Which DL techniques are used to perform the detection of Organ, Region and Landmark localization? Here we discuss about various solutions provided to address the above research question and also provided some of the important references in the Table 1.

Table 1. Summary of papers related to Organ, Region or Structure Localization.

Author Name	Imaging Modality	Method Used	Targeted Object	Remarks
Hoo-Chang Shin et al (2013)	MRI	SAE	Liver, Heart, Kidney, Spleen	Multi organ detection; probabilistic based approach on 4D DCE-MRI dataset
Sermanet P et al (2014)	General Images	CNN	-	Object Detection; Convolutional Networks for localization task using multi scale and sliding window approaches.
Su H et al (2015)	Histopathology images	Sparse Reconstruction	Lung and Brain	Sparse reconstruction with small templates along with the structured labels are used.
Bob D. de Vos et al (2016)	CT	CNN	Heart, Aortic Arch, and Descending Aorta	Automatic 3D localization of anatomical structures using CNN-based 2D image classification.
A Kumar et al (2016)	US	CNN	Abdominal	A general method for the classification of anatomical planes in fetal US images.
Sirinukunwattana et al (2016)	Histopathology Images	CNN	Colon Cancer	Nucleus detection; Spatially Constrained CNN regresses the likelihood of a pixel being the center of a nucleus
Stefano Trebeschi et al (2017)	MRI	CNN	Rectal Cancer	Developed to fully automatic localization and segmentation of locally advanced rectal tumors.
Gabriel Efrain Humpire-Mamani et al (2018)	CT	CNN	Abdomen	Efficient method for synchronized localization of numerous structures in 3D thoraxabdomen CT scans.
John D et al (2018)	Histopathology Images	CNN	Kidney	A glomerular localization channel for trichrome-stained kidney sections using an ML image classification algorithm.
Holger R.Roth et al (2018)	CT	CNN	Pancreas	Holistically-nested convolutional networks (HNNs): automated system from 3D CT volumes based on a multistage cascaded methodology.

Vaswani, et al. [20] was carried out the first GPU based implementation to accomplish fast localization of 3D anatomical structures in medical volumes using extended Adaptive Bandwidth Mean-Shift Algorithm for Object Detection (ABMSOD) algorithm applied on CT images of brain stem, eye and the parotid gland and achieves more than 90% in 40 runs and at least 50% partial structure identification in 65% runs. The winner of ICPR 2012 mitosis detection competition Ciresan [21] proposed supervised DNN for mitosis detection in breast histology images applied on publicly available MITOS dataset by achieving F-score of 0.782. Shin et al. [22] is the principal study who uses DL for organ detection in mixed MRI datasets and they have used a probabilistic based method for detection and the outcomes are most promising. Sermanet [23] conferred a multi scale, sliding window based approach which can be effectively implemented within a ConvNet and won the 2013 ILSVRC competition with 29.9% error rate. Zheng et al. [24] suggested an competen and strong landmark 3D detection in volumetric data with a mean error rate of 2.64 mm.

Chen et al. [25] proposed to recognize the fetal abdominal standard plane (FASP) from US videos. Further proposed a transfer learning strategy which reduces the overfitting problem. And also presented [26] knowledge transferred based RNN which will perceive fetal standard planes from US images discovering along with spatio-temporal feature learning. Su et al. [27] proposed sparse reconstruction method using an adaptive dictionary and insignificant templates for handling the shape variations, inhomogeneous intensity, and cell overlapping to detect the lung and brain tumor cells. De Vos [28] went a step forward and performed localized ROI on available regions of anatomy (heart, aortic arch, and descending aorta) by considering a rectangular shape 3D bounding box. Here is a kind of new strategy has started, pertained CNN architectures used for the purpose of better localization, handing the inadequacy of training data to learn better feature representation [29,30]. The other group of authors are trying to adjust the learning process of network to predict the locations directly, Payer et al. [31] suggested an approach to precisely degenerate landmark positions with the help of CNNs.

Ghesu et al. [32] specified a sparse adaptive DNN motivated by marginal space learning method, which deals with complexity of data to understand the aortic valve in 3D transesophageal echocardiogram. An exciting example is the idea from Sirinukunwattana et al. [33] outperforms the classification-based center localization by considering the center locations of nuclei. Liu et al. [34] proposed innovative algorithm for comprehensive cell detection which doesn't need the fine tuning of parameters, and this is the preliminary work to familiarize DCNN to offer weights to a graph. Trebeschi et al. [35] developed a fully automatic localization and segmentation of locally advanced rectal tumors with the help of CNN. Humpire-Mamani et al. [36] provided an decisive method for concurrent localization of several structures in 3D thoraxabdomen CT scans.

4.2 Segmentation of Organs and Other Substructures

The segmentation task in medical images permits measurable investigation of scientific specifications which are correlated with volume and shape. Besides, it is often an important first step in CADe process. The segmentation task is consistently stated as finding the group of voxels which produces either the contour/the interior of the object(s) of interest. It is the utmost common concept of papers relating DL to MI, some of them were discussed in Table 2, and also seen the broadest variability in practice, comprising the advancement of distinctive CNN involved segmentation architectures and also the broader use of RNNs to the concern (RQ2).

Table 2. Summary of papers related to Segmentation of Organ and Substructure.

Author Name	Imaging Modality	Method Used	Targeted Object	Remarks
Prasoon, A et al (2013)	MRI	CNN	Knee	It uses only 2D features at a single scale.
Guo et al (2014)	MRI	SAE	Brain	Independently learn the intrinsic feature representation for three distinct phases.
Wenlu Zhang et al (2015)	MRI	CNN	Brain	Aimed at segmenting infant brain tissue images in the isointense stage.
Wang, C et al (2015)	Wound Image	CNN	Any Body Part	Unified framework for wound segmentation and wound condition analysis.
JinzhengCai et al (2016)	MRI	LSTM	Pancreas	This accomplishes segmentation on an image by incorporating its neighboring slice segmentation predictions, in the form of a reliant sequence processing.
Jun Xu et al (2016)	Histology Images	DCNN	Epithelial (EP) and stromal(ST)	To inevitably segment or classify EP and ST regions from digitized TMAs
Youyi Song et al (2017)	PAP Smear	Learning Based Method	Cervical Cancer	A learning-based method with robust shape priors to segment separate cell in Pap smear images to maintain automatic observing of changes in cells
Kline TL et al (2017)	MRI	CNN	Kidney	Segmentation of the kidneys is desirable in order to quantify total kidney volume (TKV).
Liu, X et al (2018)	CT	CNN	Abdominal	To condense the terminated inputs, the simple linear iterative clustering (SLIC) of super-pixels and the support vector machine (SVM) classifier
Karimi, D et al (2018)	MRI	CNN	Prostate	Calculates the location of the prostate center and the constraints of the shape model, which regulate the position of prostate surface key points.

Ciresan et al. [37] presented a sliding window fashion in the microscopy imagery which can be applied to find a pixel-wise segmentation of membranes. To handle the erroneous reactions while performing voxel classification methods, many clusters have found that the association of fCNNs with graphical models resembling Markov Random Fields are the good solutions Song et al. [38]. Furthermore, in the next work by Cicek et al. [39], the U-Net, which has the limited relative 2D annotated slices, has proven to be a complete 3D segmentation. RNNs are now well developed for segmentation tasks. Xie et al. [40] used RNN to fragment the perimysium in H&E-histopathology images. It considers a former evidence against to the present patch row and the column predecessors. To integrate multi-directional information, the RNN is useful for 4 times in numerous directions and the final result is combined & served to a fully-connected layer. Lastly, Poudel, et al. [41] mixed a 2D U-net considering along with the gated recurrent unit to attain 3D segmentation. Moeskops, et al. [42] accomplished a unique fCNN to segment the MRI of brain, breast, and coronary arteries in the cardiac CT.

4.3   Computer Aided Detection

The primary objective of CADe is to discover/reduce the unusual/doubtful regions in structural images, and therefore in turn can alert experts (RQ3). While CADe intention is to reduce the false-negative rate to improve the recognition rate of unhealthy sections, it may be due to the lack of an amount of viewers involved or fatigue. Even though CADe is well recognized in MI, DL techniques has been enhancing its presentation in diverse scientific applications. Particularly, utmost methods designated in the collected works misused deep convolutional models to extremely develop essential information in 2, 2.5, or 3 dimensions. Table 3 summarizes the possible solutions to perform detection task using DL.

Table 3. Summary of papers related to Object/Lesion Detection.

Author Name	Imaging Modality	Method Used	Targeted Object	Remarks
Cruz-Roa et al (2013)	Histopathology Images	DNN	Cancer	Different image representation strategies, including bag of features (BOF), canonical (discrete cosine transform (DCT) and Haar-based wavelet transform (Haar))
Wang, H et al (2014)	Histopathology Images	CNN	Breast	Mitosis Detection; associations a CNN model and handcrafted features (morphology, color, and texture features).
Xiangyu Chen et al (2015)	MRI	CNN	Brain	Area under curve (AUC) of the receiver operating characteristic curve in glaucoma detection at 0.831 and 0.887 in the two databases, much better than state-of-the-art algorithms.
Bar et al (2015)	Histopathology Images	CNN	Chest	This is a first-of-its-kind experiment that shows that DL with large scale non-medical image databases may be sufficient for general medical image recognition tasks.
Daniel E et al (2016)	Retinal Fundus Photographs	CNN	Eye	Leading fully automated ROP detection system.
Mark Cicero et al (2016)	Chest Radiographs	DCNN	Chest	Classification of Abnormalities on Frontal Chest Radiographs
Juan Wang et al (2017)	Breast arterial calcifications	CNN	Breast	Detecting Cardiovascular Disease from Mammograms.
Hamidian S et al (2017)	CT	CNN	Lung	Automatic Detection of Lung Nodules in Chest CT.
Hailiang Li et al (2018)	US	CNN	Thyroid Papillary Cancer	An improved deep learning approach for detection of thyroid papillary cancer in ultrasound images
Hongsheng Jin et al (2018)	CT	Residual CNN	Nodule Detection	3D residual CNN (convolution neural network) to reduce false positive nodules from candidate nodules.

The leading object detection framework by using CNNs has recommended in the year of 1995, using a CNN with 4 layers to intellect nodules in x-ray's Lo et al. [43]. The integration of appropriate or 3D data is also controlled with multi-stream CNNs (for example by Roth et al. [44] and Teramoto et al. [45]). van Grinsven et al. [46] recommended a discriminating information sampling in which incorrectly categorized patterns were pumped backward to the network higher frequently to emphasis on exciting zones in retinal images. Setio et al. [47] deliberated 3 sets of rectangular perspectives for a whole of 9 perspectives from a 3D patch and considered collaborative mechanisms to integrate data from dissimilar perspectives for finding of pulmonary nodules.

4.4   Computer Aided Diagnosis

While assessing some information from an image, CADx offers an additional independent opinion. The key applications of CADx comprises of the judgment of malignant from benign lesions and the recognition of specific diseases from one or multiple images. Usually, most of the CADx systems are established to use human-intended features obtained from the domain experts. In the recent years, DL methods are productively applied to CADx systems. The concept of image classification is one of the first and foremost areas in which DL performed a major influence to MIA (RQ4). Table 4 discuss about the possible solutions to perform classification task using DL.

Table 4. Summary of papers related to Image/Exam Classification.

Author Name	Imaging Modality	Method Used	Targeted Object	Remarks
RongjianLi et al (2014)	MRI, PET	CNN	Brain	DL based framework for estimating multi-modality imaging data
Holger R. Roth et al (2015)	CT	CNN	Neck, Lungs, Liver, Pelvis and Legs	A method for organ- or body part- specific anatomical classification of medical images acquired using computed tomography (CT) with ConvNets
WeiShen et al (2015)	CT	CNN	Lung	Framework utilizes multi-scale nodule patches to learn a set of class-specific features simultaneously by concatenating response neuron activations obtained at the last layer from each input scale.
AlvinRajkomar et al (2016)	Chest Radiographs	DCNN	Chest	Classification of Radiographs Using Deep Convolutional Neural Networks
WenqingSunet al (2016)	Mammogram Images	CNN	Breast	Developed a graph based semi-supervised learning (SSL) scheme using deep convolutional neural network (CNN) for breast cancer diagnosis
Michael David Abrmoff et al (2016)	Digital Retinal Color Images	Consensus Reference Standard	Eye	used the previously reported consensus reference standard of referable DR (rDR), defined as International Clinical Classification of Diabetic Retinopathy
Hang Chang et al (2017)	Histopathology Images	Convolutional Sparse Coding	Tissue Histology	A novel multi-scale convolutional sparse coding (MSCSC) method for tissue histology.
Xiaosong Wang et al (2017)	X-ray	-	chest	A new chest X-ray database, namely ChestX-ray8
Sergey Korolev et al (2017)	MRI	CNN	Brain	The performance can be achieved skipping these feature extraction steps with the residual and plain 3D convolutional neural network architectures
Gao Huang et al (2018)	Colored natural images	CNN	-	The Dense Convolutional Network (DenseNet), which connects each layer to every other layer in a feed-forward fashion.

A timeline related to computer vision (CV) is obvious, w.r.t the kind of deep networks that are generally used in exam classification Esteva et al. [48]. The popular papers smearing these practices for image classification which seemed in 2013 and motivated on neuroimaging. Suk et al. [49], and Suk et al. [50] considered DBNs and SAEs to categorize the patients who had Alzheimer's established along with the brain MRI. A vibrant shift towards CNNs have been observed recently. Two papers has considered an architecture leveraging with limited attributes of medical data: Hosseini-Asl et al. [51] has investigated 3D convolutions irrespective of 2D to organize patients as consisting of Alzheimer; Kawahara et al. [52] smeared a CNN resembling architecture to a brain related-ness graph resulting from MRI diffusion-tensor imaging (DTI). With the help of their network they have proved that their method has outperformed the previous methods in evaluating intellectual and mechanical scores for predicting the development of a brain.

5   Discussion and Future Directions

The health sector is completely divergent in comparison with any industry. It has a top preference and customers anticipate the utmost level of care and services nevertheless of cost. Possibly, the study of medical data by medical professionals while interpreting the images is being limited due to its complexity, individuality and also having the maximum workload to the experts. In this scenario, the DL outperformed in various CV tasks nowadays successfully. Which in turn can provide the most exciting and maximum accurate solutions for MIA also and it can be seen as an important method for future applications as well.

In MIA, the deficiency of data is still two-fold and also it is most important: there is a common scarcity of freely usable data, and superior kind of labeled data is even scarcer. Data or class inequality in the training group is also a noteworthy difficulty in MIA. Currently, Variational AEs and GANS are a generative model, may avoid the data paucity problem because it can create synthetic medical data. The data imbalance outcome can be improved with the help of data augmentation process to produce maximum training images of rare/abnormal data, but still there is a possibility of overfitting. There are few important point that needs to focus: The majority of DL methods focused on supervised DL approaches; though, the annotation of medical data is not constantly achievable [53]. To eradicate the unreachability of big data, the supervised DL approach needs to shift towards unsupervised or semi-supervised systems. Thus, the usefulness of unsupervised and semi-supervised methods in medicine will be worthy.

Overall, CNNs are the most common strategy to perform localization task over 2 Dimensional image classification with decent results. Almost all of the discussed methods related to the brain are concentrating on brain MRI images [8]. But, there are other imaging modalities for brain is also aid from DL based analysis. The increasing accessibility of large scale gigapixel whole-slide images (WSI) of tissue specimen has formed digital pathology and microscopy a very prominent application area for DL techniques. In the histopathology image analysis, color normalization is a major research area. One among the major challenge is to stable the amount of imaging features for the DL network (typically thousands) with the amount of clinical features (typically only a handful) [54].

6   Conclusion

DL technology applied to MI may become the tremendous technology within the next 15 years. Applications of DL in health care will provide an extensive scope of problems extending from cancer screening and disease monitoring to personalized treatment suggestions. DL methods have regularly described as 'black boxes', specifically in medicine. Because liability is more essential which can have severe legal significance's, so it is usually not at all sufficient with a good prediction system. In this SLR, we have examined the possible state-of-the-art DL approaches used for image localization, segmentation, detection and classification tasks.

This review starts with the discussion on the evolution of MIA and the standards it follows. The process of research methodology considered for this SLR is elaborately discussed. Then, presented a brief summary of more than 130 contributions to the field is discussed and provided insights through discussion along with the future directions. As the communities of engineers, computer scientists, statisticians, physicians, and biologists continue to integrate; this holds great promise for the development of methods that combine the best elements of modeling and learning approaches for solving new technical challenges. The potential of this field is most promising, so there is a lot of scope for budding researchers.

References

1. Litjens, G., et al.: A survey on deep learning in medical image analysis. Med. Image Anal. **42**, 60–88 (2017)
2. Qayyum, A., Anwar, S.M., Majid, M., Awais, M., Alnowami, M.: Medical image analysis using convolutional neural networks: a review. Comp Vis. Pattern Recogn. ArXiv: 1709.02250 (2017)
3. Suzuki, K.: Overview of deep learning in medical imaging. Radiol. Phys. Technol. **10**, 257 (2017)
4. Shortliffe, E.H.: Computer-Based Medical Consultations: MYCIN, vol. 2. Elsevier, New York (1976)
5. Lawrence, S., Giles, C.L., Tsoi, A.C., Back, A.D.: Face recognition: a convolutional neural-network approach. IEEE Trans. Neural Netw. **8**(1), 98–113 (1997)
6. McCulloch, W.S., Pitts, W.: A logical calculus of the ideas immanent in nervous activity. Bull. Math. Biol. **5**(4), 115–133 (1943)
7. Ker, J., Wang, L., Rao, J., Lim, T.: Deep learning applications in medical image analysis. IEEE Access **6**, 9375–9389 (2018)
8. Rosenblatt, F.: The perceptron: a probabilistic model for information storage and organization in the brain. Psychol. Rev. **65**(6), 365–386 (1958)
9. Fukushima, K., Miyake, S.: Neocognitron: a self-organizing neural network model for a mechanism of visual pattern recognition. In: Amari, S., Arbib, M.A. (eds.) Competition and Cooperation in Neural Nets. LNBM, vol. 45, pp. 267–285. Springer, Berlin (1982). https://doi.org/10.1007/978-3-642-46466-9_18
10. LeCun, Y., et al.: Backpropagation applied to handwritten zip code recognition. Neural Comput. **1**(4), 541–551 (1989)

11. Rumelhart, D.E., Hinton, G.E., Williams, R.J.: Learning representations by back-propagating errors. Nature **323**, 533–536 (1986)
12. Digital Imaging and Communications in Medicine (DICOM). https://www.dicomstandard.org/
13. Health Level Seven (HL7). http://www.hl7.org/index.cfm
14. Integrating the Healthcare Enterprise (IHE). http://www.ihe.net/
15. Picture Archiving and Communications Systems (PACS). http://www.pacshistory.org/index.html
16. Dallora, A.L., Eivazzadeh, S., Mendes, E., Berglund, J., Anderberg, P.: Prognosis of dementia employing machine learning and microsimulation techniques: a systematic literature review. Procedia Comput. Sci. **100**, 4808 (2016)
17. Pai, M., McCulloch, M., Gorman, J.D., Pai, N., Enanoria, W., Kennedy, G., et al.: Systematic reviews and metaanalyses: an illustrated, step-by-step guide. Natl. Med. J. India **17**(2), 8695 (2004). PMID 15141602
18. Sharma, K., Mediratta, P.: Importance of keywords for retrieval of relevant articles in medline search. Indian J. Pharm. **34**, 369–371 (2002)
19. Moher, D., Liberati, A., Tetzlaff, J., Altman, D.G.: The PRISMA group: preferred reporting items for systematic reviews and meta-analyses: the PRISMA statement. PLoS Med. **6**(7), e1000097 (2009). https://doi.org/10.1371/journal.pmed1000097
20. Vaswani, S., Thota, R., Vydyanathan, N., Kale, A.: Fast 3D structure localization in medical volumes using CUDA-enabled GPUs. In: 2nd IEEE International Conference on Parallel, Distributed and Grid Computing, Solan, pp. 614–620 (2012)
21. Cireşan, D.C., Giusti, A., Gambardella, L.M., Schmidhuber, J.: Mitosis detection in breast cancer histology images with deep neural networks. In: Mori, K., Sakuma, I., Sato, Y., Barillot, C., Navab, N. (eds.) MICCAI 2013. LNCS, vol. 8150, pp. 411–418. Springer, Heidelberg (2013). https://doi.org/10.1007/978-3-642-40763-5_51
22. Shin, H.C., Orton, M.R., Collins, D.J., Doran, S.J., Leach, M.O.: Stacked autoencoder for unsupervised feature learning and multiple organ detection in a pilot study using 4D patient data. IEEE Trans. Pattern Anal. Mach. Intell. **35**, 1930–43 (2013)
23. Sermanet, P., Eigen, D., Zhang, X., Mathieu, M., Fergus, R., Le Cun, Y.: OverFeat: integrated recognition, localization and detection using convolutional networks. ArXiv: 1312.6229 (2014)
24. Zheng, Y., Liu, D., Georgescu, B., Nguyen, H., Comaniciu, D.: 3D deep learning for efficient and robust landmark detection in volumetric data. In: Navab, N., Hornegger, J., Wells, W.M., Frangi, A.F. (eds.) MICCAI 2015. LNCS, vol. 9349, pp. 565–572. Springer, Cham (2015). https://doi.org/10.1007/978-3-319-24553-9_69
25. Chen, H., et al.: Standard plane localization in fetal ultrasound via domain transferred deep neural networks. IEEE J. Biomed. Health Inform. **19**(5), 1627–1636 (2015)
26. Chen, H., et al.: Automatic fetal ultrasound standard plane detection using knowledge transferred recurrent neural networks. In: Navab, N., Hornegger, J., Wells, W.M., Frangi, A.F. (eds.) MICCAI 2015. LNCS, vol. 9349, pp. 507–514. Springer, Cham (2015). https://doi.org/10.1007/978-3-319-24553-9_62
27. Su, H., Xing, F., Kong, X., Xie, Y., Zhang, S., Yang, L.: Robust cell detection and segmentation in histopathological images using sparse reconstruction and stacked denoising autoencoders. In: Navab, N., Hornegger, J., Wells, W.M., Frangi, A.F. (eds.) MICCAI 2015. LNCS, vol. 9351, pp. 383–390. Springer, Cham (2015). https://doi.org/10.1007/978-3-319-24574-4_46

28. De Vos, B.D., Wolterink, J.M., de Jong, P.A, Viergever M.A., Isgum I.: 2D image classification for 3D anatomy localization: employing deep convolutional neural networks. In: Medical Imaging, Proceedings of the SPIE, vol. 9784, p. 97841Y (2016)

29. Cai, Y., Landis, M., Laidley, D.T., Kornecki, A., Lum, A., Li, S.: Multi-modal vertebrae recognition using transformed deep convolution network. Comput. Med. Imaging Graph. **51**, 11–19 (2016)

30. Kumar, A., et al.: Plane identification in fetal ultrasound images using saliency maps and convolutional neural networks. In: IEEE International Symposium on Biomedical Imaging, pp. 791–794 (2016)

31. Payer, C., Štern, D., Bischof, H., Urschler, M.: Regressing heatmaps for multiple landmark localization using CNNs. In: Ourselin, S., Joskowicz, L., Sabuncu, M.R., Unal, G., Wells, W. (eds.) MICCAI 2016. LNCS, vol. 9901, pp. 230–238. Springer, Cham (2016). https://doi.org/10.1007/978-3-319-46723-8_27

32. Ghesu, F.C., Georgescu, B., Mansi, T., Neumann, D., Hornegger, J., Comaniciu, D.: An artificial agent for anatomical landmark detection in medical images. In: Ourselin, S., Joskowicz, L., Sabuncu, M.R., Unal, G., Wells, W. (eds.) MICCAI 2016. LNCS, vol. 9902, pp. 229–237. Springer, Cham (2016). https://doi.org/10.1007/978-3-319-46726-9_27

33. Sirinukunwattana, K., Raza, S.E.A., Tsang, Y.W., Snead, D.R.J., Cree, I.A., Rajpoot, N.M.: Locality sensitive deep learning for detection and classification of nuclei in routine colon cancer histology images. IEEE Trans. Med. Imaging **35**, 1196–206 (2016)

34. Liu, F., Yang, L.: A novel cell detection method using deep convolutional neural network and maximum-weight independent set. In: Lu, L., Zheng, Y., Carneiro, G., Yang, L. (eds.) Deep Learning and Convolutional Neural Networks for Medical Image Computing. ACVPR, pp. 63–72. Springer, Cham (2017). https://doi.org/10.1007/978-3-319-42999-1_5

35. Trebeschi, S., van Griethuysen, J.J.M., Lambregts, D.M.J., et al.: Deep learning for fully-automated localization and segmentation of rectal cancer on multiparametric MR. Sci. Rep. **7**, 5301 (2017)

36. Humpire Mamani, G.E., Setio, A.A.A., van Ginneken, B., Jacobs, C.: Efficient organ localization using multi-label convolutional neural networks in thorax-abdomen CT scans. Phys. Med. Biol. **63**(8), 085003 (2018)

37. Ciresan, D., Giusti, A., Gambardella, L.M., Schmidhuber, J.: Deep neural networks segment neuronal membranes in electron microscopy images. In: Proceedings of Advances in Neural Information Processing Systems, pp. 2843–2851 (2012)

38. Song, Y., Zhang, L., Chen, S., Ni, D., Lei, B., Wang, T.: Accurate segmentation of cervical cytoplasm and nuclei based on multi-scale convolutional network and graph partitioning. IEEE Trans. Biomed. Eng. **10**, 2421–2433 (2016)

39. Çiçek, Ö., Abdulkadir, A., Lienkamp, S.S., Brox, T., Ronneberger, O.: 3D U-Net: learning dense volumetric segmentation from sparse annotation. In: Ourselin, S., Joskowicz, L., Sabuncu, M.R., Unal, G., Wells, W. (eds.) MICCAI 2016. LNCS, vol. 9901, pp. 424–432. Springer, Cham (2016). https://doi.org/10.1007/978-3-319-46723-8_49

40. Xie, Y., Zhang, Z., Sapkota, M., Yang, L.: Spatial clockwork recurrent neural network for muscle perimysium segmentation. In: Ourselin, S., Joskowicz, L., Sabuncu, M.R., Unal, G., Wells, W. (eds.) MICCAI 2016. LNCS, vol. 9901, pp. 185–193. Springer, Cham (2016). https://doi.org/10.1007/978-3-319-46723-8_22

41. Poudel, R.P.K., Lamata, P, Montana, G.: Recurrent fully convolutional neural networks for multi-slice MRI cardiac segmentation. ArXiv: 1608.03974 (2016)

42. Moeskops, P., et al.: Deep learning for multi-task medical image segmentation in multiple modalities. In: Ourselin, S., Joskowicz, L., Sabuncu, M.R., Unal, G., Wells, W. (eds.) MICCAI 2016. LNCS, vol. 9901, pp. 478–486. Springer, Cham (2016). https://doi.org/10.1007/978-3-319-46723-8_55

43. Lo, S.-C., Lou, S.-L., Lin, J.-S., Freedman, M.T., Chien, M.V., Mun, S.K.: Artificial convolution neural network techniques and applications for lung nodule detection. IEEE Trans. Med. Imaging **14**, 711–718 (1995)

44. Roth, H.R., et al.: Improving computer-aided detection using convolutional neural networks and random view aggregation. IEEE Trans. Med. Imaging **35**(5), 1170–1181 (2016)

45. Teramoto, A., Fujita, H., Yamamuro, O., Tamaki, T.: Automated detection of pulmonary nodules in PET/CT images: ensemble false-positive reduction using a convolutional neural network technique. Med. Phys. **43**, 2821–2827 (2016)

46. van Grinsven, M.J.J.P., Ginneken, V., Hoyng, C., Theelen, B., Sanchez, C.: Fast convolutional neural network training using selective data sampling: application to hemorrhage detection in color fundus images. IEEE Trans. Med. Imaging **35**(5), 1273–1284 (2016)

47. Setio, A.A., et al.: Pulmonary nodule detection in CT images: false positive reduction using multi-view convolutional networks. IEEE Trans. Med. Imaging **35**(5), 1160–1169 (2016)

48. Esteva, A., et al.: Dermatologist-level classification of skin cancer with deep neural networks. Nature **542**, 115–118 (2017)

49. Suk, H.-I., Shen, D.: Deep learning-based feature representation for AD/MCI classification. In: Mori, K., Sakuma, I., Sato, Y., Barillot, C., Navab, N. (eds.) MICCAI 2013. LNCS, vol. 8150, pp. 583–590. Springer, Heidelberg (2013). https://doi.org/10.1007/978-3-642-40763-5_72

50. Suk, H.I., Lee, S.W., Shen, D.: Hierarchical feature representation and multimodal fusion with deep learning for AD/MCI diagnosis. NeuroImage **101**, 569–582 (2014)

51. Hosseini Asl, E., Gimelfarb, G., El-Baz, A.: Alzheimer's disease diagnostics by a deeply supervised adaptable 3D convolutional network. arXiv: 1607.00556 (2016)

52. Kawahara, J., et al.: BrainNetCNN: convolutional neural networks for brain networks; towards predicting neurodevelopment. NeuroImage **146**, 1038–1049 (2017)

53. Lee, J.G., et al.: Deep learning in medical imaging: general overview. Korean J. Radiol. **4**, 570–584 (2018)

54. Shen, D., Wu, G., Suk, H.I.: Deep learning in medical image analysis. Ann. Rev. Biomed. Eng. **19**, 221–248 (2017)

Cross-Domain Spam Detection in Social Media: A Survey

Deepali Dhaka[(⊠)] and Monica Mehrotra

Department of Computer Science, Jamia Millia Islamia,
New Delhi 110025, India
deepali.dhaka@gmail.com, drmehrotra2000@gmail.com

Abstract. Social media is now an integral part of everyone's life. Due to its exponential growth with rising interest of users, they have become the source of the abundant amount of data prevailing on the internet. This tremendous amount of data is not only useful for researchers but also fascinates spammers. To reach more users, to increase monetary gain and to disseminate malicious activities, spammers are using multiple content sharing platforms. Conventional spam detection techniques have focused more on spam detection on one platform only. This paper discusses cross-domain detection techniques of email and web spams, social spams, opinion spams, and their comparisons. This is an attempt to provide various challenges in this area. As far as our knowledge concerned, this is the first detailed literature study in the field of cross-domain spam detection.

Keywords: Cross-domain · Content-based · Spàm · Social media ·
Spams detection

1 Introduction

Social media are interactive technologies that facilitate dissemination of information, ideas, thoughts, reviews etc., through networks or virtual community over the internet. It is a collective of websites and applications. They provide alluring features to users and the ease to communicate and share information. With their increasing popularity, they have become breeding ground for spamming activities also. Spams are unwanted content sent to a number of users with malicious intentions. Though there are different kinds of spam prevalent over the internet, their evolving nature and evolving detection techniques result in more intricate forms of spam such as email spam, blog spam, microblog spam, bookmarking spam, social networking spam, review spam, location search map, cross-media spam [25, 26].

Earlier spam was limited to email spam and web spam only. But now it has evaded OSNs (online social networks), due to their growing popularity among users. Email spam has adversely affected the user messaging experience over electronic mail communication. Web spam degrades the quality of search of users over the World Wide Web. These spams aim to generate traffic and monetary gain. According to [1] social spam is more harmful than email and web spam as they exploit the trust relationship between users. Social networks are more prone to spam attacks, they pose a great threat to the security and privacy of user's online data.

© Springer Nature Singapore Pte Ltd. 2019
A. K. Somani et al. (Eds.): ICETCE 2019, CCIS 985, pp. 98–112, 2019.
https://doi.org/10.1007/978-981-13-8300-7_9

To combat social spam various spam detection strategies have been proposed in the literature. According to [2] anti-spam strategies can be classified as (1) detection based, (2) demotion based, (3) prevention based. But these strategies should have scope to evolve as spams are volatile in nature. Spamming strategies and social media have evolved a lot in the last few years. A survey of various spam detection approaches has been given by [3]. They categorized them as (1) honey profiles, (2) clustering, (3) supervised, (4) URL/blacklist, (5) incremental learning.

Spams also exist on the e-commerce sites as a review spam or opinion spam. Malicious reviews aim to attract a number of customers for monetary gain or to degrade the image of a company. There are many challenges in spam detection techniques, few of them are listed here-

- With increasing globalization, OSNs are available in different languages to facilitates interaction of people worldwide example LinkedIn which is available in 20 languages [4]. Therefore we can find malicious content also in different languages. So for OSNs, an intelligent, scalable, multilingual spam detection framework is required.
- The humungous amount of data is prevailing over the internet which is not homogeneous as a user may share videos, photos with text on social media. So spam could also prevail in any such form, which implies an analysis of multimedia data is also important with textual data.
- To make it more flexible and attractive for users, the social networks will interact shortly, which implies there would be features which would be common among different networks. This cross-domain interaction might also facilitate similar spamming activities across domains. So cross domain spam detection is one of the emerging areas in OSNs.

In this manuscript, we have focused on cross-domain detection of email spam & web spam, social spam, and opinion spam as shown in Fig. 1.

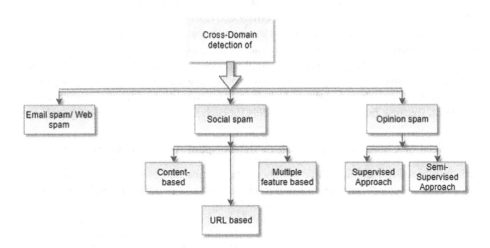

Fig. 1. Taxonomy of cross-domain span detection

The remainder of this manuscript proceeds as follows. Section 2 discusses the motivation behind cross domain spam detection and what is the need to study cross domain spam detection. Section 3 discusses spam detection across email and web. Section 4 discusses work done in cross-domain social spam detection. Section 5 discusses cross-domain opinion spam detection. Section 6 lists down the challenges. Section 7 concludes this work.

2 Motivation Behind Cross-Domain Spam Detection

The motivation of Cross-domain spam detection comes from the fact that traditional machine learning techniques to detect spam use those source of training and testing data, which have the same input feature space and same data distribution. But in the realworld, the scenario is different training and test data does not come from the same domain. They may have different data distribution and different input feature space, though may share some features. Also, it is sometimes very expensive and difficult to collect training data. So there is a need to have a learner built by easily available dataset and which can be efficiently used to test dataset of some other domain for which getting training data is not so easy.

Different OSNs have the ability to interact with each other. To provide more flexible features, they will develop more connections among themselves. Similar kind of spam can attack different social networks like the phishing campaign affecting Facebook, Twitter, and Hotmail[1]. Therefore there is a need to do a cross-domain analysis of data from different networks having a different distribution of data. If different social networks and their spam detection model, cooperate with each other, it will speed up the spam detection by leveraging information from one social network and incorporating into the other [5].

Few such pieces of evidence of cross-platform communication or information sharing are mention here. Twitter bots are encouraging SoundCloud content [6]. A real-time system called monarch, crawls URLs coming from different web services to check whether they land to spam content or not irrespective of the context associated with that URL [7]. So it might be helpful in accurately identifying campaigns across web services. It has also been observed that bots that are part of campaigns over Twitter are also observed on Facebook [8]. The aforementioned work gives the pointer to think how the same kinds of spammers are spread across the social networks and since they have things in common, a single framework might help in combating them efficiently across social networks. Because building separate models for each OSNs would be very expensive.

2.1 Need of Common Framework to Detect Spam Across Social Media

Existing detection techniques like collaborating filtering and behavioral analysis filters are already reducing spam significantly. There are various different and independent

---

[1] https://www.lastwatchdog.com/unstoppable-phishing-attacks-blanket-facebook-twitter/.

strategies adopted by different social media to combat spam. But what if there is a common framework for all the social media to detect spam, then

1. Spam recognized on one social media would be automatically recognized on other social media.
2. It would decrease the cost of building independent frameworks for every social media.
3. It would give more accurate results.
4. If other existing techniques get integrated with the common framework it will boost the performance further, especially speed and accuracy.

3 Spam Detection Across Email and Web Spam Corpus

For cross-domain classification across different domains, we need to extract those features which are common between them. These features are acting as a bridge between two domains [25]. The values of these features from one domain can be used to train a classifier and that would be further used to test dataset of another domain, which overcomes the problem of scarcity of labeled training data for a particular domain and requires single classifier which would work for both the domains instead of multiple frameworks. According to [9] URL is considered as a bridge across email spam and web spam to combat spam in both domains as URL advertise by email spam message correspond to a spam web page. They build a large collection of web spam samples which can be helpful in improving the traditional email classification algorithm by incorporating information from web pages advertise by email spam messages through embedded URLs.

4 Cross-Domain Social Spam Detection

We have classified the various detection techniques for cross-domain social spam detection into three categories as mention below. Comparison of these techniques is given in Table 1.

- Content-based
- URL based
- Multiple features based

4.1 Content-Based

In 2011 [10] proposed a framework which they called SPADE in 2014 [11] a social spam analytics and detection framework which can do two types of classification cross-domain and associative. Domains on which they evaluated their model are MySpace, Twitter, Email and Webb Spam Corpus. Their model comprises of – First Mapping and assembly, the incoming data is mapped into message model, profile model, web page model, and associated objects are also taken care. Then Pre-filtering is done to check if the new object is already there in the existing spam pool then it is filtered, finally leads to the classification of incoming objects and objects associated with them. They have used common features across domains like features related to about me section for profile object, the body of the message for message model and content for web page model. Cross-domain classification is done by building a classifier using a dataset from one domain and testing it on a dataset of other domain. They have shown that cross-domain classification gives a better result for web page classification and the poor result for message and profile related dataset as this kind of dataset has different distribution across the network due to different lexicons and format. In associative classification, it extracts the object associated with the income one and classifies them and combines the result at the end. They have shown that the cross-domain classifications can enhance the performance of the objects having common features across domains and associative classification would improve the result of other object classification.

Limitation: There is the scope of using more critical features. Also, behavior analysis of spammer can be taken into consideration as spammers are evolving in nature
Advantage: Their Pre-filtering module keeps account of all the new spams which may help in blacklisting the incoming spam by comparing the already existing one in this pool

In 2014 [14] made use of content analysis instead of using network related information as the former could outperform the later. They believed that spammers can evade network related information easily. They perform a linguistic variation analysis of spam resources from different social media using different metrics. They learn the information from an external source at the topic level instead of word level. They defined a novel framework which incorporates information from existing social media to aid spammer detection in microblogging. So there are two constrained on the proposed model one is lexicon information from other media and other is Laplacian regularization learned from microblogging. They evaluated their proposed framework on, SMS, Twitter, Web and Email datasets. The base of their proposed model is ONMTF orthogonal nonnegative matrix tri-factorization model.

Their result shows that by incorporating information from Email and SMS, the performance of spammer detection framework is enhanced in Twitter, as their linguistic analysis shows that language used in microblogging is similar to email and SMS. If we remove any one of the external source the performance decreases.

Limitations: Hybrid features set may also be evaluated for comparison
Advantage: Instead of word-level, they capture information from other media from topic level, as information at word level leads to high dimensional feature space

As the social media is growing with time, a variety of real-time data is streaming over the internet which is different in distribution because coming from different domain having multidomain (Flicker, Google, YouTube) property and heterogeneous having different modality i.e. multimodal (text, image, audio, video) property. In 2015 [15] proposed a novel algorithm called CDFL cross-domain feature learning algorithm by using mSDA marginalized stacked denoising autoencoder. It is the first work which studied both the multimodal and multidomain property simultaneously and applied them to autoencoder. To minimize the effect of different data distribution across domains and to learn domain invariant features, the cross-domain constraint is introduced in auto encoders using MMD maximum mean discrepancy. To increase the correlation among different modalities of data CCA canonical correlation analysis is used. They evaluate their algorithm on–spam filtering, sentiment classification, and event classification. For spam filtering, they split UCI spam dataset as-. One set is used as source domain data and other two samples are used as target samples one as unlabeled training data and other as a test sample of the target domain. Gaussian noise is added in target samples to simulate domain discrepancy.

Limitation: They may evaluate their algorithm for spam detection by working on live data coming from different social media. Also for spam detection, they did not use the multimodal constraint to avoid complexity

Advantage: It is the first study which considered both the multimodal and multidomain property of data together in a framework of deep learning and evaluates it by applications like sentiment classification and event classification

4.2 URL Based

To gain more and more profit and to reach as many users as they can, spammers are disseminating malicious information to multiple platforms. In 2012 [12] focused to study spam on multiple platforms at the same time period. They did combine analysis of data from Yahoo emails and one-month tweets from Twitter. They worked on the basis that spammers mostly use URL links to direct users to a malicious site. They analyze the tweets and emails to extract all the URLs and links to their final landing page and compare them with URL blacklists and spamtrap emails to find malicious URLs. They state that there is a need for a common methodology for detecting quick spam across domains with better accuracy.

Limitation: The yahoo dataset comprises of pre-filtered emails, but as Twitter has its own spam filters so tweets collected are after spam filtering. So all the common features are not covered in detecting spam across platforms

Advantage: They concluded that because of data limitation they are not able to know the way how spammers are using Twitter and email, but their result shows that incorporating information from multiple platforms at the same time may be helpful in getting more accurate results. In a way, they inspire us to build a common framework to detect spam across multiple platforms

To give users more alluring features, different social media may communicate with each other. So the possibility of having the same users and same kind of information prevailing across the network increases. This fascinates spammers also to spread the same malicious content to multiple platforms. So if spam detection frameworks across different domain have the capability to interact then it will significantly reduce the spam action across the different domains. In 2016 [16] came up with a similar thought, they collected posts and tweets from Facebook and Twitter respectively. They collected this data over the same time period and with similar keywords from both the platforms, to have common features. They manually labeled the dataset whether it is spam or ham on the basis of URL link they are pointed to. They split the Twitter spam dataset and Facebook spam dataset into training and test set respectively. To show how the spam on one social network influences the other social network, they combine the twitter spam dataset with Facebook training dataset, trained the classifier and test the rest of the dataset. Then they do the same with Twitter training dataset i.e. combine it with Facebook spam dataset then trained the classifier and test the original test set. They did the classification through WEKA using different classifiers. Random Forest gives the best performance for both Twitter and Facebook spam dataset.

Limitation: They did not work on extracting effective features which are domain invariant to remove domain discrepancy

Advantage: They have considered the time factor in collecting data, most of the earlier work has not considered it

4.3   Multiple Features Based

Twitter and Facebook are the sources of ample data for research, as millions of users are members of these networks, with popularity vulnerability grows. So these networks are also on the hit list of spammers.

In 2013 [13] a generic statistical approach was used to find 14 features which are common to both Twitter and Facebook. The approach is novel in the sense that they have used such features which are generic to both Twitter and Facebook to detect spam profiles and applied them to train classifiers independently as well as jointly and compare the result. Earlier work has focused more on wall post contents and tweets. This is the first attempt especially in case of Facebook to identify features for spam profile detection. They have found 7 most significant features to discriminate spammers and legitimate users and give a clustering-based approach using Markov Clustering Algorithm to identify spam campaigns on both Twitter and Facebook.

Their result shows that the J48 algorithm achieves a better result when it is trained with combined dataset.

Limitation: More features can be incorporated for better detection accuracy, as these features are not complete

Advantage: The first study who focused on profile related features rather than only wall post features especially in the case of Facebook. Also, this work throws some light on how to extract common features across domains, which might be useful for cross-domain data analysis

Table 1. Comparison of cross-domain social spam detection techniques

	Pros	Cons
Content-based	They are easy to compute and easily available in form of tweets, posts, etc.	By augmenting contextual features with another category of features, the performance of a framework can be further improved. Since they are not a complete set of features
URL-based	As per the current literature, this is the most common way to spread spam, also it is independent of the context	Time lag problem, detection after clicking the link, also spammer use URL shortening services to evade detection
Multiple features based	Using different features together may increase the chance to get a more accurate result	As spammers are evolving in nature, static set of features is never complete and sound. They need to be evolving with time

Summarization of existing work done in cross-domain social spam detection is given in Table 2.

Table 2. Summarization of existing work done in cross-domain social spam detection

Reference	Year	Social networking site	Data statistics	The time during which data is collected	Ground Truth	Approach	Features	Observation
[12]	2012	Twitter & Yahoo	Messages with URLs- 47% Yahoo only, 64% common. 1% twitter only, 99% common. Domains - > 99% Yahoo only, < 0% Yahoo. 45% Twitter only, 55% common	One month in March 2011. (focus on emails and tweets sent during the same time period)	Manually Messages with URLs-Blacklists -75% Yahoo only, 47% common. < 1% Twitter only, > 99% common. Domains – Blacklists- 99%yahoo only, 1% common. < 1% Twitter only, > 99% common	Blacklisting URL and domains	URL	They have shown spam that is common across multiple platforms has higher exposure and is able to spread more malicious content than spam that advertises on only one platform
[13]	2013	Facebook, Twitter	320 Facebook profiles, 305 Twitter profiles. ((java API)	Not available	Manually identified 165 spam profiles and 155 normal user profiles (Facebook), and 160 spam profiles and 145 normal user profiles (Twitter)	Machine learning	Multiple features like - post driven, Interaction driven, tag driven, URL driven	The J48 algorithm performs best with combined dataset

(*continued*)

Table 2. (*continued*)

Reference	Year	Social networking site	Data statistics	The time during which data is collected	Ground Truth	Approach	Features	Observation
[14]	2014	Twitter, SMS, Email and web datasets	TAMU Social Honeypots Dataset - 41,499. Enron email dataset 200000 emails. Web & SMS dataset already constructed in previous work	TAMU Social Honeypots Dataset - December 30, 2009 to August 2, 2010. Twitter Suspended Spammers Dataset - July to September 2012 via the Twitter Search API	Manual identification of spam & ham messages	ONMTF	Content-based	With the use of information from different media, the proposed method performs effective spammer detection
[11]	2014 (In 2011 this work was proposed then extended in 2014)	MySpace, Email, Twitter, Webb spam corpus & WebBase dataset.	1.8 million MySpace profiles, 50,000 spam emails from email honeypot and over 25,000 legitimate emails, 900,000 Twitter users, over 2.4 million Tweets, 350,000 spam web pages, crawled from spam links. 392,000 legitimate web pages	MySpace- June to September 2006, Email – not mentioned, Twitter - November 2009 to February 2010. Webb spam corpus & WebBase dataset November 2002 to January 2006	1,500 spam MySpace profiles from the previous study.26,000 Twitter users and 138,000 Tweets suspended by Twitter violating their norms	Machine learning	Content-based – include textual attribute like -About me section of the profile, the body of the message, the content of web page model	Cross-domain classification does not perform well if objects have nothing in common and associative classification would improve the result of other object classification
[15]	2015	Email (Gaussian noise is added in the sample set to introduce data discrepancy)	UCI spam dataset 4601 emails	Not available	2,788 non-spam and 1,813 spam instances (given in the UCI spam dataset)	CDFL based on stacked denoising autoencoder. The maximum mean discrepancy is introduced in each layer of autoencoder to reduce domain discrepancy	Content-based	CDFL against state of art approaches performs best
[16]	2016	Facebook, Twitter	Twitter - 12879 tweets. (Twitter API) Facebook – 10623 posts. (Facebook API)	Dataset collected with keyword "Taylor Swift" on Twitter June to August (2015). Facebook - Dataset collected with keyword "World of Taylor Swift" July to August (2015)	Manually labeled Twitter - 1937 spam tweets and 10942 genuine tweets. Facebook - 1338 spam posts and 9285 genuine posts	Machine learning	URL	Random forest performs best for classification across different OSNs

5 Cross-Domain Opinion Spam Detection

Opinion spam is a kind of spam which refers to illicit activities like writing fake reviews to gain incentives. In e-commerce sites reviews about a product play a major role, it is an unbiased opinion of an individual about a product, which may influence the opinion of other users about that product. Spammers are exploiting this platform also by writing fake reviews. Cross-domain analysis of opinion spam has been done by few researchers. We have classified their work to detect opinion spam into 2 categories as mention below. Comparison of these approaches is given in Table 3.

- Supervised approach
- Semi-supervised approach

5.1 Supervised Approach

In 2014 [17] gave cross-domain results on the multi-domain public dataset created by them, including reviews of Restaurants, Doctors, Hotels by taking help of domain experts and through crowdsourcing. They were the only one who builds this standard dataset, which was further used by many researchers. A sparse additive generative model (SAGE) which is a generative Bayesian approach was used for cross-domain classification. The model was trained on one domain to test the dataset of other domains. Unigram features among all perform the best but SAGE does not outperform SVN. [18] evaluated the effectiveness of SVN (support vector networks) as domain independent deception detection. Features were generated using various methods like LDA (latent Dirichlet allocation), LIWC (linguistic inquiry and word count), WSM (a word space model). The cross-domain classification has been done to detect deception in that domain that has not been used in training. The best result was obtained by using a combination of LDA and WSM with SVN.

Detecting Cross-Domain Opinion Spam using Neural Networks

Neural network approach outperforms the existing approaches of handcrafted features. They are able to handle high dimensional feature space. Their hidden layers are responsible for automatic combinations of features to figure out the other hidden features [19]. Presented a sentence weighted neural network (SWNN) based model to learn a representation of reviews, in form of short documents. The model comprised of two convolution layer one for sentence compositions and another one to transform sentence to document vector. Each sentence has a weight associated with it, as a sentence is comprised of different review's words each having different importance to review, some may be more important to identify spam review some are not. Two features are added to the model POS and first-person pronouns. The cross-domain experiment was done on dataset released by [17] and compared with SVM as it outperforms SAGE (also mentioned above). It was concluded from the analysis that SWNN does not perform well as sentence weight are domain dependent. For mixed domain experiment, Unigram feature is the most robust one [20]. Also made use of neural network to learn document level representation. For sentence composition, convolution layer is used and to transform sentence to document vector bi-directional gated recurrent neural network is used as the second layer. The model was assessed on the dataset build by [17]. In domain and cross-domain tests outperform the state of art methods. For the Cross-domain test, this model performs better than [17].

5.2 Semi-supervised Approach

Plenty of unlabeled data is available over the internet. Opinion spam was first studied in 2008 [24]. But very limited work has been done in using approaches other than supervised. To recognize ground truth spam data for the supervised approach is an expensive task and require domain expertise. A semi-supervised multi-task learning model via Laplacian regularized logistic regression (SMTL-LLR) was developed, which make use of unlabeled data available online to increase the learning for one review spam detection task by incorporating information from another related task [21]. A stochastic alternating optimization method is developed for the optimization problem. Evaluation of the model was done on the multi-domain dataset released by [17]. Though this work is slightly different from domain adaption task as here we are interested in making a generalized model which will improve the learning of all the task simultaneously and in the domain adaption task, the performance of the target task is expected to increase by leveraging knowledge from source task.

Table 3. Comparison of cross-domain opinion spam detection approaches

	Pros	Cons
Supervised approach	It is the most common and easy way to detect if labeled data is available	Getting Ground truth for labeled data is not easy, Suffers from class imbalance problem, may be domain specific
Semi-supervised approach	It fits in well where plenty of unlabeled data is present and labeled data is limited	Maybe domain specific

Summarization of existing work done in cross-domain opinion spam detection is given in Table 4.

Table 4. Summarization of existing work done in cross-domain opinion spam detection

Ref	Year	Domains	Statistics of data	Ground Truth	Technique	Features	Observations
[17]	2014	Hotel, Restaurant, Doctor.	Hotel(p/n): Turker 400/400, Expert 140/140, Customer 400/400; Restaurant(p/n): Turker 200/0, Expert 120/0, Customer 200/200; Doctor(p/n): Turker 200/0, Expert 32/0, Customer 200/0	Customer generated truthful reviews, Turker& employee generated deceptive	SAGE and SVM are compared (supervised)	LIWC, Unigram,and POS	Unigram feature performs the best for cross-domain classification, though SAGE doesnotoutperform SVN.
[21]	2016	Hotel, Restaurant, Doctor.	Doctor: Spam/non spam 400/400, unlabelled 10,000; Hotel: Spam/non spam 200/0, unlabelled 10,000; Restaurant: Spam/non spam 200/0, unlabelled 10,000	Used already existing ground truth datasets	SMTL-LLR(semi-supervised)	Text unigram and bigram term-frequency	Learning of one task can be enhanced by incorporating information from related task's training signal.
[18]	2017	DeRev corpus opinion about books, OpSpam (opinion about hotels), Opinions dataset about abortion, death penalty, and a best friend.	DeRev corpus - 118 are truthful texts and 118 are deceptive texts. OpSpam corpus – 400 deceptive texts and 400 truthful texts. Opinions dataset - 100 deceptive texts and 100 truthful texts.	AMT is used to generate deceptive reviews for OpSpam and opinion dataset. Trip advisor is used for OpSpam truthful review. Rest reviews are extracted manually keeping some aspects in mind.	SVN (supervised)	LDA, LIWC, WSM	Accuracy - in one-domain 86%, in mixed-domain 75%, and in cross-domain 52 to 64%.
[19]	2017	Hotel, Restaurant, Doctor.	Hotel: Turker 800, Expert 280, Customer 800; Restaurant: Turker 200, Expert 0, Customer 200; Doctor: Turker 356, Expert 0, Customer 200	AMT &employee collect deceptive reviews from online workers. Truth reviews are from customers.	SWNN (sentence weighted neural network model) (supervised)	POS, First person pronouns	SWNN is domain-specific, so it does not give the expected result.
[20]	2017	Hotel, Restaurant, Doctor.	Hotel(p/n): Turker 400/400, Expert 140/140, Customer 400/400; Restaurant(p/n): Turker 200/0, Expert 120/0, Customer 200/200; Doctor(p/n): Turker 200/0, Expert 32/0, Customer 200/0	Customer-generated truthful reviews, Turker& employee generated deceptive reviews.	Neural network model (consists of both CNN and GRNN). (supervised)	Document representation is used as features.	For cross-domain classification,ac curacies can be improved by combining discrete and neural network.

6   Challenges Corresponding to Cross-Domain Spam Detection

- Different social networks have different characteristics, so having a common detection framework is a tedious job. Like on Twitter a user can follow anyone else without the followee's consent to build strong social relations with legitimate users, this is not the case on Facebook. So one of the major challenges in cross-domain classification is limited by common features across the different online social network.

- Data collected during the same time may have some common context like the discussion on social media about a trending topic or some issues. To collect data during the same time period from different OSNs is a challenging task.
- Limited accessibility to get the private data makes the task less transparent.
- As spammers behavior is dynamic in nature i.e. with evolving detection techniques their evading techniques are also evolving. So the traditional ways of classification i.e. a classifier trained on some static features set are not enough. To capture this evolving nature of spam we need classifiers which are trained on source data at time (t), would also classify target data at time (t + 1) efficiently, for that we need to figure out the change in context with time in source and target data, which is called concept drift [22]. Concept drift arises due to evolving features.
- Most of the existing work relies on supervised machine learning, which requires labeled data which is limited as spams are volatile in nature and it is not easy to find the ground truth. So we can switch to a semi-supervised or unsupervised learning approach. Also, the existing supervised approaches are domain specific so it is still unsure whether they would work well in other domains or not.
- To extract handcrafted features from a huge dataset is a complex task. With the help of deep learning technology, high-level features could learn automatically making the feature learning task less complex. So switching from traditional machine learning based spam detection to deep learning is required.
- There is a need to improve training time with the increasing volume of data over the social networks. Techniques which give the same accuracy but improved training time must be adopted, like Extreme Learning Machines ELM a new scheme of feedforward neural networks [23].

7 Discussion and Conclusion

With a growing user base on social media, spammers are also proliferating and using multiple content sharing platforms to reach as many users as possible. To detect them, techniques which can work across domains to give more accurate and cost-effective results are required while keeping all the above challenges in mind.

This manuscript gives a detailed study of cross-domain spam detection and its challenges. In future researchers will focus more on analyzing the activities of spammers, their behavior in different OSNs and intend to construct a common framework for spam detection. Online cross-domain spam detection tools are required because most of the present detection tools are offline and they cannot work well online due to long lag time and their inefficiency to handle huge live streaming data. We think this area has great potential in future research as it has been scantily studied yet.

References

1. Grier, C., Thomas, K., Paxson, V., Zhang, M.: @ spam: the underground on 140 characters or less. In: Proceedings of the 17th ACM Conference on Computer and Communications Security, pp. 27–37. ACM (2010)
2. Mathur, A., Prachi, G.: Spam detection techniques: issues and challenges. Int. J. Appl. Inf. Syst. (IJAIS) (2013). ISSN 2249-0868
3. Kaur, R., Singh, S., Kumar, H.: Rise of spam and compromised accounts in online social networks: a state-of-the-art review of different combating approaches. J. Netw. Comput. Appl. **112**, 53–88 (2018)
4. Sun, Q., et al.: Transfer learning for bilingual content classification. In: Proceedings of the 21th ACM SIGKDD International Conference on Knowledge Discovery and Data Mining, pp. 2147–2156. ACM (2015)
5. Ramalingam, D., Chinnaiah, V.: Fake profile detection techniques in large-scale online social networks: a comprehensive review. Comput. Electr. Eng. **65**, 165–177 (2018)
6. Bruns, A., et al.: Detecting Twitter bots that share SoundCloud tracks. In: Proceedings of the International Conference on Social Media + Society, vol. 8, pp. 251–255. ACM Press (2018)
7. Thomas, K., Grier, C., Ma, J., Paxson, V., Song, D.: Design and evaluation of a real-time URL spam filtering service. In: 2011 IEEE Symposium on Security and Privacy (SP), pp. 447–462. IEEE (2011)
8. Stringhini, G., Kruegel, C., Vigna, G.: Detecting spammers on social networks. In: Proceedings of the 26th Annual Computer Security Applications Conference, pp. 1–9. ACM (2010)
9. Webb, S., Caverlee, J., Pu, C.: Introducing the web spam corpus: using email spam to identify web spam automatically. In: Proceedings of the Third Conference on Email and Anti-Spam (CEAS) (2006)
10. Wang, D., Irani, D., Pu, C.: A social-spam detection framework. In: Proceedings of the 8th Annual Collaboration, Electronic Messaging, Anti-Abuse and Spam Conference, pp. 46–54. ACM (2011)
11. Wang, D., Irani, D., Pu, C.: Spade: a social-spam analytics and detection framework. Soc. Netw. Anal. Min. **4**(1), 189 (2014)
12. Lumezanu, C., Feamster, N.: Observing common spam in Twitter and email. In: Proceedings of the 2012 Internet Measurement Conference, pp. 461–466. ACM (2012)
13. Ahmed, F., Abulaish, M.: A generic statistical approach for spam detection in Online Social Networks. Comput. Commun. **36**(10–11), 1120–1129 (2013)
14. Hu, X., Tang, J., Liu, H.: Leveraging knowledge across media for spammer detection in microblogging. In: Proceedings of the 37th International ACM SIGIR Conference on Research & Development in Information Retrieval, pp. 547–556. ACM (2014)
15. Yang, X., Zhang, T., Xu, C.: Cross-domain feature learning in multimedia. IEEE Trans. Multimed. **17**(1), 64–78 (2015)
16. Xu, H., Sun, W., Javaid, A.: Efficient spam detection across online social networks. In: 2016 IEEE International Conference on Big Data Analysis (ICBDA), pp. 1–6. IEEE (2016)
17. Li, J., Ott, M., Cardie, C., Hovy, E.: Towards a general rule for identifying deceptive opinion spam. In: Proceedings of the 52nd Annual Meeting of the Association for Computational Linguistics (Volume 1: Long Papers), vol. 1, pp. 1566–1576 (2014)
18. Hernández-Castañeda, Á., Calvo, H., Gelbukh, A., Flores, J.J.G.: Cross-domain deception detection using support vector networks. Soft. Comput. **21**(3), 585–595 (2017)

19. Li, L., Qin, B., Ren, W., Liu, T.: Document representation and feature combination for deceptive spam review detection. Neurocomputing **254**, 33–41 (2017)
20. Ren, Y., Ji, D.: Neural networks for deceptive opinion spam detection: an empirical study. Inf. Sci. **385**, 213–224 (2017)
21. Hai, Z., Zhao, P., Cheng, P., Yang, P., Li, X.L., Li, G.: Deceptive review spam detection via exploiting task relatedness and unlabeled data. In: Proceedings of the 2016 Conference on Empirical Methods in Natural Language Processing, pp. 1817–1826 (2016)
22. Henke, M., Souto, E., dos Santos, E.M.: Analysis of the evolution of features in classification problems with concept drift: application to spam detection. In: 2015 IFIP/IEEE International Symposium on Integrated Network Management (IM), pp. 874–877. IEEE (2015)
23. Zheng, X., Zhang, X., Yu, Y., Kechadi, T., Rong, C.: ELM-based spammer detection in social networks. J. Supercomputing **72**(8), 2991–3005 (2015)
24. Jindal, N., Liu, B.: Opinion spam and analysis. In: Proceedings of the 2008 International Conference on Web Search and Data Mining, pp. 219–230. ACM (2008)
25. Jain, G., Sharma, M., Agarwal, B.: Spam detection in social media using convolutional and long short term memory neural network. Ann. Math. Artif. Intell. **85**, 21–44 (2019). https://doi.org/10.1007/s10472-018-9612-z
26. Jain, G., Sharma, M., Agarwal, B.: Spam detection on social media using semantic convolutional neural network. Int. J. Knowl. Discov. Bioinf. (IJKDB) **8**(1), 12–26 (2018)

A Comparative Study on k-means Clustering Method and Analysis

Rajdeep Baruri[1(✉)], Anannya Ghosh[2], Saikat Chanda[2], Ranjan Banerjee[1(✉)], Anindya Das[1], Arindam Mandal[1], and Tapas Halder[3]

[1] Department of Computer Science and Engineering,
Jadavpur University, Kolkata, India
rajdeepbaruri2602@gmail.com, rbkpccst@gmail.com,
blueanindyadas@gmail.com, arrrindam@gmail.com
[2] Department of Computer Science and Engineering,
Institute of Engineering and Management, Kolkata, India
anannyaghosh2@gmail.com, saikatcd5@gmail.com
[3] Cyber Patrol Cell, Kolkata Police, Kolkata, India
tapas.cs.ju@gmail.com

Abstract. A study of three clustering methods using four different cluster validity metrics is being presented here. We have discussed the clustering methods and made an analysis. We have given the mathematical formation of four cluster validity measures. From the experimental outcomes, indications regarding the optimal validation method, as well as, optimal clustering method are being presented. Choice of preferable clustering technique is presented after getting outcomes using real-world data sets.

Keywords: Data analytics · Machine learning · Algorithm analysis · Clustering validity

1 Introduction

Clustering is one kind of unsupervised learning technique in machine learning. Clustering may be useful when we do not have labeled data. K-means clustering is one of the many clustering algorithms. Clustering is the process of partitioning a set of data objects into subsets of clusters so that objects within the same cluster are similar to one another yielding dissimilarities with objects within other clusters.

Clustering is widely used in many applications including business intelligence, DNA analysis in computational biology, security [10,14], intrusion detection [7], intelligent transportation system, music sound features analysis [13], social studies [22]. Partitioning method, hierarchical method, density-based method, and grid-based method are mainly four categories of clustering. Clustering helps us to discover information hidden in the given data sets [10]. In this research work, we are interested in only k-means clustering which is a partition-based method.

© Springer Nature Singapore Pte Ltd. 2019
A. K. Somani et al. (Eds.): ICETCE 2019, CCIS 985, pp. 113–127, 2019.
https://doi.org/10.1007/978-981-13-8300-7_10

This paper is divided into mainly seven parts. In the second part, we have discussed the recent current research interest about clustering. In the third part, we have outlined the k-means algorithm and analyzed the time complexity and then we have investigated two improvements over the traditional k-means algorithm. In the fourth part, we have given a brief outline of our experimental setup process followed by the data sets used in our evaluation. In the sixth part, we have discussed about the performance evaluation criteria. In the next two section, we have described our results and analysis over our experiments.

2   Related Work

Though k-means is a simple clustering technique and easy to implement, there are certain factors upon which it depends heavily [20]. Choice of initial centroids and the value of k are the two main drawbacks of k-means. Some of the common limitations are discussed below.

- **Effects of outliers.** Outliers may be defined as those data points which are present in the data set but do not result in the clusters formed from the classification algorithm. Outliers can increase the sum of squared error within clusters. Researchers have identified that the presence of outliers results in unstable outputs while different executions are executed on same data set. Preprocessing technique is a popular method to remove outliers.
- k, **the number of clusters.** It may be worthwhile if we know, in advance, the number of clusters for a particular dataset. Finding the number of clusters is still a challenging task for researchers [1, 17].
- **Null set of clusters.** A bad choice of initial centroids may results in null set of clusters. Thus choice of initial centroids always play an important role in k-means clustering. Our experimental results agree with this facts, as we have observed.
- **Convex shapes of clusters.** The choice of clustering method drastically affects the shapes of cluster, scalability of cluster, and quality of resulting clusters. Different clustering algorithms perform different task in order to make cluster more dynamic and effective [19].

Several efforts have been applied to overcome these challenges. Here we mention some of them which works as our motivation factor.

A heuristic method has been proposed in [16] that find better initial centroids as well as more accurate clusters with lesser computational time.

Another method of refinement of initial point is being described in [3]. The refinement algorithm operates over small subsamples of a given data set. It is suitable mostly for large-scale clustering problems.

Another brilliant effort has been applied in [23] where a greedy methodology-based constructive approach is performed to reduce the clustering cost [12]. Empirical results on biological data show the accuracy of their implementation is quite well.

Again another effort has been accomplished on real world data where the analysis of k-means is done carefully. They have provided an improved version of traditional k-means clustering algorithm in [18]. They have used two special data structure to improve the time complexity of k-means.

3 Clustering Techniques

Researchers have experimented with several distance metrics for the purpose of evaluation of clustering algorithms. We have used standard Euclidean distance between two points for this purpose [8].

3.1 Lloyd's k-means Clustering

Suppose we are given a dataset of n data points where each data point is d-dimensional. And we need to organize these points into k clusters so that some the objective function tries to minimize the sum of all squared distances within a cluster, for all k clusters. The objective function is defined as

$$\arg \min \sum_{i=1}^{k} \left(\sum_{x_j \in S_i} ||x_j - \mu_i||^2 \right)$$

where S_i is a cluster, μ_i is cluster mean or centroid and x_j is a data point in the data set.

The procedure of standard k-means clustering procedure is given in Algorithm 1 [8,15]. The output is the set of clusters.

Algorithm 1. LLOYDKMEANS(D, k)

Input: $D = \{d_i : 1 \leq i \leq n\}$ is a database containing n datapoints and c_j is the center of j-th cluster, $1 \leq j \leq k$.
Output: a set of k clusters
1: select k random data objects from D as initial centroids
2: **repeat**
3:    calculate the distance between all d_i and all c_j
4:    assign d_i to the nearest cluster
5:    **for** $j = 1$ to k **do**
6:        recompute the j-th centroid
7:    **end for**
8: **until** convergence criteria is met
9: **return** k clusters

Objective Function Requirement. We need to find a method to minimize the objective function iteratively by choosing a new set of clusters and centroids so that the value of the function decreases in each iteration.

Time Complexity. Let p is the dimension of data set *i.e.* we are given n data points, and each of them has p attributes. Let us assume that the loop at line number 2 of Algorithm 1 repeats t times in order to meet the required centroids or clusters according to the convergence criteria. Then the complexity of Algorithm 1 is thus $\mathcal{O}(nkpt)$ [11]. Practically $k \ll n$ and $t \ll n$.

3.2   Greedy k-means Clustering

The implementation part of Algorithm 1 is quite easy in practice. It is clear from the loop begins at line number 5 of Algorithm 1 that every cluster is being recomputed in each iteration. That means we have to move a lot of data points in each and every iteration. It may be preferable if we move just single data point among the clusters in each iteration. What we need to find is which point makes the clustering quality best. We call that point P_{best}. A more constructive approach may be to move only and only P_{best} [12]. The idea is presented in Algorithm 2 [12,23].

Algorithm 2. GREEDYKMEANS(D, k)

Input: $D = \{d_i : 1 \leq i \leq n\}$ is a database containing n datapoints and c_j is the center
    of j-th cluster, $1 \leq j \leq k$.
Output: a set of k clusters
 1: choose an initial partition P of k clusters randomly
 2: **repeat**
 3:    PROFIT $= 0$
 4:    **for** $j = 1$ to k **do**
 5:       find the d_i for which PROFIT of C_j is maximum
 6:    **end for**
 7:    **if** PROFIT > 0 **then**
 8:       update partition by moving d_i to C_j
 9:    **else**
10:      **return** k clusters
11:    **end if**
12: **until** convergence criteria is met

Close Observation. Intuitively greedy method should give us a better execution time as indicated in [23]. But our empirical results, in Sect. 7, does not support this fact. The clustering produced by greedy version is not better than that of the original version. We are not yet sure the reason behind this. This may be the choice of dataset we have taken, or may be types of data we have worked with. The reason behind this is still unknown to us. What we can comment from our evaluation, in Sect. 7, is that this greedy version performs worst among all the three version we have taken care of.

Method 1. PROFITCALCULATION(C_j)

1: **for** $s = 1$ to k **do**
2:     **if** $s \neq j$ **then**
3:        find the cost $Cost_i$ after moving $d_i \in C_s$ to C_j
4:     **end if**
5:     find $Profit_i = CurrentCost - Cost_i$
6: **end for**
7: **return** $Profit$

3.3 Improved k-means Clustering

While Algorithm 1 iteratively finds the partition of datapoints into a predefined number of clusters, it needs to calculate the distance between each and every data points in every iteration. If there are n datapoints, the it takes almost $\binom{n}{2}$ pairs and for every pair it recompute the distances again and again. As mentioned in [18], our empirical results also suggests that it is not necessary to recompute all of the $\binom{n}{2}$ pairs which takes a considerable amount of time. An improvement may be to use two data structures – one to store the label of centroid within which d_i exists, another to store the distance to the nearest centroid from d_i, which we call $Distance_{old}$. Algorithm 3 presents the improved version of Algorithm 1 with the help of two data structures DIST and CLUSTER [18].

Algorithm 3. IMPROVEDKMEANS(D, k)

Input: $D = \{d_i : 1 \leq i \leq n\}$ is a database containing n datapoints and c_j is the center of j-th cluster, $1 \leq j \leq k$.
Output: a set of k clusters
1: INITIALCLUSTER = INITIALIZATION(D, k)
2: recalculate the centroids
3: **repeat**
4:     **for** $i = 1$ to n **do**
5:        find distance between d_i and its new centroid
6:        **if** NEWDISTANCE $>$ DIST[i] **then**
7:           compute newDistance among d_i and all centroids
8:           move d_i to nearest cluster c_x
9:           update CLUSTER[i]
10:          update DIST[i]
11:        **end if**
12:     **end for**
13: **until** convergence criteria is met
14: **return** k clusters

Intuitive Idea. The main idea behind Algorithm 3 is that before we (re)calculate the centroids, we shall check the old distance with the new distance. If new distance is less than the old distance, then we do not need to calculate the pairwise distance for that d_i. If new distance is greater than the old distance, then we shall calculate the pairwise distance for that d_i as we were doing it same for Algorithm 1. Thus we shall be able to save execution time of original algorithm to some extent.

Working Procedure. While Algorithm 3 is an improved version of Algorithm 1, the improvement is a trade off between a little space. Before the program control executes the loop at line number 3 of Algorithm 3, two data structures are already being computed. If the new distance within the new cluster is less than the previous distance stored at DIST, the computation for that point is omitted. Otherwise the pairwise distance computation is done for that point. At this point the new nearest distance and the corresponding centroid is updated in the corresponding data structures.

Method 2. INITIALIZATION(D, k)

1: choose k data points randomly as initial centroids
2: **for** $i = 1$ to n **do**
3:    calculate the distance between d_i and all c_j
4:    set DIST$[i]$ = minimum value in the previous step
5:    assign d_i to the nearest cluster, c_g
6:    set CLUSTER$[i] = g$
7: **end for**
8: **return** INITIALCLUSTER

Time Complexity. Suppose p is the dimension of data set *i.e.* we are given n data points, and each of them has p attributes. The for loop at line number 2 of Method 2 executes one time for the n data points. So for the INITIALIZATION(D, k) procedure it takes only $\mathcal{O}(nkp)$. At this point some datapoints will move to another clusters whereas others will remain at their existing clusters. If the datapoints moves, then it takes $\mathcal{O}(kp)$. If the datapoints do stay at the old cluster, then it takes $\mathcal{O}(1)$ time. According to the convergence of clustering algorithms, the number of datapoints moved from some cluster to some other cluster will decrease. If half of the datapoints move, then the time complexity being $\mathcal{O}(\frac{nkp}{2})$.

Difference Between Algorithms 1 and 3. The first difference between Algorithms 1 and 3 lies in the initialization step. Here the distance between all datapoints and all centroids is calculated along with some extra computation to calculate the distance and labels using two data structures. This extra computation can be done in $\mathcal{O}(1)$ time.

The second difference lies in the line number 6. The conditional statements saves lot of execution time omitting some the calculation of distances between datapoints and centroids.

If the program control enters into the else block at line number 8, then it is usually doing the calculation of Algorithm 1. The updating method at line number 11 and line number 12 is the third difference. This updating step can be done in $\mathcal{O}(1)$ time.

4 Experimental Setup

We have implemented the above mentioned clustering methods on a Lenovo ThinkPad E460 Ultrabook running the Windows 10 professional 64-bit operating system and intel core i7 processor. The clock speed of the processor being 2.5 GHz with 8G bytes DDR3 memory size.

The experiments were conducted in 3 phases. In the first set of experiment, we have implemented Algorithms 1, 2 and 3. We have used python programming language version 2.7.14 with pandas 0.21.0 version and NumPy 1.15.1 version. In the second phase we have computed the clusters generated from the corresponding algorithm and validate against cluster validate metrics. We have mentioned details about the validation metrics in Sect. 6. In the third phase we have compared the running time of the algorithms (Table 1).

Table 1. Algorithms we have experimented with

Sl. no.	Name of the algorithm	Nature
A01	Algorithm 1	Iterative
A02	Algorithm 2	Greedy
A03	Algorithm 3	Improved

5 Datasets

We have performed our experiments on seven real-world datasets available from UC Irvine Machine Learning Repository [5]. The following data sets have been used:

- **Iris Dataset** [5]. The class of iris plant is one of the many multivariate real-world datasets used in the pattern recognition. This dataset contains three classes of fifty instances each, where each class refers to one type of iris plant. It contains 150 instances and 4 numeric attributes.
- **Wine Dataset** [5]. This dataset comes from the analysis of origins of wines. It is a multivariate dataset consisting of real and integer attributes. This wine recognition data has 178 instances divided into three classes, and each data item is 13-dimensional.
- **Glass Identification Dataset** [5]. This dataset comes from USA Forensic Science Services. It has six different types of glass defined in terms of their oxide contents. This dataset contains all over 214 instances and each of them is 10-dimensional. There are mainly two distinguishable classes each of them has their own subdivisions.

- **Ecoli Dataset** [5]. This dataset is from Protein subcellular localization. It has 336 instances where each of them is 8-dimensional. Eight different classes are used to distribute the dataset.

Table 2 represents the datasets in ascending order of their sizes. The first column represents the symbolic names.

Table 2. Description of real world datasets

Sl. no.	Name of the dataset	Size in KB	Number of instances	Number of attributes
D01	Iris	4.4	150	4
D02	Wine	10.5	178	13
D03	Glass	11.6	214	10
D04	Ecoli	19	336	8

6   Performance Evaluation Criteria

In order to evaluate the quality of the clustering, we introduce four basic coefficients, namely silhouette index, sum of squared error, Davies-Bouldin index, and in the last but not the least is Dunn index, in this paper.

6.1   Silhouette Index

In our current research we have conducted our experiment on the clustering quality with respect to variety of parameter settings of cluster number. We have taken the value of $k = i$, for $3 \leq i \leq 20$

We have used silhoutte index (SI) for our purpose of evaluation. It helps us to be independent from the number of cluster produced.

For a given cluster C_j, for $1 \leq j \leq k$, the *silhouette width*, s_i, for $1 \leq i \leq m$, to the ith sample of C_j is defined as

$$s_i = \frac{b_i - a_i}{\max\{a_i, b_i\}}$$

where a_i is the average distance between the ith sample and all of the samples included in C_j.

$$a_i = \frac{1}{|C_j|} \sum_{x_i \in C_j, y_j \in C_j} d(x_i, y_j)$$

b_i is the minimum average distance between the ith sample and all of the samples clustered in C_y, for $y = 1, 2, \cdots, k; y \neq j$.

$$b_i = \min_k \left\{ \frac{1}{|C_k|} \sum_{x_i \in C_j, y_j \in C_k, k \neq j} d(x_i, y_j) \right\}$$

The *global silhouette index* [21], denoted by GS_U, can be used as a validity index for a partition U.

$$GS_U = \frac{1}{k} \sum_{j=1}^{k} s_j$$

Requirement. A high value of GS_U indicates that partition U is better or optimal cluster.

6.2   Davies-Bouldin Index

Apart from SI, we have used Davies-Bouldin Index (DBI) [4] as a validity metric. DBI is a measure that helps us to estimate the ideal number of clusters in a given data set. DBI assumes that we have a predefined number of clusters, for our experimental evaluation we have taken $k = i$, for $3 \leq i \leq 20$. DBI is based upon the average ratio between the within cluster scatter (S_l) and the inter-cluster distance.

As we mentioned in earlier that the value of k has a deep impact on the quality of cluster, Davies-Bouldin Index [4] is a validity metric which helps us to find the ideal value of k. A lower DBI indicates a better cluster configuration.

Let us assume that k be the predefined cluster number. For our experiments we have started with $k = 3$ and gradually increased it. Let v_l and v_m be the centroids for l-th and m-th cluster, respectively. Let x_l denotes any arbitrary point within l-th cluster. Figure 1. represents the intuitive idea of centroid v_l, and N_l points inside l-th cluster.

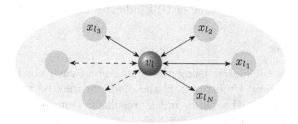

Fig. 1. Centroid and general points in l-th cluster

Then the intra-cluster scatter, denoted by S_l, is defined mathematically as:

$$S_l = \sqrt{\frac{1}{N_l} \sum_{j=1}^{N_l} ||x_j - v_l||^2}$$

Let D_{lm} be the distance between two clusters whose centroids are v_l and v_m. Let N_l and N_m denotes the number of datapoints in cluster l and l respectively. Then

$$D_{lm} = \sqrt{\frac{1}{N_l N_m} \sum_{l \in C_l} \sum_{m \in C_m} ||x_l - x_m||^2}$$

Figure 2 represents the constructive idea behind D_{lm}.

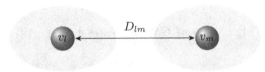

Fig. 2. Two centroids and distance between corresponding clusters

Let R_{lm} denotes the joint cluster scatter over the distance between cluster l and m. R_{lm} can be calculated as:

$$R_{lm} = \frac{S_l + S_m}{D_{lm}}$$

The Davies-Bouldin index is defined as

$$\text{DBI} = \frac{1}{K} \sum_{l=1}^{K} R_l,$$

where

$$R_l = \max_{l \in K, l \neq m} R_{lm}$$

Requirement. A minimal value of DBI indicates an optimal k [4].

6.3 Dunn Index

Dunn index [6] is calculated using the diameter of each cluster and the distance between clusters. Unfortunately the diameter may be grievously affected by noise, which affects the Dunn index, resulting Dunn index's performance undesirable. That is why Dunn index may not perform well and good in a cluster validity index. This observation has been mentioned in [2].

Let l be the label of an arbitrary cluster C_l. Let ΔS denotes the maximum distance between two datapoints in C_l. Figure 3 represents the basic idea behind ΔS.

$$\Delta S = \max_{x,y \in S} \{d(x,y)\}$$

Let C_k be another cluster. The inter-cluster distance between C_l and C_k, denoted by $\delta(k, l)$, is the smallest distance between all pair of points where one

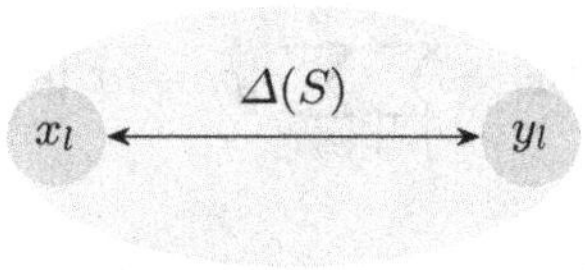

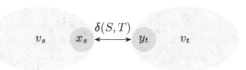

Fig. 3. Computing $\Delta(S)$

point belongs to l and the other point belongs to k. Figure 4 represents the basic idea behind $\delta(S,T)$.

$$\delta(S,T) = \min_{x \in S, y \in T} \{\delta(x,y)\}$$

Fig. 4. Computing $\delta(S,T)$

Dunn index [6] is defined as

$$\frac{\min\{\Delta S\}}{\max\{\delta(S,T)\}}$$

It may be noted that DI, along with SI and DBI, is commonly used as inter-cluster validity metrics. We have computed all these metrics to validate the clusters generated from A01, A02, A03 although, in some literature, it may be observed that DI may not be perform well in validity.

Requirement. A higher DI is preferable over a lower value, indicating a better cluster formation.

6.4 Sum of Squared Error

Let p_i be a datapoint inside j-th cluster C_j. The centroid of C_j, denoted by m_j, is computed as below:

$$m_j = \frac{\sum_{p_i \in C_j} p_i}{|C_j|}$$

Now distance between point p_i and cluster C_j is defined as $||p_i - m_j||$. The sum of squared error (SSE) [9] is defined as

$$\text{SSE} = \sum_{j=1}^{k} \sum_{i=1}^{n} ||p_i - m_j||^2$$

Requirement. A smaller SSE, corresponding to one cluster, is preferred to a larger SSE, corresponding to another cluster.

7   Results of Evaluation and Analysis

In this section we present our results of experiments. We have implemented the previous algorithms and validate them against the metrics.

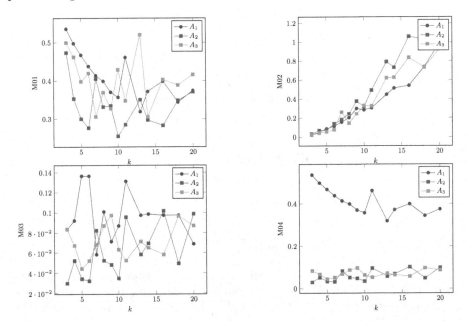

Fig. 5. Experimental results on D01 (Color figure online)

Algorithms 1, 2, and 3 is shown in blue, red and green line respectively. Algorithm 1 performs better than Algorithm 2, as indicated by Fig. 5. Figures 6, 7, 8 also agrees the same. Algorithm 2 performs worst. A lower k value gives us a better cluster formation, as these results demand. Some time Algorithm 3 gives better results than Algorithm 1, but Algorithm 2 never does so. We are yet unaware of the reason behind why algorithm behaves so. We shall study that in our next research.

8   Future Research Directions

Bayesian Information Criterion [24] is an alternative method for detecting the number of optimal number clusters. It may an interesting choice to study the

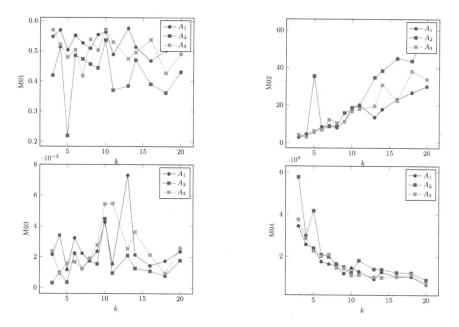

Fig. 6. Experimental results on D02 (Color figure online)

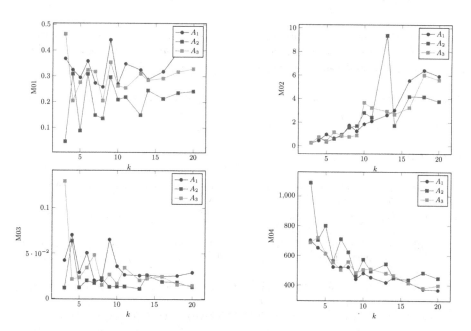

Fig. 7. Experimental results on D03 (Color figure online)

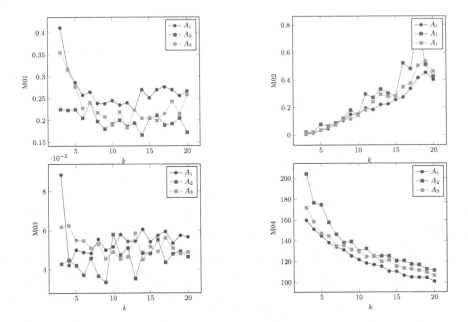

Fig. 8. Experimental results on D04 (Color figure online)

behavior of Bayesian information criterion and compare the results with k-means clustering methods. Moreover k-Strange point [11] algorithm is another choice instead of traditional k-means algorithm.

Acknowledgment. This research is funded by Jadavpur University (UGC-UPE, Phase-II, grant no. P-1/RS/115/13).

References

1. Abbas, O.A.: Comparisons between data clustering algorithms. Int. Arab J. Inf. Technol. **5**, 320–325 (2008)
2. Bezdek, J.C., Pal, N.R.: Some new indices of cluster validity. IEEE Trans. Syst. Man Cybern. **28**, 301–315 (1998)
3. Bradley, P.S., Fayyad, U.M.: Refining initial points for k-means clustering. In: Proceedings of the Fifteenth International Conference on Machine Learning, pp. 91–99 (1998)
4. Davies, D.L., Bouldin, D.W.: A cluster separation measure. IEEE Trans. Pattern Anal. Mach. Intell. **2**, 224–227 (1979)
5. Dheeru, D., Taniskidou, E.K.: UCI Machine Learning Repository (2017)
6. Dunn, J.C.: Well-separated clusters and optimal fuzzy partitions. J. Cybern. **4**, 95–104 (1974)
7. Eslamnezhad, M., Varjani, A.Y.: Intrusion detection based on MinMax K-means clustering. In: 7th International Symposium on Telecommunications, pp. 804–808 (2014)

8. Han, J., Kamber, M., Pei, J.: Data Mining: Concepts and Techniques, 3rd edn. Morgan Kaufmann, Burlington (2011)
9. Hand, D., Smyth, P.: Principles of Data Mining. MIT Press, Cambridge (2001)
10. Huang, Z.: Extensions to the k-means algorithm for clustering large data sets with categorical values. Data Min. Knowl. Discov. **2**, 283–304 (1998)
11. Johnson, T., Singh, S.K.: K-strange points clustering algorithm. In: Jain, L.C., Behera, H.S., Mandal, J.K., Mohapatra, D.P. (eds.) Computational Intelligence in Data Mining - Volume 1. SIST, vol. 31, pp. 415–425. Springer, New Delhi (2015). https://doi.org/10.1007/978-81-322-2205-7_39
12. Jones, N.C., Pevzner, P.A.: An Introduction to Bioinformatics Algorithms. The MIT Press, Cambridge (2004)
13. Krey, S., Ligges, U., Leisch, F.: Music and timbre segmentation by recursive constrained K-means clustering. Comput. Stat. **29**, 37–50 (2014)
14. Li, W.: Modified K-means clustering algorithm. In: 2008 Congress on Image and Signal Processing, pp. 618–621 (2008)
15. Lloyd, S.P.: Least squares quantization in PCM. IEEE Trans. Inf. Theory **28**(2), 129–137 (1982)
16. Mahmud, M.S., Rahman, M.M., Akhtar, M.N.: Improvement of k-means clustering algorithm with better initial centroids based on weighted average. In: International Conference on Electrical & Computer Engineering, pp. 647–650 (2012)
17. Maulik, U., Bandyopadhyay, S.: Performance evaluation of some clustering algorithms and validity indices. IEEE Trans. Pattern Anal. Mach. Intell. **24**, 1650–1654 (2002)
18. Na, S., Xumin, L., Yong, G.: Research on k-means clustering algorithm: an improved k-means clustering algorithm. In: Proceedings of the Third International Symposium on Intelligent Information Technology and Security Informatics, pp. 63–67 (2010)
19. Patil, Y.S., Vaidya, M.B.: A technical survey on cluster analysis in data mining. Int. J. Emerg. Technol. Adv. Eng. **2**, 503–513 (2012)
20. Peña, J.M.S., Lozano, J.A., Larrañaga, P.: An empirical comparison of four initialization methods for the k-means algorithm. Pattern Recogn. Lett. **20**, 1027–1040 (1999)
21. Rousseeuw, P.J.: Silhouettes: a graphical aid to the interpretation and validation of cluster analysis. J. Comput. Appl. Math. **20**, 53–65 (1987)
22. Wasserman, S., Faust, K.: Social Network Analysis: Methods and Applications. Cambridge University Press, Cambridge (1994)
23. Wilkin, G.A., Huang, X.: K-means clustering algorithms: implementation and comparison. In: Proceedings of the Second International Multi-Symposiums on Computer and Computational Sciences, pp. 133–136 (2007)
24. Zhao, Q., Hautamaki, V., Fränti, P.: Knee point detection in BIC for detecting the number of clusters. In: Blanc-Talon, J., Bourennane, S., Philips, W., Popescu, D., Scheunders, P. (eds.) ACIVS 2008. LNCS, vol. 5259, pp. 664–673. Springer, Heidelberg (2008). https://doi.org/10.1007/978-3-540-88458-3_60

Smart Judiciary System: A Smart Dust Based IoT Application

Shelendra Kumar Jain$^{(\boxtimes)}$ and Nishtha Kesswani

Central University of Rajasthan, NH-8, Bandar Sindri,
Ajmer 305817, Rajasthan, India
Shelendra23@hotmail.com, nishtha@curaj.ac.in

Abstract. Due to the popularity of IoT, applications related to IoT
have gained focus. In this paper, we have proposed a Smart Dust based
Smart Judiciary System. We have also discussed the merits and demerits
of the proposed system. And have also illustrated the risks and challenges
related to Smart Dust. We have also briefly highlighted some of the
emerging domains of IoT.

Keywords: Smart Dust · Internet of things · IoT applications

1 Introduction

Internet of things (IoT) has the potential to contribute to the economic growth of
a country. IoT and its variants such as Internet of battlefield Things, Internet of
everything, Internet of Military Things [1] technologies are based on sensor networks
and its usage is increasing day by day [7]. It leverages several application
sectors in this current era. This technology is enhancing accuracy, transparency
and smartness of applications. Currently it is being used and many more applications
are developing in many sectors such as transportation, industrial, health,
agriculture, smart city, supply chain management etc. One of the emerging and
interesting application areas of IoT is detecting criminal activities. IoT and cheap
sensors can be used for detection and reduction of crimes [10]. Information and
communication technologies (ICT) are playing important role in detecting several
type crimes like a networked camera based surveillance system to detect
crimes. IoT based system can be very useful for crime detection.

Our proposed applications are Smart Dust based surveillance system in IoT
ecosystem. Major contributions and novelties in this paper are as follows:-

1. We have proposed Smart Dust based potential applications of IoT.
2. We have discussed challenges and risks related to Smart Dust based Systems.
3. Merits & demerits of the proposed application are discussed.

The remainder of this paper is organized as follows. Related work is given in
Sect. 2. An overview of the basic building block of the proposed work is described

© Springer Nature Singapore Pte Ltd. 2019
A. K. Somani et al. (Eds.): ICETCE 2019, CCIS 985, pp. 128–140, 2019.
https://doi.org/10.1007/978-981-13-8300-7_11

in Sect. 3. Architecture of the Smart Judiciary System is presented in Sect. 4. Description of the other proposed emerging applications is given in Sect. 5. In the Sect. 6, Merits and demerits of the proposed applications are given. Issues related to the proposed applications are highlighted in the Sect. 7. Finally, we conclude the paper in Sect. 8.

2   Background and Related Work

Among the Vision and Objectives of CCTNS (a Project of Govt. of India), some are to provide the tools, technology and information to related officers for investigation of crime, enhancing services for citizen through effective usage of ICT [8].

Dickson [17] mentioned that US military must spend energy and money today for the research work on technologies (like micro-electromechanical systems (MEMS)) for development of the surveillance applications such as Smart Dust for the future. Smart dust can help in border surveillance against terrorism [3,17,18]. In near future, clouds of several tiny sensors will be moved in the air [3].

Micro Electro Mechanical Systems (MEMS), digital circuitry are reducing in the size, consumption of power and cost due to the advances in the hardware technology and engineering design. There are several application areas of Smart Dust such as, In deployment of Bridges, Tracking the Climate, Traffic Monitoring, Biological Studies etc. [4].

As dust particles, MEMS based devices can be suspended in the environment. The optical lenses based miniaturized sensors can capture the finest quality images. Smart Dust can measure anything and enable wireless monitoring of people [11].

Development in technologies has made possible devices having one or more sensors, computation and communication capabilities and very small in the size and low cost. Smart Dust can be used in civil and military applications for notification of any poison or hazardous biological agent in the air [16].

Thiele et al. [5] have presented a highly miniaturized camera by 3D-printing technology and CMOS image sensor. The footprint of the optics on the chip is less than 300 μm × 300 μm and height is less than 200 μm. This innovation can lead to many application sectors like optical sensing, surveillance systems.

Motlagh et al. [9] have presented a case of Unmanned aerial vehicles (UAVs) based crowd surveillance applying facial recognition tools. In this article authors describe how UAVs can be used for crowd surveillance.

Byun et al. [10] presented the design of IoT based smart crime detection system. Proposed system consists of modules for sensing and recording of emotions, crime prediction, visualization and detection. For monitoring purpose, proposed system uses CCTV cameras.

Several IoT based crime detection systems have been proposed but to the best of our knowledge none of research work has proposed framework for Smart

Dust based Systems likewise Smart Judiciary System, Meta Security & Privacy Protection and Microbe Detection System. In this paper we have proposed advanced and new IoT based potential applications that can benefits the society by providing direction and envision for future research and developments.

3   Smart Dust

The complex digital IC's like, dimensions of a microprocessor shrinks to less than 1 mm (i.e., a dust particle size) due to the development in CMOS scaling in the last few decades. As a result, Smart Dust can be used to enhance the capabilities of IoT devices and lead to reduction in the cost [2]. A typical smart dust mote (Fig. 1) is made up of different components in a single package. These different components are MEMS sensor(s), semiconductor laser diode and MEMS beam steering mirror, MEMS corner cube retroreflector, an optical receiver, signal processing & control circuitry, and a power source based on thick film batteries and solar cells [6,12,16]. Data is collected by the Smart Dust Mote and transmitted to control base. Smart Dust Motes communicates using radio frequency transmission or optical transmission [16].

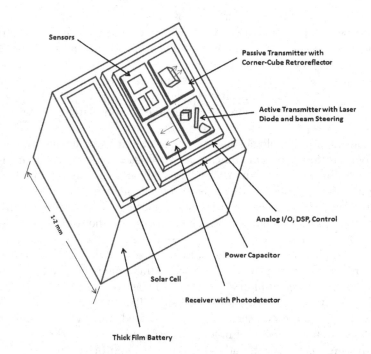

Fig. 1. A comprehensive block diagram of a Smart Dust Mote for overview (Constructed using source [6])

3.1   Benefits of Smart Dust Mote

Smart Dust Mote is small in the size, comparable to the size of dust particle. It is difficult to detect the presence of the Smart Dust and it also decreases the cost of system and infrastructure [16].

3.2   Applications of Smart Dust

The scope of Smart Dust based potential application is becoming hot research area due to characteristics of smart dust (i.e., size, cost effective, difficult to detect). Smart Dust has potential to strengthen several applications areas like, it can leverage robotics, sensing different environmental parameters, regional & national security, military, effective perimeter surveillance, healthcare, agriculture, transportation, parking, biological research, planetary research, machinery, recording geographical data etc. [3,4,6].

4   Proposed Architecture

In this section, we propose an application that can exploits best features of the Smart Dust Mote, such as very small size, cost effectiveness and difficult to detect the presence of Smart Dust Mote. Basic idea behind the proposed application is sensing different types of criminal activities by distributing potential Smart Dust in the surroundings. A detailed description of the proposed system is given in the Subsect. 4.2. This type of application can bring a big revolution in the society by reducing the crime rate.

Our proposed Smart Judiciary System is Smart Dust based IoT application. In this section we have described different ways in which Smart Dust can be distributed in the environment and a detailed architecture of the Smart Judiciary System.

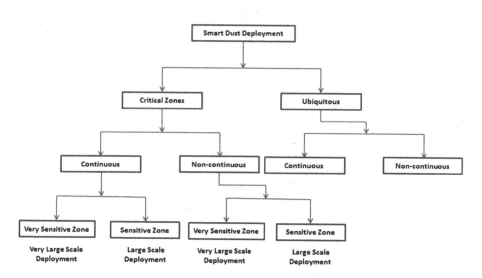

Fig. 2. Strategies to deploy smart dust in smart city

4.1   Strategies to Deploy Smart Dust in Smart City

There can be two basic ways to distribute Smart Dust in a smart city, namely Critical Zones and Ubiquitous Fig. 2. Critical Zone strategy can be used when monitoring is required in only some specific zones which are prone to be sensitive for the criminal activities. This strategy can help in reducing installation cost as it causes only selective areas. Further, continuous and non-continuous monitoring can be selected according to the probability of the occurrence of different activities. A continuous monitoring is regular sensing with the time and non-continuous monitoring is irregular sensing of the surrounding parameters. If a zone is very sensitive to occurrence of crimes then very large scale deployment of the smart dust can result in more accuracy in sensing, storing, processing and communicating the events. Ubiquitous strategy can be used to deploy Smart Dust to cover whole region of a smart city and further it can be divided into continuous and non-continuous monitoring. As compared to the Critical strategy, Ubiquitous strategy is more expensive as we need to deploy more devices in order to cover the entire area.

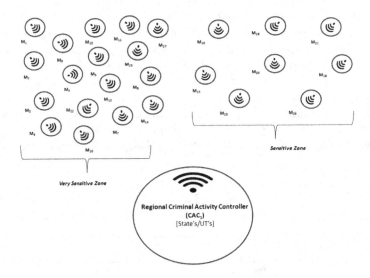

Fig. 3. Deployment of smart dust in the environment and communication with criminal activity controller

4.2   Smart Judiciary System Using Smart Dust in IoT

Our proposed Smart Judiciary System Fig. 4 has four basic components as mentioned below:-

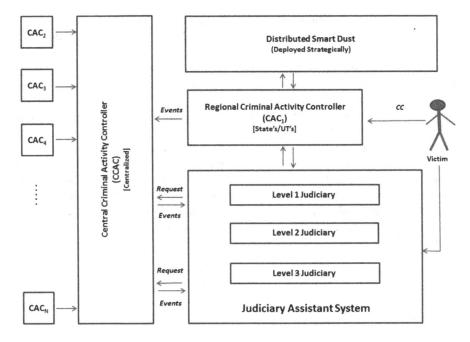

Fig. 4. A framework of smart judiciary assistant system using IoT

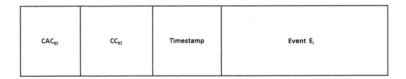

Fig. 5. An example record sent to the CCAC

Distributed Smart Dust: It consists of several motes. A comprehensive block diagram of a Smart Dust Mote is given in Fig. 1. Each mote is very small in size and has computational and communication capabilities. It is hard to detect its presence. We are assuming that there exist a miniaturized image sensor and MEMS microphone which can be integrated with a Smart Dust Mote which freely flow in the air and are invisible to the human eyes. Each mote senses the event from the surrounding environment and transmit sensed event to the Regional Criminal Activity Controller (CAC). Smart Dust consisting of 25 Motes are shown in Fig. 3 Each mote can communicate to CAC through wireless channels.

Regional Criminal Activity Controller (CAC): Its basic function is to collects and save all the data which is transmitted by the Smart Dust Motes. It monitors and keeps maintenance of all the deployed motes. A victim can approach to respective regional CAC for the registration of the Crime Request (CC). Each CAC belongs to a state or union territory and has its unique identification.

Judiciary Assistant System: Each CAC is always connected to the concerned Judiciary. For example, let us assume CAC_1 is implemented in a region R_1 then CAC_1 will be connected to that regional level's court. If any CC is registered then it can be forwarded to Judiciary Assistant System of the concerned regional level's court.

Central Criminal Activity Controller (CCAC): It is a national level central system that acts as a repository for all the data forwarded through the Regional Criminal Activity Controller. A CAC forwards an event E_i related each record to the CCAC. A typical record structure is shown in Fig. 5. The entire nation's CAC's are linked with the CCAC. It provides services to the national Judiciary System by providing necessary data for solving many registered cases.

Table 1. Summary of notations

Symbol	Meaning
L	Location where event occured
E_i	An Event
T_{E_i}	Timestamp of event E_i
ET_{E_i}	Expected time of occurance of event E_i, that is mentioned in CC
T_{Limit}	Maximum time limit to store a event E_i
T_P	Additional time required to fetch event E_i
CC	Crime Complain
V_i	Victim of a crime
E_v	Event present in CC which registered by V_i
AR_{V_i}	All records related to an incident with V_i
CAC_{ID}	Unique Identification number of Regional Criminal Activity Controller (CAC)
CC_{ID}	Unique CC number

4.3 Working of Smart Judiciary System

The basic functionalities of the proposed system are based on the Algorithm 1 and Algorithm 2 as described below. Summary of notation used in algorithms is given in the Table 1.

Algorithm 1. Operational algorithm for Regional Criminal Activity Controller

Input Data: *CC, Location (L), Timestamp T_{E_i}*
Output: *Matched Event E_i saved at CCAC*

1: Start
2: **if** $ET_{E_i} < T_{Limit}$ **then**
3:     **for** $ET_{E_i} - T_P$ to $ET_{E_i} + T_P$ **do**
4:         Fetch Event E_i
5:         **if** $E_i == E_v$ **then**
6:             Forward record to CCAC
7:         **end if**
8:     **end for**
9: **end if**
10: Stop

Description of Algorithm 1: CAC is informed for an event of incident E_i with an expected timestamp ET_{E_i} ($\approx T_{E_i}$) and location L by registering the CC. Next, if timestamp ET_{E_i} is less than the maximum limit time (T_{Limit}) to register CC then event E_i is searched within a defined range of time to be found E_i. Now if E_i matched with a victim's incident E_v (it can be checked manually by the CAC authority on a display) then record R_{E_i} of event E_i is sent to the Central Criminal Activity Controller (CCAC) for a permanent storage till registered case is solved. T_{Limit} is used to manage high volume of big data, i.e., an event recorded with timestamp T_{E_i} saved in datastore and if it exceed the storage time duration to T_{Limit}, it will be removed automatically.

Algorithm 2. Operational algorithm for Judiciary Assistant System

Input Data: *Request all records AR_{V_i} associated to Victim V_i, CAC_{ID}, CC_{ID}*
Output: *AR_{V_i}*

1: Start
2: **if** CAC_{ID} Exist **then**
3:     **if** CC_{ID} Exist **then**
4:         Sent AR_{V_i} to Judiciary Assistant System
5:     **end if**
6: **end if**
7: Stop

Description of Algorithm 2: If Judiciary Assistant System requires records related to a victim V_i then Judiciary Assistant System sends a request to CCAC which is the central repository for records of criminal activities. On the receiving request, if CCAC found CAC_{ID} and CC_{ID} in the database then it will forward all related records to the Judiciary Assistant System.

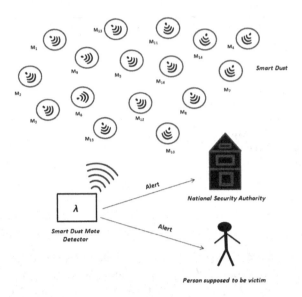

Fig. 6. Overview of meta security and privacy protection system

5   Emerging Application Domains of IoT

5.1   Meta Security and Privacy Protection

Privacy leakage of a citizen is a serious issue. One of the major issues with the Smart Dust is the privacy concern due to smaller size [3] of the Smart Dust Motes. This subsection describes the concept and importance of Meta security and privacy protection system.

Privacy Issue. Let us assume that Smart Dust is distributed surrounding the individual's private places quietly and it is hard to detect due to very small size of the mote. If the Smart Dust Mote is equipped with the image sensor and sound sensor then these sensors can easily sense private activities and send to controller of the Smart Dust. In the next upcoming decades these types of issues can emerge faced due to fast growth of IoT related technologies.

Meta Security and Privacy Protection System. Its functionality is similar to an Intruder Detection System (IDS) i.e., detecting malicious activities. But it detects privacy invasion in the user's surrounding environment i.e., detects malicious physical objects in the user's surrounding environment which is suspicious (In fact, for the non-suspicious condition also) for privacy invasion. In this proposed system Fig. 6 we are assuming that there exist such a mechanism

'λ' that can detect presence or absence of a Smart Dust Mote in the surrounding environment. If Smart Dust exists in the surrounding environment then λ mechanism can send alert to person(s) which can be a victim(s) of the privacy invasion. This alert will include location of the Smart Dust Mote(s).

5.2   Microbe Detection

In this section, we illustrate how smart dust can be used for microbe detection. This proposed system can be useful for public in a smart city to reduce different types of diseases from spreading due to microbes like bacteria, fungi etc. Proposed system is based on the Microbe Detection Sensor(s). Microbe Detection Sensor such as Microbe sensor (BM-300C) [13,14] (and other efficient miniaturized Microbe Detection sensor if currently present or developed in future) can be used to detect microbes. Useful sensed data such as; number of microbes counted can be sent to the City Central Data Analyzer Fig. 7 for analysis and can extract useful patterns. Further, City Central Data Analyzer will send alerts to the public to avoid visiting locations where high concentration of the microbes exists. Public healthcare authority can initiate steps to control concentration of microbes in the concerned areas. For the real time notification, real time microbe detection sensor is necessary and for Smart Dust based Microbe Detection System, research and development of a very small size Microbe Detection Sensor is essential (i.e., development of a sensor which can be integrated with the Smart Dust Mote and which can freely flow in the air to detect microbe's presence and their types).

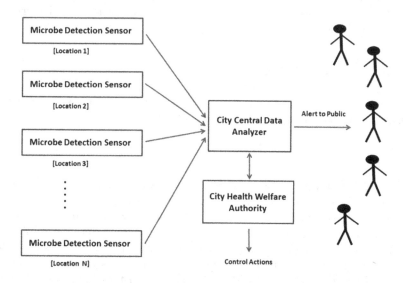

Fig. 7. Overview of an IoT based microbes detection system

6   Merits and Demerits

Table 2. Merits and demerits of the proposed smart dust based IoT applications

Merits	Demerits
• Helps to reduce criminal rate • Empower administration • Empower Citizens • Strengthen economic growth • Helps to reduce diseases due to microbes	• Increased privacy risks • Increased environmental issues • Increased health issues due to pollution generated by e-waste and other causes

With the suspense and reality of invisible surveillance due to smart dust based IoT applications; it can lead to drastic fall in the crime rates. The reduced rate of crime directly empower the regional administration that results in increasing the economic growth of country (i.e., reduced rate of the crimes, many national resources will be utilized effectively). Smart Judiciary System will help to empower citizens due to the availability or tracing of the proofs of any incident that occurs. Other smart dust based IoT applications such as Microbe Detection System can help in preventing or spreading many types of diseases due to microbe infection. Along with these merits, smart dust based IoT applications demerits include increased privacy risks, environmental and health issues due to e-waste or other forms of pollutions. All of these merits and demerits are summarized in Table 2.

7   Risks and Challenges

Though useful, Smart Dust based IoT applications can lead to several risks and challenges. Some which are discussed in this Section.

7.1   Risks

Security and Privacy Risks. A Smart Dust Mote (specially, a Mote which includes image sensor/sound sensor) can pose high risks by privacy invasion. These two types of information (Image and Sound) may contain very sensitive data. The continuous monitoring of unaware public or individual can result losses in any form such as mental, financial, health and/or social respect. Privacy leakage due to Smart Dust based IoT system will be a major challenge in near future. We believe that as the size of the Smart Dust Mote decreases, it will result in fair chances of privacy issues if proper measure to ensure privacy are not taken into consideration.

Environmental and Health Risks. Very fast expansion of the microelectronic components contributes to worrying effects on the environment such as pollution and deposition of the e-waste [15]. We believe that in the near future smart dust and other likewise technologies will be very common in many potential applications, resulting in the increased volume of such devices. The semiconductors used in construction of Smart Dust Motes can adversely affect the environment through e-waste. Due to increasing level of pollution and toxics health issues will be increased.

7.2 Challenges

Smart Dust based IoT applications are based on Motes, which is one of the basic building blocks. Development of the technologies to identify Smart Dust Mote(s) in the environment, miniaturized microbe detection sensor which can be integrated with the Smart Dust Mote can be a challenging task. If we suspend these motes in the environment then there can be weather based challenges like how the rainy weather will affect functioning of these smart dust motes.

8   Conclusion

In this paper, we proposed Smart Dust based Smart Judiciary System. We have discussed how the proposed system can aid to the functioning of judiciary making it more 'Smart' and more beneficial for the citizens. In the emerging world of connected devices, the proposed system can be a breakthrough in detection and reduction of crimes and strengthen the economic growth of the nation. We have also discussed some emerging application domains such as Meta security & privacy protection system and Microbe detection system.

References

1. Gai, K., et al.: Privacy-preserving content-oriented wireless communication in internet-of-things. IEEE Internet Things J. **5**(4), 3059–3067 (2018)
2. Sadana, D.K., Li, N., Bedell, S.W., Shahidi, G.S.: "Smart Dust" & Internet of Things (IoT): progress & challenges. J. Lasers Optics Photonics **4**(2), 1–2 (2017). https://doi.org/10.4172/2469-410X.1000160
3. Sathyan, S., Pulari, S.R.: A deeper insight on developments and real-time applications of smart dust particle sensor technology. In: Hemanth, D.J., Smys, S. (eds.) Computational Vision and Bio Inspired Computing. LNCVB, vol. 28, pp. 193–204. Springer, Cham (2018). https://doi.org/10.1007/978-3-319-71767-8_16
4. Mannir, M., Getso, A., Ismail, M.: Internet of things and smartdust: the future of wireless internet of things and smartdust. Int. J. Inf. Syst. Eng. **5**(1), 13–23 (2017). https://doi.org/10.24924/ijise/2017.04/v5.iss1/13.23
5. Thiele, S., et al.: 3D-printed eagle eye: Compound microlens system for foveated imaging. Sci. Adv. **3**(2), e1602655 (2017)
6. Kahn, J.M., Katz, R.H., Pister, K.S.J.: Next century challenges: mobile networking for "Smart Dust". In: Proceedings of the 5th Annual ACM/IEEE International Conference on Mobile Computing and Networking, pp. 271–278. ACM (1999)

7. Menaria, V.K., Jain, S.C., Nagaraju, A.: A fault tolerance based route optimisation and data aggregation using artificial intelligence to enhance performance in wireless sensor networks. Int. J. Wireless Mobile Comput. **14**(2), 123–137 (2018)

8. NCRB.: Crime and Criminal Tracking Network & Systems (CCTNS). CCTNS Branch, http://ncrb.gov.in/BureauDivisions/cctnsnew/index.html. Accessed 24 Nov 2018

9. Motlagh, N.H., Bagaa, M., Taleb, T.: UAV-based IoT platform: a crowd surveillance use case. IEEE Commun. Mag. **55**(2), 128–134 (2017)

10. Byun, J.-Y., Nasridinov, A., Park, Y.-H.: Internet of things for smart crime detection. Contemp. Eng. Sci. **7**(15), 749–754 (2014)

11. Marr, B.: Smart dust is coming. are you ready?. Forbes Media LLC, 16 September 2018. https://www.forbes.com/sites/bernardmarr/2018/09/16/smart-dust-is-coming-are-you-ready/#27df1cb15e41. Accessed 23 Nov 2018

12. Warneke, B., Bhave, S.: Smart dust mote core architecture. Project report, Berkeley Sensor and Actuator Center, Berkeley, CA (2000). http://bwrc.eecs.berkeley.edu/Classes/CS252/Projects/Reports/warnke.pdf

13. Sharp Corporation: Sharp Introduces Microbe Sensor for Fast[*1], Automatic Measurement and Visualized Count of Airborne Microbes. Sharp, 26 September 2013. http://www.sharp-world.com/corporate/news/130926.html. Accessed 25 Nov 2018

14. Abbasian, F., Ghafar-Zadeh, E., Magierowski, S.: Microbiological sensing technologies: a review. Bioengineering **5**(1), 20 (2018)

15. Villard, A., Lelah, A., Brissaud, D.: Drawing a chip environmental profile: environmental indicators for the semiconductor industry. J. Clean. Prod. **86**, 98–109 (2015)

16. Rishika Poojari, Snehal Shah. Smart Dust Technology. Int. J. Recent Innovation Trends Comput. Commun. (IJRITCC) **5**, 65–68 (2017). ISSN 2321–8169. http://www.ijritcc.org/download/conferences/ICEMTE_2017/Track_6_(Others)/1489906857_19-03-2017.pdf

17. Dickson, S.A.: Enabling Battlespace Persistent Surveillance: The Form, Function, and Future of Smart Dust. AIR WAR COLL MAXWELL AFB AL CENTER FOR STRATEGY AND TECHNOLOGY (2007)

18. Mohan, S.C., Arulselvi, S.: Smartdust network for tactical border surveillance using multiple signatures. IOSR-JECE **5**(5), 1–10 (2013)

Performance Evaluation of Tree Ensemble Classification Models Towards Challenges of Big Data Analytics

Hanuman Godara[1(✉)], M. C. Govil[1,2], and E. S. Pilli[2]

[1] National Institute of Technology Sikkim, Ravangla 737139, India
hanuman_godara@live.in, govilmc@gmail.com
[2] Malaviya National Institute of Technology Jaipur,
Jaipur 302017, Rajasthan, India
espilli.cse@mnit.ac.in

Abstract. Big Data Analytics poses challenges like effective and accurate real-time data mining, lack of suitable tools & techniques and in-memory processing problem. Tree-based ensemble methods (machine learning models) are able to perform such kind of large-scale analytical processing in combination with high-performance cluster computing (special kind of distributed computing) using parallel processing. Random Forest (forest of randomized trees, a tree ensemble) algorithm is considered for the performance evaluation, as tree model supports concurrency and all trees are grown simultaneously in it, so it is a suitable parallel approach with good accuracy, noisy & imbalance dataset handling capability and also it never overfit unlike a single tree model for large dataset. However significant notable improvement over the original approach is available, but some limitation still exists regarding performance and streaming dataset such that performance rate decreases on increasing the compute nodes due to a redundant allocation of feature subsets in the hybrid approach of task & data parallelization and inability to handle stream data. So these performance issues are identified and a problem statement is formulated with an objective to achieve the linear scalable speedup and incremental processing capability of random forest algorithm to perform predictive analytics over massive datasets in the cluster environment.

Keywords: Tree ensemble models · Big Data Analytics · HPC

1 Introduction

This paper aims to improve the performance of existing techniques like tree ensembles, for large-scale data processing by applying appropriate parallel processing concepts and using suitable tools in a distributed environment having powerful configurations such as a cluster. Overview of involved fields, concepts, tools and techniques in this work, is such as: Big Data concern large-volume, complex, growing data sets with multiple; autonomous sources like WWW, YouTube, Blogs, Social Networking & E-commerce sites e.g. Twitter, Flicker, Facebook, Amazon, and Flipkart [1]. Big Data is a collection of datasets consisting of massive unstructured, semi-structured and

© Springer Nature Singapore Pte Ltd. 2019
A. K. Somani et al. (Eds.): ICETCE 2019, CCIS 985, pp. 141–154, 2019.
https://doi.org/10.1007/978-981-13-8300-7_12

structured data. The four main characteristics (dimension) of Big Data are as [2]: Volume (amount of data), Variety (range of data types and sources), Velocity (speed of incoming data) and Veracity (data quality) [29, 30]. The HPC (High-Performance Computing) equips with commodity clusters, which breaks the barrier of processing power (with many logical & physical processing cores), memory & storage constraints (with RAID storage), with high-speed intranet (with InfiniBand). Virtual clusters (HPCaaS) is also found similarly capable as actual physical clusters. Parallel processing in the form of task & data parallelization reduces the execution time and increases performance with various supporting tools and libraries like Spark, Hadoop, MapReduce, and Python. Random Forest as an ensemble of several tree models provides more accurate, fast method to process such a huge Big Data, it never over fit and also capable of handling the missing, noisy and imbalanced dataset. It is a powerful machine learning algorithm used in all areas of Data Science like Artificial Intelligence, Data Mining, Machine Learning, Big Data, Pattern Recognition, statistics, and Deep Learning.

In current scenario with the fast development of networking, data storage & collection capacity and the emergence of Internet of Things (IoT) & Cloud Computing technologies, Big Data is growing day by day in such a way that it doubles on every two years [1]. In present time Big Data have following challenges [1, 3]:

- Extractions of useful information (Data Mining) from massive data efficiently & accurately in real time, while this Big Data is high dimensional, complex and noisy.
- To choose effective techniques and tools for fast (performance) and time-bound analysis of Big Data.
- In-memory processing is not feasible for Big Data.

Common or usual computing systems like desktop PCs (Personal Computers), workstations and servers have limitations in terms of processing power, storage & memory capacity, tools & techniques and can't process such a large volume Big Data (in Terabytes). So, HPC (High-Performance Computing) or HPCaaS (HPC as a Service, Cloud) is the only way for it. Tree-based ensemble models are capable to perform such kind of processing (analytical) for Big Data in combination with HPC technology, as both supports parallel processing to boost the execution speed. Hence handling of this Big Data is beyond the capabilities of usual computing systems & techniques, so main motivation behind this work is Big Data processing (analytical) and mitigating its challenges by recognizing & enhancing suitable tools, techniques, and methods for it. Initial studies toward the motivation results with the identification of High-Performance Cluster, Parallel Processing & Tree ensemble methods such as Random Forest as robust, efficient & accurate viz. tools, techniques & methods for such a terabyte Big Data processing.

Rest of the paper is organized such as: Sect. 2 includes a background study of performance oriented parallel processing along with machine learning tree models and their evaluation followed by Sect. 3 which involves a review of recent improvements over existing approaches. Section 4 describes the issue of scalable speedup for such a processing with a proposed methodology to be implemented for resolving it. Future work with experimental setup of proposed work is given in Sect. 5 with a conclusion of presented work.

2 Background Studies

2.1 Parallel Processing

High-Performance Computing allows solving complex data and computing intensive problems, here key factor is performance which is focused more by supporting parallel processing. HPC systems have faster computing resources as the latest many integrated core accelerators in combination with multicore multiple processors, high speed interconnect, parallel file system and storage, large amount of memory. The major concepts behind HPC are cluster computing and parallel processing. Cluster computing is a special kind of distributed computing (with shared & distributed memory both). A cluster consists of a set of tightly connected compute nodes like one master and 32 slave-compute nodes in PARAM-Kanchenjunga [4], which work together so that, in many respects, they can be viewed as a single system. Compute nodes may be of the same configuration with equal processing power or may have different-different computing capability on this basis it can be categorized into homogeneous and heterogeneous. Clusters have each node set to perform the same task, controlled and scheduled by cluster scheduler like TORQUE (Terascale Open-source Resource and QUEue Manager). In parallel processing tasks are divided into subtasks and allocated on multiple processors to run simultaneously to reduce the execution time of the whole program. Thus speedup of program execution is achieved by allocating the task to the different-different processor or many cores within a processor or both. Several types of parallelization used in parallel processing as Task Parallelization and Data Parallelization. HPC clusters are very expensive and not available to all, so moving to HPC systems on the cloud environment can be cost effective for many organizations. In HPCaaS (HPC as a service, virtual cluster in the cloud), a virtual cluster (a group of VM's, virtual machines) is provided to the user on demand pay per use basis by implementing cloud setup on actual clusters using cloud tools.

2.2 Decision Tree Models

Machine learning is used to build an analytical model by which it can be found the hidden insights or patterns from given datasets. Mostly there are two types of machine learning methods: supervised and unsupervised as given in Fig. 1. An analytical model is created by labeled (known or past) training data in supervised and from unlabeled data (no train data, no response variable) in unsupervised methods. Classification and regression are the examples of supervised while clustering is categorized under unsupervised learning. Classification algorithms forecast (predict) the target class (like Yes/No, Fail/Pass). If the trained model is for predicting any of two target class, it is known as binary classification. Like predicting student's Pass or Fail, Customer will buy or not buy. If it is to predict more than two target classes it is known a multi-classification. Like in which subject student will score more. Regression algorithms forecast continuous value like predicting student's marks or percentage. In machine learning, all datasets are divided into training & testing data set according to a predefined ratio.

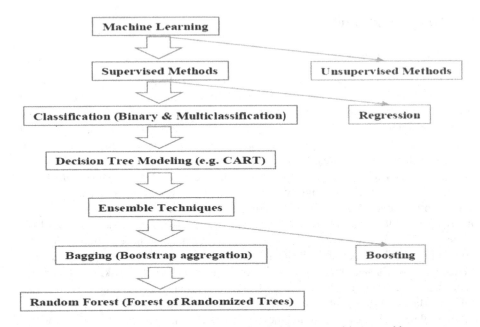

Fig. 1. Machine learning hierarchy for decision tree and its ensembles

To inferring a function (rule) or creating a model, from training data (data that has the correct answers attached or past data), called training. Applying the inferred rule (function) on test data that has no answers (class label) to forecast (predict) their class label called testing. How well an algorithm can forecast new answers (labels) for testing data based on its training show its accuracy. Decision tree learning model is used for classification as well as a regression task; here decision means where to split (partition or branch) in a tree model. In decision tree classification task begins from the root node to the leaf or terminal node involving greedy, recursive partition in gapproach. Terminal node shows output for response variable and a branch indicates partition of attributes or variables means each branch corresponds to an attribute value. CART (Classification and Regression Tree) is an example of the Decision Tree creation algorithm by Breiman [5]. CART uses the Gini method (Best Split, a greedy approach) to create binary splits, which is a measure of homogeneity of attributes. For a higher value of Gini, homogeneity will be higher in sub-nodes.

Evaluation of Tree Models. Overfitting and bias-variance are the major issues during building decision tree models. In the case of overfitting, a decision tree model fails to provide exact accuracy, for every observation it delivers 100% accuracy as it ends up with one leaf node. To prevent overfitting stopping criteria (setting a limit on tree size) and pruning are suitable ways where leaves with negative returns are removed in pruning which is an opposite process of splitting. Bias reflects the difference between prediction and actual values on an average basis, while variance shows different-different predictions of a model at the same point with the different samples. Bias is high for building a simple decision tree model but it shows low variance. Making complex decision tree losses accuracy due to the lower bias of model, on making more complex decision tree model will show overfitting due to high variance. So a proper balance of these factors is to be maintained during building a decision tree model for good accuracy and to reduce such errors.

2.3   Tree Ensemble Models

Making ensemble of decision tree models is a good solution for bias-variance related errors of individual tree models. In an ensemble model, a group of tree models is built together instead of a single large or complex tree model for a given data which provides an improvement in accuracy and more stable powerful model than a single weak model. On increasing numbers of tree models in an ensemble model will enhance its prediction accuracy because variance will be reduced in 1/n manner. Two popular methods of making ensembles are Bagging (Bootstrap Aggregation) and Boosting.

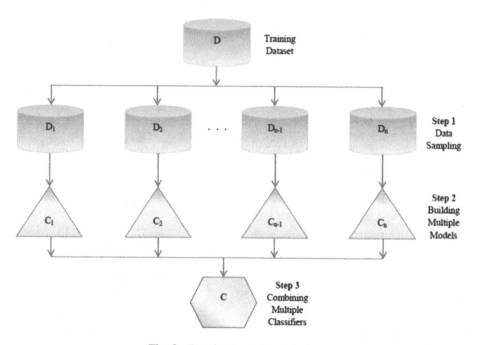

Fig. 2. Bagging ensemble technique

Boosting is a sequential approach where the next stage depends on earlier one while in Bagging models can be created independently so it leads to a faster model in term of training time. The idea of Bagging [6] ensemble models was given by Breiman, in it after the creation of separate models from subsamples of the whole dataset, the results of these models are combined by averaging these as shown in Fig. 2. Random Forest or forest of randomized trees is an example of implementation of Bagging techniques.

Random Forest Algorithm. Random Forest or Random Forest or Random Forests or RF are same and has a trademark of Breiman and Cutler [7], is considered panacea for all data science problems. Breiman presented random forest, which is an ensemble model of random, independent decision trees for classification or regression tasks here final classification result is decided on the basis of majority votes of results from individual trees. It involves the key concept of bagging with the random feature selection instead of considering all features as bagging thus it involves two kinds of randomization, one is random sampling and another is random feature selection so random forest can be seen as "forest of randomized trees". Each tree in it can be developed using CART like tree classification algorithms where branching is based on the Gini Index or any other functions like Information entropy/gain ratio etc. Model building process using the random forest method can be understood as the flowchart given in Fig. 3 and steps provided in Table 1.

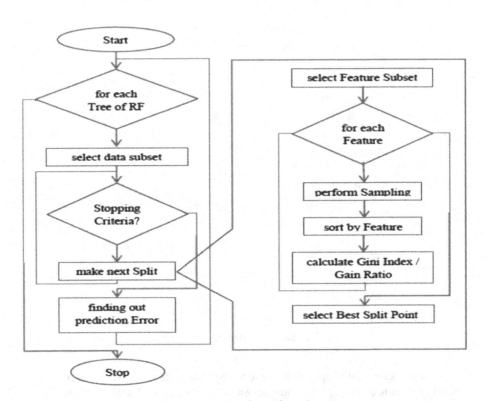

Fig. 3. Random forest flow chart

Table 1. Random forest algorithm.

Steps for building a random forest model:
1. Take N number of samples from a dataset, where M is the number of features/variables
2. Decide m number of features for each split in such a way that m2 = M (in bagging m = M)
3. Perform random sampling by choosing one sample at a time with replacement from N
4. Use the rest of samples (N − 1) for testing (no separate test data required) and find the error
5. Random Feature selection (m strong features from M) for each split or branch node
6. Find out best split point on basis of a suitable method (Gini Index/Entropy)

The complexity of a random forest model depends upon the number of samples, variables, trees, and average tree height. Tree models support concurrency and all trees are grown simultaneously in a random forest so it is considered a suitable parallel approach for fast big data analytics with good accuracy.

Random forest provides better prediction accuracy with faster training (learning) [8–10], large-scale data processing, missing value handling and model reusability. If enough trees are built in this ensemble model then it doesn't show the issue of over-fitting. Variable interaction can also be seen in it during model building. But it has some limitations as it may over fit for noisy datasets and can show bias toward variables (attributes) with more levels. Application areas where random forest algorithm is widely used are E-mail spam filtering, remote sensing, image processing, pattern recognition, banking (credit risk analysis), medicine (to identify the correct combination of the components to validate the medicine and to identifying the disease), stock market (to identify the stock behaviour) and E-commerce (in recommendation engine for identifying the likely hood of customer).

3 Review of Relevant Work

Further study of literature in the combined perspective of Big Data, Parallel Processing, HPC, and HPCaaS & Random Forest shows the current related work. It involves research papers on pre-processing of data like dimension reduction, variable importance measures for Random Forest, combination of boosting and bagging approaches, analysis of boosting & bagging for noisy & imbalanced data, mega-ensemble methods of Random Forest with different sampling & voting approaches, histogram-based parallelization of training for tree models, dynamic resource allocation & scheduling in cloud along with job & execution graph based task parallelization in distributed environment, resource allocation & scheduling of stochastic tasks on heterogeneous parallel system & elastic cloud with workload balance for data processing, horizontal & vertical data partitioning and multiplexing in task parallelization.

3.1 Pre-processing of High Dimensional and Noisy Data

Xu [11] proposed a PCA (Principal Component Analysis) based dimensionality reduction algorithm for remote sensing image registration with high dimensional datasets. By this algorithm registration specific metric is minimized, which results to obtain more information from the datasets but this algorithm is applicable only when CRCB(Cramer Rao Lower Bound) of transformation parameter estimates can be obtained, which is difficult to derive in some more complicated misalignment cases (for non-rigid body). Tao et al. [12] presented a multilevel maximum margin feature extraction approach called RSVM (Recursive Support Vector Machine). In comparison to PCA and LDA (Linear Discriminant Analysis), the proposed method has no singularity (undersampled) problems and provides efficient dimensionality reduction. But this method works for binary classification problems only not for multi-classification. Lin et al. [13] introduced MKL-DR (Multiple Kernel Learning-Dimension Reduction) approach, which extends basic dimension reduction approaches like LDA, LDE (Local Discriminant Embedding) used for supervised, PCA used for unsupervised models and SDA (Semi-supervised Discriminant Analysis) used for semi-supervised model, with significant improvement over these basic dimension reduction approaches. But it is suitable only for graph based and binary class data.

3.2 Variable Importance Measures

Strobl et al. [14] worked on variable importance measures of Random Forest, which show bias towards correlated predicator variables, so a new, conditional permutation scheme for the computation of the variable importance measure proposed, which uses the partition that is automatically provided by the fitted model as a conditioning grid and reflects the true impact of each predicator variable better than the original, marginal approach, but it cannot entirely eliminate the preference for correlated predictor variables.

3.3 Combination of Bagging and Boosting

Bernard et al. [15] introduced Dynamic Random Forest (DRF) algorithm based on adaptive tree induction procedure, in which each tree will complement the existing trees in the ensemble through resampling of the training data, inspired by boosting algorithms, and combined with other randomization processes used in Random Forest method. The DRF algorithm shows a significant improvement in terms of accuracy compared to standard static Random Forest algorithm but without the pre-required condition of independence of trees, it is no more obvious to demonstrate the generalization error convergence always reached high, while trees are added to the ensemble, as a consequence, DRF is not proved to overfitting.

3.4 Analysis for Noise and Data Imbalance

Khoshgoftaar et al. [16] compared the performance of several boosting and bagging techniques in the context of learning from imbalanced and noisy binary class data.

After testing the results of parameter-based experiments on these classifiers, it is proved that bagging techniques generally outperform boosting, and hence in noisy data environments, bagging is the preferred method for handling class imbalance. Biau [17] presented an in-depth analysis of Random Forest model, which show in particular that the procedure is consistent and adapts to sparsity, in the sense that its rate of convergence depends only on the number of strong features and not on how many noise variable are present.

3.5 Mega-Ensemble

Basilico et al. [18] proposed COMET (Cloud of Massive Ensemble Trees) algorithm, in which each compute node constructs a Random Forest and it merges them into a mega-ensemble. COMET uses I Voting (Importance-sampled voting) instead of usual bagging. This approach is appropriate when learning from massive-scale data that is too large to fit on a single machine and provided good accuracy with less time consumption but this accuracy may be affected, as all data are not available for every tree. Also, it uses only a single pass of map-reduce to avoid overhead which results in less performance than multiple pass approaches like Google's PLANET. Panda et al. [19] introduced PLANET (Parallel Learner for Assembling Numerous Ensemble Trees) motivated by a real application in sponsored search advertising, for learning tree models over large datasets in a distributed environment. It is a MapReduce based tree learner, MapReduce [20] effectively address many scalability issues in this learning but as MapReduce was not intended to be used for a highly iterative procedure like tree learning, so it creates a primary performance bottleneck. It uses bagging and boosting techniques both for construction of trees ensembles which provides good accuracy. Cost of traversing model in order to determine a split point in parallel turned out to be higher. It doesn't handle multiclass classification and incremental learning.

3.6 Resource Allocation and Task Parallel Execution

Tyree et al. [21] proposed a novel method for parallelizing the training of GBRT (Gradient Boosted Regression Trees) for web search ranking. This approach is based on data partitioning using histograms, provides impressive (almost linear) speedups on several large scale web-search data set without any sacrifice in accuracy on shared memory machines, but on distributed memory machines it fails to get a linear speedup with accuracy. Warneke et al. [22] designed Nephele, which is a data processing framework to explicitly exploit the dynamic resource allocation offered by today's IaaS cloud for both, task scheduling and execution. It is based on job graph (DAG, Directed Acyclic Graph, by user submitted jobs) and execution graph (DAG in term of processing time). This models task parallelization, resulting to improve overall resource utilization and consequently, reduce the processing cost in comparison to Hadoop. But it is not able to handle load balancing in case of overload or underutilization during job execution automatically. Briceno et al. [23] presented a resource allocation scheme for satellite weather data processing on a heterogeneous parallel system. It is robust to the uncertainty in the arrival of new datasets to be processed and also balances the overloaded and underutilized machine resources, but it does not support multithreaded environment. Zhang et al. [24] proposed eOO (Evolutionary Ordinal Optimization) method based on the concept of workload pattern, to partition and merge the scheduling periods considering fluctuation within the workload, for Big Data analytics in the elastic cloud. It handles dynamic scenarios but obtains only a suboptimal solution. Simulation is limited to virtual clusters, not for an actual cluster. Li et al. [25] proposed a stochastic dynamic level scheduling (SDLS) for precedence constrained stochastic tasks (task processing time is not known in advance, assumed as random variable with probability distribution function) on heterogeneous cluster systems (computation capabilities of processors are different, so they calculate different-different execution time for the same task), which integrated variances with expected values of random variables, hence outperform to other existing stochastic scheduling algorithms like Rob-HEFT and SHEFT in term of make span (completion time, schedule length) and speedup. But it deals with normal distributed random variable only and not suitable for other probability distribution. Apache Mllib (Machine Learning Library) [26] introduced a data parallelization method for Random Forest algorithm using horizontal data partitioning, which provides increased performance but not so better, as it takes data-communication of gain ratio (GR) computing as global communication. Its best split method reduces data transmissions but it results into degraded accuracy of the algorithm. Chen et al. [27] presented parallelization of original random forest algorithm by applying a task and data parallel hybrid approach for big data processing in spark cluster environment, in which task parallelization involves dual parallel optimization of training phase and then performed DAG (Directed Acyclic Graph) based task scheduling for reduced execution time, where DAG creation depends upon tasks in training and RDD (Resilient Distributed Dataset); data parallelization contains data partitioning in a vertical manner for reduced communication and multiplexing for data reuse purpose. Before this parallelization, approaches to reduce dimension of big data having noise; and weighted voting instead of usual majority approach is performed for increased prediction accuracy. However these approaches show results of increased accuracy and less execution time then relevant algorithms and implementations as MLlib, but this algorithm is not applicable to streaming dataset. Linear speedup is not achieved as on adding or increasing compute nodes, performance increment rate decreases slightly.

4 Performance Issue

Somehow these research approaches (mentioned in the above literature review) achieved their goal, but limitations are also present with each of them. Some limitations are resolved by their successive work but few of them still exists regarding performance & streaming datasets. In this direction latest work is illustrated by Chen et al. using a hybrid approach of data parallelization with method for dimension reduction & weighted voting over Random Forest for Big Data Processing. It achieved good accuracy with better performance, but the performance rate decreases on increasing the no. of compute nodes and can't handle streaming datasets. So concluding the above review it is found out that "Linear Scalable Speedup & Streaming Data Handling" for

Big Data processing through Random Forest in an actual and virtual cloud environment, is an issue and research problem, which is yet to be resolved and we have considered it for our future research work.

4.1 Methodology

This work is focused towards the issue of linear speedup (Constant performance increment rate with increasing no. of compute nodes) and datastream handling capability of Random Forest algorithm for Big Data analytical processing in the actual and virtual cluster (HPC and HPCaaS). At this initial stage, the possible reason of this non-linear behaviour of speedup in cluster environment, is found (on basis of analysis) that the high data communication operations during task parallelization approach of Chen et al. (due to redundant allocation of feature subsets to multiple compute nodes) increases the communication overhead and cause the decrease in performance rate. So to avoid it we have to set constraints on its existing resource allocation cum scheduling scheme in such a way that redundant allocation may be avoided, or search for other suitable resource allocation cum scheduling approach as presented by Li et al. which can also handle streaming datasets with stochastic approach for Random Forest. To reduce communication overhead among feature subsets allocated on different-different compute nodes, process level (MPI provides process-level parallelization) and thread level (OpenMP provides thread-level parallelization) parallelization should be used rather than existing task parallelization, which are more fine grain levels of parallelization with better performance, supporting within-compute-node parallel processing (local processing on many cores), so inter-compute node communication (global) will be reduced which will result into increased performance. To maintain constant performance rate we will apply tiling concept [28] of parallel processing with process level and thread level parallelization which is one of the most important strategies for distribution of data and computations in parallel programs and for locality enhancement and Tiles are of great importance for the implementation of iterative and recursive, parallel algorithms which surely boost the speedup and maintain its linearity while increasing no. of compute nodes. So our objective includes developing new algorithms or modifying existing methods which can contribute to performance enhancement for massive data processing with following methodology:

- Adapting a particular data pre-processing (dimension reduction, missing value & imbalanced data handling) method (on the basis of literature analysis or with some modifications according to the scenario).
- Parameter tuning (no. of trees in forest N, no. of feature variables in each sample m, the ratio of subset and OOB for training & testing, random sampling & random variable selection using various probability distribution function instead of by default in Random Forest).
- Applying best split method (based on compute gain ratio, information gain & entropy instead of Gini Index).
- Lookup for other tree creation algorithm (ID, C4.5 instead of CART) for the optimal tree (in the manner of left and right skew and handling multi-classification).

- Setting constraint for stopping criteria (for tree height or fully grown tree).
- Checking scope to apply tree pruning or not, on the basis of practical analysis.
- Combining of bagging with boosting technique and validating effects.
- Making mega ensemble (more than one forest) of Random Forests.
- Constructing an algorithm for a hybrid approach of process-level and thread-level parallelization with tiling concept to minimize communication overhead.
- Creating an algorithm for stochastic aware resource allocation and job and execution graph based scheduling scheme.

5 Conclusion

In this paper an overview of challenges in big data analytics, background study of decision tree and ensemble models of machine learning with their problems are discussed; and provided an evaluation of random forest tree ensemble model in the combined perspective of big data, HPC and parallel processing, followed by tracing out the performance issue of scalable speedup; with help of reviewing involved approaches as dimension reduction of big data, variable importance measures, boosting bagging combination, mega ensembles, histogram or graph based horizontal/vertical task/data portioning and different allocation cum scheduling approaches. Possible cause of the problem and its mitigation by developing performance oriented localization aware tiling concept based hybrid parallel approaches using process and thread level parallelization to original random forest algorithm to reduce communication overhead and avoid redundant allocation with appropriate new allocation strategies are suggested, which can contribute toward our objective of performance enhancement for massive data processing. Further, we will design feasible approaches with an experimental setup in distributed cluster environment as well as also in virtual cluster environment by deploying apache spark, code through MPI and OpenMP and will execute& validate the results in term of performance and accuracy.

References

1. Wu, X., Zhu, X., Wu, G., Ding, W.: Data mining with big data. IEEE Trans. Knowl. Data Eng. **26**(1), 97–107 (2014)
2. Kuang, L., Hao, F., Yang, L., Lin, M., Luo, C., Min, G.: A tensor-based approach for big data representation and dimensionality reduction. IEEE Trans. Emerg. Topics Comput. **2**(3), 280–291 (2014)
3. Rajaraman, A., Ullman, J.: Mining of Massive Data Sets. Cambridge University Press, Cambridge (2011)
4. PARAM-Kanchenjunga. http://www.nitsikkim.ac.in/research/hpc/HPC.php
5. Breiman, L.: Classification and Regression Trees. Chapman & Hall, London (1984)
6. Breiman, L.: Bagging predictors. Mach. Learn. **24**(2), 123–140 (1996)
7. Breiman, L.: Random Forests. https://www.stat.berkeley.edu/~breiman/RandomForests/

8. Chand, N., Mishra, P., Krishna, C.R., Pilli, E.S., Govil, M.C.: A comparative analysis of SVM and its stacking with other classification algorithm for intrusion detection. In: Proceedings of International Conference on Computing Communication & Automation, pp. 1–6. IEEE, Dehradun (2016)
9. Chauhan, H., Kumar, V., Pundir, S., Pilli, E.S.: Comparative study of classification techniques for intrusion detection. In: Proceedings of International Symposium on Computational and Business Intelligence, pp. 40–43. IEEE, New Delhi (2013)
10. Mishra, P., Varadharajan, V., Tupakula, U., Pilli, E.S.: A detailed investigation and analysis of using machine learning techniques for intrusion detection. IEEE Commun. Surv. Tutorials (2018). https://doi.org/10.1109/comst.2018.2847722
11. Xu, M., Chen, H., Varshney, P.K.: Dimensionality reduction for registration of high-dimensional data sets. IEEE Trans. Image Process. **22**(8), 3041–3049 (2013)
12. Tao, Q., Chu, D., Wang, J.: Recursive support vector machines for dimensionality reduction. IEEE Trans. Neural Netw. **19**(1), 189–193 (2008)
13. Lin, Y., Liu, T., Fuh, C.: Multiple kernel learning for dimensionality reduction. IEEE Trans. Pattern Anal. Mach. Intell. **33**(6), 1147–1160 (2011)
14. Strobl, C., Boulesteix, A., Kneib, T., Augustin, T., Zeileis, A.: Conditional variable importance for random forests. BMC Bioinf. **9**(14), 1–11 (2008)
15. Bernard, S., Adam, S., Heutte, L.: Dynamic random forests. Pattern. Recog. Lett. **33**(12), 1580–1586 (2012)
16. Khoshgoftaar, T.M., Hulse, J.V., Napolitano, A.: Comparing boosting and bagging techniques with noisy and imbalanced data. IEEE Trans. Syst. Man Cybern. **41**(3), 552–568 (2011)
17. Biau, G.: Analysis of a random forests model. J. Mach. Learn. Res. **13**(1), 1063–1095 (2012)
18. Basilico, J.D., Munson, M.A., Kolda, T.G., Dixon, K.R., Kegelmeyer, W.P.: COMET: a recipe for learning and using large ensembles on massive data. In: Proceedings of 11th International Conference on Data Mining, pp. 41–50. IEEE, Washington (2011)
19. Panda, B., Herbach, J.S., Basu, S., Bayardo, R.J.: PLANET: massively parallel learning of tree ensembles with MapReduce. In: Proceedings of VLDB Endowment, pp. 1426–1437. ACM, Lyon (2009)
20. Dean, J., Ghemawat, S.: MapReduce: simplified data processing on large clusters. Commun. ACM **51**(1), 107–113 (2008)
21. Tyree, S., Weinberger, K.Q., Agrawal, K.: Parallel boosted regression trees for web search ranking. In: Proceedings of International Conference on World Wide Web, pp. 387–396. ACM, Hyderabad (2011)
22. Warneke, D., Kao, O.: Exploiting dynamic resource allocation for efficient parallel data processing in the cloud. IEEE Trans. Parallel Distrib. Syst. **22**(6), 985–997 (2011)
23. Briceno, L.D., et al.: Heuristics for robust resource allocation of satellite weather data processing on a heterogeneous parallel system. IEEE Trans. Parallel Distrib. Syst. **22**(11), 1780–1787 (2011)
24. Zhang, F., Cao, J., Tan, W., Khan, S.U., Li, K., Zomaya, A.Y.: Evolutionary scheduling of dynamic multitasking workloads for big-data analytics in elastic cloud. IEEE Trans. Emerg. Topics Comput. **2**(3), 338–351 (2014)
25. Li, K., Tang, X., Veeravalli, B., Li, K.: Scheduling precedence constrained stochastic tasks on heterogeneous cluster systems. IEEE Trans. Comput. **64**(1), 191–204 (2015)
26. Spark mllib-random forest. http://spark.apache.org/docs/latest/mllib-ensembles.html
27. Chen, J., Li, K., Tang, Z., Bilal, K., Yu, S., Weng, C., Li, K.: A parallel random forest algorithm for big data in a spark cloud computing environment. IEEE Trans. Parallel Distributed Syst. **28**(4), 919–933 (2017)

28. Guo, J., Bikśhandi, G., Fraguela, B., Garzaran, M., Padua, D.: Programing with tiles. In: Proceedings of of 13th SIGPLAN Symposium on Principles and Practice of Parallel Programming, pp. 111–122. ACM, Salt Lake City (2008)
29. Garg, R., Mittal, M., Son, L.H.: Reliability and energy efficient workflow scheduling in cloud environment. Cluster Comput. (2019). https://doi.org/10.1007/s10586-019-02911-7
30. Shastri, M., Roy, S., Mittal, M.: Stock price prediction using artificial neural model: an application of big data. EAI Endorsed Trans. Scalable Inf. Syst. (2019). https://doi.org/10.4108/eai.19-12-2018.156085

The Survival Analysis of Big Data Application Over Auto-scaling Cloud Environment

R. S. Rajput[1]([✉]), Dinesh Goyal[2], and Anjali Pant[3]

[1] Centre of Cloud Infrastructure and Security,
Suresh Gyan Vihar University, Jaipur, India
rajpoot.rs@gmail.com
[2] Department of Computer Engineering,
Poornima Institute of Engineering and Technology, Jaipur, India
[3] Government Polytechnic College, Sankifarm, Sitarganj, Uttarakhand, India

Abstract. The cloud resource provisioning is a mechanism of cloud resources allocation to cloud customers, and cloud customers have to interact with cloud resources using any cloud data center. The workload of the cloud environment consists of the significance of computing resources running situation in the cloud data centers. Cloud resource provisioning has a signification relation with cloud workload. The workload of cloud data centers is not the same all the time. For smooth and effective working of cloud resources at the cloud customer end, scaling of cloud resources required at cloud data center end. The scaling is a primary plan that to manage the extended work-load of the cloud data center. Scaling is implemented by adding additional or increasing computing power and memory capacity. Auto-scaling is one of an essential attribute of cloud computing that facilitates automatic provisioning of computing resources like add, remove, scale-up or scale-down resources depending upon workload. Big data applications associated with the large storage capacity and high processing power, cloud environment is suitable for fulfilling big data application requirement using auto-scaling of resources. In the present study, we have estimated the survival probability of auto-scaled cloud environment in the context of big data applications. Further, we investigated in this paper the importance of cloud resources that are used to build auto-scaling based cloud computing environment.

Keywords: Cloud computing · Big data · Survival analysis · Auto-scaling · Mean time to failure · Importance of cloud resources

1 Introduction

1.1 Cloud Computing System

Cloud computing bases consist of services that are granted and allotted through a cloud service center, such as data center, services can be accessed through the cloud customers' side interface such as the web browsing, remote desktop, thin client, mobile application, terminal emulator, etc., anywhere in the world. A cloud computing system divides it into two primary segments. The front end segment and the back end segment.

© Springer Nature Singapore Pte Ltd. 2019
A. K. Somani et al. (Eds.): ICETCE 2019, CCIS 985, pp. 155–166, 2019.
https://doi.org/10.1007/978-981-13-8300-7_13

Both segments connected through a communication network, usually the Internet. The front end segment is the part of the cloud computing model seen by the cloud customers, i.e., the cloud user. That includes the cloud customers' network and the cloud resources accessed as a service. Backend segment of the cloud computing model is cloud itself; it is a pooling of cloud resources like computing servers, storage, and network, memory in the virtual or physical forms. Availability Zones of AWS is one of an example of the cloud backend segment [10].

Cloud computing system has four basic deployment models, three fundamental service models, and four key attributes. Deployment models of the cloud computing system are defined by the type of access to the cloud resources, i.e., how the cloud located? Cloud computing systems' access models are public access, private access, hybrid access, and community access. Service model of the cloud computing system is classified into three fundamental service models namely infrastructure as a service (IaaS) model, platform as a service (PaaS) model and software as a service (SaaS) model. The primary attributes of the cloud computing system are broad network access, rapid elasticity, measured service, and on-demand self-service respectively [1].

1.2 Big Data

Big data is data that concerned with large volume, heterogeneous types, growing nature, and generating with multiple resources [17]. Big data has five primary characteristics i.e. volume, variety, velocity, value, and variability. Variety includes the data coming from different sources with three different categories i.e. structured, semi-structured and unstructured sources. Volume means the enormous amount of data. The extremely fast rate at which the data is generated is termed as velocity. The relevance of the important information which is taken out by applying queries over data means the value. The inconsistencies which arise during the data flow are known as variability. Many techniques are implemented to find solution with big data. It is surrounded by issues related to the storage, processing, and management of big data. Healthcare, Administration in Public sector, Retail marketing, Manufacturing, Personal location data, Fact-based decision making, Improvement in customer experience, Improvement in sales, Innovation of new products, Reducing risk in various sectors, Quality improvement of product and services are some significance fields of big data applications [16]. As big data associated with the large storage capacity and high processing power, Rapid elasticity, on-demand self service and resource pooling are some key characteristics of cloud computing, those are suitable for fulfilling issues related with big data implementation.

1.3 The Auto-Scaling of a Cloud Computing Environment

The auto-scaling cloud computing system is designed with the ability to increase or withdraw the number of running resources as the demands of application over time. Auto-scaling is usually used for the application tier for enhancing the performance of software applications. Scaling actions focused on horizontal scaling, i.e., combining new Virtual Machine (VM) or vertical scaling, i.e., allocating extended physical CPU or memory to a running VM. Unfortunately, the conventional operating system does not support action required vertical scaling without rebooting so that performing vertical scaling, in this case, it is feasible to change VM with new configured VM The Figs. 1 and 2 describes auto-scaling architecture of cloud computing system [14].

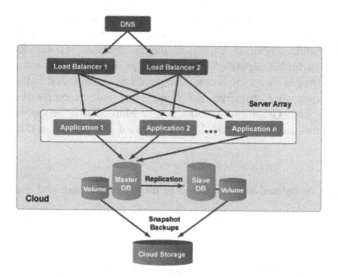

Fig. 1. Auto-scaling architecture

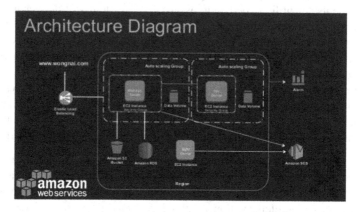

Fig. 2. Auto-scaling architecture of AWS

1.4   Cloud Work Load

Lorido-Botran et al. [15] described the term workload as a sequence of users' job requests with arrival timestamp. Abawajy [5] explained workload has some features like it has arrival rate, execution time, memory usages, Input/output operations, communication and numbers of processors (VMs) required. We can categorize the workload of the cloud into two board categories, static workload, and dynamic workload. A static workload in which a certain amount of workload is given one time and, in the dynamic workload arrives all the time. Based on workload execution platform it can be further separated into two groups, server-centric and client-centric. Websites, Web Applications, Scientific Computations, Enterprises Software Applications, Database query, e-Commerce, and Storage, are some examples of server-centric workload. Computer Graphics, Mobile app, e-Mail, Word processing are some examples of client-centric workload. Workload-based job requests and defined as

$$J = \{j_1, j_2, \ldots j_n\} \tag{1}$$

The job can be defined as a collection of the sequential task and defined as

$$j_i = \{T_1, T_2, \ldots T_m\} \tag{2}$$

1.5   Reliability Study for Cloud Computing System

The Reliability Society of Institute of Electrical and Electronics Engineers (IEEE) declares that reliability or reliability engineering is a discipline of design engineering which implements scientific expertise to assure that a system will deliver its assigned function for the required duration within a given setting, including the capacity to test and support the system through its entire lifecycle. For software, it represents reliability as the probability of failure-free software operation for a particularized period in a specified situation [7, 8].

All items used in the scenario of the cloud are resources. Resources include computing, storage, networking and energy, resources directly or indirectly associated with the set of cloud applications. A cloud resource is any physical or virtual component of limited availability within a cloud system or environment [11–13].

Survival Probability. Reliability is the strength of an item to achieve expected functions under given environmental and operational situations and for a declared period. Survival probability is the probability that the item does not fail in the time interval (0, t] as the implementation of cloud computing, customers' demand 24/7 access to their services and data. Reliability remains a difficulty for cloud service providers everywhere. As per Marvin Rausand and Arnljot Heyland, the survival probability of component i R_i (t) at time t.

$$R_i(t) = \int_t^\infty f(u)du = e^{-\lambda_i t} \, for \, t > 0 \tag{3}$$

where

λ_i Failure rate of an ith component of the system
$R_i(t)$ Survival probability of component i at time t
$f(u)$ Probability density function

Failure Analysis of Cloud Service in Context of Big Data Application. Various types of possible failure may be affected by the reliability of a cloud service. Some well-known are overflow of job requests, maximize service availability, minimize the impact of any failure on cloud customers, minimize the impact of any failure on customers, session timeout, data resource missing, Maximize business continuity, hardware failure, computing resource missing, software failure, database failure, and network failure [2, 3, 9].

Importance of Component. Some resources in the system are more critical for system reliability than other resources [4]. The importance of a resource depends on two factors. Location of resource in the system, the reliability of the resource in question.

Some well-known measures are Birnbaum's measure, Improvement potential, Risk achievement worth, Risk reduction worth and Critical importance. We are here worked only on Birnbaum's measure. We considered a system of n components with component reliability p_i (t), for $i = 1, 2, \ldots n$. The system reliability concerning a specific system function is denoted h $(p$ $(t))$. Birnbaum (1969) proposed a measure of the reliability importance of component i at time t is

$$I^B(i|t) = \frac{\partial h(p(t))}{\partial p_i(t)} \, for \, i = 1, 2, \ldots, n. \tag{4}$$

2 Methodology

2.1 Analytical Model

The exponential distribution, a particular case of the Weibull distribution family, is the most widely used distribution in reliability and survival studies. In the study of continuous-time stochastic processes, the exponential distribution is usually used to model then the time until something happen in the process. The exponential model can be used as the primary form of the reliability growth model of software. In this study, we will refer to applications related with big data, VM's, kernel, data storages, and hypervisor as software application as they all follow an exponential distribution time till failure distribution [9].

In Fig. 3, represents the system structure of auto-scaling architecture of cloud computing system model. The system structure of the proposed model is a mixed

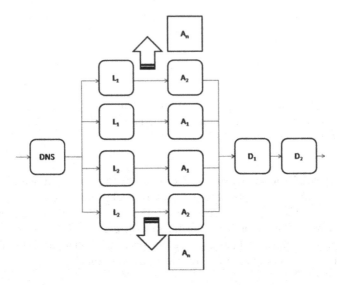

Fig. 3. The system structure of auto-scaling cloud environment

system structure of series structure, parallel structure, and k-out-of-n structure. This model has a single entry point DNS of job request for service. The service requests can the form of http, https, ftp, telnet, database query, remote-desktop, or any other I/O request. The job loads arrive according to a Poisson process. L_1, L_2 denoted Load Balancers. Load balancer forwards the user request to one of an application server. A_1, A_2, A_3 ... A_n. denoted application servers. An Application server can be a node like AWS EC2 instance, core or processor representing the actual computing resources of cloud architecture where the services like MapReduce are computed. D_1, D_2 denoted Database instances like AWS RDS or AWS DynamoDB. D_1 accounts for the master database server and use to access files, directories, database or any I/O. Data inconsistencies are avoided by choosing only one database server in our design. D_2 represents a slave database server, use to store details data and accessed by D_1. According to Marvin Rausand and Arnljot Heyland, the survival probability of system $R_S(t)$ at time t for n application servers is defined in by the Eq. 5 [6].

$$R_s(t) = R_{DNS}(t)\left[1 - (1 - R_L(t).R_A(t))^{2n}\right]R_{D1}(t) \cdot R_{D2}(t) \tag{5}$$

Equations 3 and 5 give

$$R_s(t) = e^{-\lambda_1 t}\left[1 - \left(1 - e^{-\lambda_2 t}e^{\lambda_3 t}\right)^{2n}\right]e^{-\lambda_4 t}e^{-\lambda_5 t}$$

The failure rates of all components are almost the same, consider all are same then Survival probability of $R_S(t)$ at time t is

$$R_s(t) = e^{-3\lambda t}\left[1 - \left(1 - e^{-2\lambda t}\right)^{2n}\right] \tag{6}$$

Mean time to failure (MTTF) is defined as

$$MTTF = \int_0^\infty R_s(t)dt \tag{7}$$

Equations 6 and 7 give

$$MTTF = \int_0^\infty \left[e^{-3\lambda t}\left[1 - \left(1 - e^{-2\lambda t}\right)^{2n}\right]\right]$$

$$MTTF = \frac{1}{3\lambda} - \frac{1}{\lambda}\frac{\frac{1}{2}\sqrt{\pi}(2n)!}{(2(n+1))!} \tag{8}$$

The failure rate function defined as

$$z_s(t) = \frac{-R'_s(t)}{R_s(t)} \tag{9}$$

Equations 9 and 6 give

$$z_s(t) = \frac{-\frac{d}{dt}\left[e^{-3\lambda t}\left[1 - \left(1 - e^{-2\lambda t}\right)^{2n}\right]\right]}{\left[e^{-3\lambda t}\left[1 - \left(1 - e^{-2\lambda t}\right)^{2n}\right]\right]}$$

$$Z_s(t) = \frac{-\left[-3\lambda e^{-2\lambda t} - e^{-3\lambda t}\left(1 - e^{-2\lambda t}\right)^{2n-1}\left[4n\lambda e^{-2\lambda t} - 3\lambda + 3\lambda e^{-2\lambda t}\right]\right]}{\left[e^{-3\lambda t}\left[1 - \left(1 - e^{-2\lambda t}\right)^{2n}\right]\right]} \tag{10}$$

Birnbaum's measures for components DNS is

$$I^B(DNS|t) = \frac{\partial(Rs(t))}{\partial R_{DNS}(t)} \tag{11}$$

Birnbaum's measures for components Load Balancer L is

$$I^B(L|t) = \frac{\partial(Rs(t))}{\partial R_L(t)} \tag{12}$$

Birnbaum's measures for components Application Server A is

$$I^B(A|t) = \frac{\partial(Rs(t))}{\partial R_A(t)} \tag{13}$$

Birnbaum's measures for components Database Server D_1 is

$$I^B(D1|t) = \frac{\partial(Rs(t))}{\partial R_{D1}(t)} \tag{14}$$

Birnbaum's measures for components Database Server D_2 is

$$I^B(D2|t) = \frac{\partial(Rs(t))}{\partial R_{D2}(t)} \tag{15}$$

From Eqs. 11 to 15 if failure rates of all components are almost the same then $I^B(DNS|t)$, $I^B(D_1|t)$ and $I^B(D_2|t)$ are equal, and $I^B(L|t)$, $I^B(A|t)$ and equal. Value of $I^B(L|t)$, $I^B(A|t)$ is larger than $I^B(DNS|t)$, $I^B(D_1|t)$ and $I^B(D_2|t)$.

2.2 Simulation

A simulation program is developed using Scilab 6 mathematical software. Data generated by this program can be helpful in the investigation of the model. In the current study, we assume the failure rate of all resources is the same. The algorithm of a program to simulate the proposed model is as under.

ALGORITHM FOR THE SIMULATION PROGRAM

Set $\lambda=0$, $\lambda_{max}=1$, $\lambda_{inr}=0.1$, $t=0$, $t_{max}=25$, $t_{inr}=1$, $n=1$, $n_{max}=10$, $i=1$
While performing until $n < n_{max}$
Set $\lambda=0$
While performing until $\lambda < \lambda_{max}$
Set $t=0$
While performing until $t < t_{max}$
$R_S(t):= e^{-3\lambda t}[1-(1-e^{-2\lambda t})^{2n}]$
$MTTF_S:=[(1/3\lambda) - (1/\lambda)[(0.5*sqrt(22/7)*(2n)!) / ((2n+2)!)]$
$Z_S(t)=[-[-3\lambda e^{-2\lambda t}-e^{-3\lambda t}(1-e^{-2\lambda t})^{2n-1}[4n\lambda e^{-2\lambda t}-3\lambda+3\lambda e^{-2\lambda t}]]]/R_S(t)$
$R_S(t)$, $MTTF_S(t)$, $Z_S(t)$, t, λ, n and i, insert its values in the ith row of data table
$i=i+1$
$t=t+t_{inr}$
end
$\lambda=\lambda+\lambda_{inr}$
end
$n=n+1$
end

Result: Database table i rows and 6 columns for relation ($R_S(t)$, $MTTF_S(t)$, $Z_S(t)$, t, λ, n)

3   Results and Discussion

We have drawn diagrams as shown in the Figs. 4, 5 and 6 using data which have been generated through a simulation program.

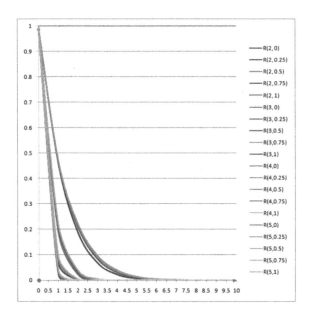

Fig. 4. Survival probability w.r.t. time for different scaling servers and failure rates

The Analysis of Survival Probability of the System. One can easily observe from Fig. 4 that the survival probability of system decreases as increases failure rates of cloud resources. The survival probability of system also decreases as increases system time. The survival probability of system increases as increases number of scaling servers, the changes in survival probability of system follow an exponential decreasing pattern.

The Analysis of Failure Rate of the System. In Fig. 5, the failure rate of system increases as the increases failure rate of cloud resources. The failure rate of the system also increases as increases system time. The Failure rate of system decrease as the increases scaling servers of cloud resources, the changes in the failure rate of the system follow an exponentially increasing pattern.

The Analysis of MTTF. We observed from Fig. 6, the MTTF of system decrease as the increased failure rate of cloud resources. The MTTF of system increase as increases scaling servers. The pattern of changes also fellow an exponential pattern, change of MTTF of the system divided into two sections. First MTTF quickly decreases and the second section decreases slowly.

The Analysis of Importance of Component. From Eqs. 11 to 15, the failure rates of all components are almost the same then $I^B(DNS|t)$, $I^B(D_1|t)$ and $I^B(D_2|t)$ are equal, and $I^B(L|t)$, $I^B(A|t)$ and equal. Value of $I^B(L|t)$, $I^B(A|t)$ is larger than $I^B(DNS|t)$, $I^B(D_1|t)$ and $I^B(D_2|t)$. We can conclude that the parallel components are more reliable compared to series.

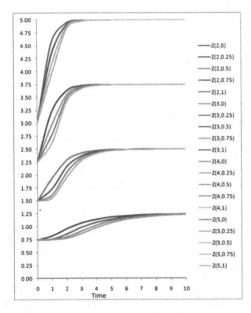

Fig. 5. Failure Rate w.r.t. time for different scaling servers and failure rates

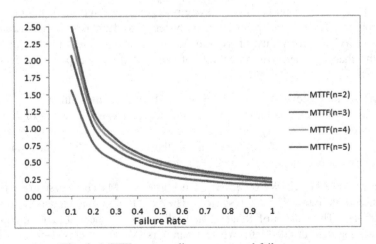

Fig. 6. MTTF w.r.t. scaling servers and failure rates

4 Conclusion

After critical examination of results, the following points are discovered for cloud resources in the context of big data applications.

- The failure rate of cloud resources increases then the survival probability of system or cloud environment decrease. When the system time increase then the survival probability of system or cloud environment also decreases. An increasing number of scaling units then increases survival probability. The changes follow an exponential decreasing pattern.
- The failure rate of cloud resources increases then the failure rate of a system or cloud environment increase. An increasing number of scaling units decreases failure rate. The system time increase then the failure rate of a system or cloud environment also increases. Changes follow an exponentially increasing pattern,
- When the failure rate of cloud resources increases, then the MTTF of a cloud system or cloud environment decreases. The increasing number of scaling units then increases MTTF. The pattern of changes is an exponential pattern, changes of MTTF of the system divided into two sections, first MTTF quickly decreases and the second section decreases slowly.
- We also concluded that the parallel components are more reliable compared to series based on Birnbaum's measures.

5 Future Work

In this study, we have considered the failure rate of all cloud resources in the context of big data application as the same failure rate. Furthermore, nonidentical cloud resources having different failure rates may be included in the future study. We also used an exponential distribution which is a particular case of Weibull distribution other distributions may also be applied. The present study also focused on horizontal scaling vertical scaling may be topic of future study.

References

1. Sosinsky, B.: Cloud Computing Bible. Wiley Publishing Inc., Indianapolis (2011)
2. Bills, D.: Fundamentals of Cloud Service Reliability. https://cloudblogs.microsoft.com/microsoftsecure/2014/03/24/reliability-series-1-reliability-vs-resilience/. Accessed 29 Nov 2018
3. Kaur, G., Kumar, R.: A review on reliability issues in cloud service. In: Proceeding of International Conference on Advancements in Engineering and Technology (ICAET 2015), pp. 9–13 (2015). Int. J. Comput. Appl.
4. Dui, H.: Reliability optimization of automatic control systems based on importance measures: a framework. Int. J. Performability Eng. 12(3), 297–300 (2016)
5. Abawajy, J.: What is workload (cloud data center service provisioning: theoretical and practical approaches). https://www.jnu.ac.in/content/LAB05/presentation/gian2018/day2.pdf. Accessed 9 Sept 2018

6. Rausand, M., Hayland, A.: System Reliability Theory Models, Statistical Methods, and Applications, 2nd edn. Wiley, Hoboken (2004)

7. Adams, M.: An Introduction to designing reliable Cloud Services. https://chapters. cloudsecurityalliance.org/seattle/files/2013/08/An-Introduction. Accessed 8 Aug 2018

8. Sah, N., Singh, S.B., Rajput, R.S.: Stochastic analysis of a Web Server with different types of failure. J. Reliab. Stat. Stud. 3(1), 105–111 (2011)

9. Yadav, N., Singh, V.B., Kumari, M.: Generalized reliability model for cloud computing. Int. J. Comput. Appl. 88(14), 13–16 (2014)

10. Nabeela, N.: All you need to know about cloud computing. http://eid100nujhatn.blogspot.in/ 2015/10/all-you-need-to-know-about-cloud.html. Accessed 20 Oct 2018

11. Rajput, R.S., Pant, A.: Optimal resource management in the cloud environment - a review. Int. J. Converging Technol. Manag. (IJCTM) 4(1), 12–24 (2018)

12. Rajput, R.S., Goyal, D., Singh, S.B.: Study of performance evolution of three-tier architecture based cloud computing system. In: Proceeding of Third International Conference on Internet of Things and Connected Technologies (ICIoTCT) (2018). http:// dx.doi.org/10.2139/ssrn.3166719

13. Rajput, R.S., Goyal, D., Pant, A.: The survival analysis of three-tier architecture based cloud computing system. Int. J. Adv. Stud. Sci. Res. 3(11), 300–305 (2018). http://ssrn.com/ abstract=3320440

14. RightScale Docs: Cloud Computing System Architecture Diagrams. http://docs.rightscale. com/cm/designers_guide/cm-cloud-computing-system-architecture-diagram.html. Accessed 20 Oct 2018

15. Lorido-Botran, T., Miguel-Alonso, J., Lozano, J.A.: Auto-scaling techniques for elastic applications in cloud environments. Technical report, Department of Computer Architecture and Technology University of the Basque Country (2012)

16. Arora, Y., Goyal, D.: Big data technologies: brief overview. Int. J. Comput. Appl. 131(9), 1–6 (2015)

17. Agarwal, B., Ramampiaro, H., Langseth, H., Ruocco, M.: A deep network model for paraphrase detection in short text messages. Inf. Process. Manag. 54(6), 922–937 (2018)

Rating Prediction by Combining User Interest and Friendly Relationship

H. P. Ambulgekar[1(✉)] and Manesh B. Kokare[2]

[1] Department of Computer Science and Engineering,
SGGS IE & T, Nanded, India
ambulgekar@sggs.ac.in
[2] Department of Electronics and Telecommunication Engineering,
SGGS IE & T, Nanded, India

Abstract. Due to the popularity of social media, users of these sites are sharing what they're doing with their friends within numerous social sites. Now, we've an enormous quantity of explanations, ratings and comments for native facilities. The data is effective for fresh users to evaluate whether or not the facilities meet their needs before its. During this paper, suggest an approach for rating prediction by combining user interest, friendly relationship info along with item reputation factor to improve the prediction accuracy. So as to predict ratings, we tend to concentrate on users' rating behaviors and reputation similarity between items. Within the proposed approach for rating prediction, we tend to fuse five factors like personal interest (items and user's topics related), similarity of social interest (user interest related), social rating behavior similarity (users' rating pattern habits related), social rating pattern diffusion (behavior of users diffusions related), and item similarity, this can be deduced by distributing the rating of a user set that represent customers evaluation—into a combined framework of matrix factorization. We tend to perform a number of experiments with the Yelp dataset. Figures in the results show that our approach is having good performance.

Keywords: Data mining · Recommender system · Social user behavior · Collaborative filtering

1 Introduction

Modern consumers are flooded with choices. Users are having huge choices. Providing a matching product to the choice of customer is now became priority of retailers. This is possible with the use of recommender system. Recommender Systems (RS) manage data overload by suggesting to users the items that are probably of their interests. In Amazon-the Ecommerce domain, it's necessary to handle bulk of knowledge, like recommending user most popular things and product [1]. A survey demonstrate that, in Amazon, a minimum of twenty percent of the sales is because of use of recommenders. First generation recommenders [2] using conventional algorithms of collaborative were used [3–14] to forecast interest of user. Though, with the apace growing variety of

© Springer Nature Singapore Pte Ltd. 2019
A. K. Somani et al. (Eds.): ICETCE 2019, CCIS 985, pp. 167–178, 2019.
https://doi.org/10.1007/978-981-13-8300-7_14

listed users as well as numerous product, the matter of cold start begin for users of the system and therefore the poorness of datasets are progressively unmanageable. Social platforms collect volumes of knowledge added by users from all over the world. This data is flexible. It invariably contains information about the products or services (including matter descriptions, emblems and photos), user's observations, attitudes and user's social groups, charges, and locations. It's extremely accepted for recommending users' preferred services from wide-source contributed data.

To resolve cold start begin and drawback of sparseness, social relationship, particularly in group of friends of social networks are used. Social media provide some precious hints to suggest user preferred things like music, video, most popular make/products [15], user's most popular tags once sharing a photograph to social media networks [16], and user fascinated travel places by searching photos on social media [17]. However, is it needed by all users to connect to social sites for suggesting items? Will such connection conceal personality of user, particularly for tough users? It's still a good trial to express personality of user in recommender, and it's still an open problem that the way to create social features be successfully unified in model of a recommender system for enhancing the exactness of recommender.

In this paper, we predict ratings by combining user interest and relationship among the friends. Four social factors like individual interest, relational interest similarity between users, rating behavior similarity between users, and relational rating behavior diffusion combine into a recommendation model which is built on matrix factorization. The interpersonal relationship of social network contains factors like: social concern connection, social rating behavior likeness, and social rating behavior distribution. Personal interest establishes link among user and product latent feature vectors with the help of three different factors that relates socially. For such social interest connection, we tend to understand interest circle to reinforce key connection of user latent feature. For social rating behavior similarity, conception of rating timetable is employed to denote users regular rating actions. We tend to leverage the correspondence between schedules of user rating for social rating action similarity. For social rating behavior diffusion, social circle of user is divided into three parts such as indirect, direct, and mutual friends, to deep perceive rating actions diffusions of social users.

A personalized recommendation system is proposed in this paper by adding new item similarity factor along with personal interest, similarity of users relational interest, similarity between relational rating actions and relational rating action diffusion factor to get an accurate results.

The paper is organized as follows. Section 2 is about the brief literature survey of the existing work. We define problem which is discussed in this paper is also described. Section 3 is about the work that we have proposed. Section 4 is about dataset, experimentation and result. Conclusions are drawn in Sect. 5.

2 Earlier Work

Recommender systems are one of the widely researched area in recent times. Most of the work is related to the problem predication of rating for the items that is to be purchased by user. This estimation is generally based on the rating given by the user for other product or item and other information. Once we succeed in estimating the rating for items which are not rated, we can endorse items having maximum rating to users [38].

Many authors proposed recommendation models based on social network [18, 19, 27, 28, 37] to increase performance of recommenders. Yang et al. [29] proposed to use the idea of 'domain-obvious deduced friends trust group on social networks to advocate user most desired things. This approach enhances social faith within massive complicated network. This also minimizes the load of massive knowledge. For the time being, further the social impact, Jiang et al. [24] proved that a single inclination is additionally an important aspect in social network. Rather similar to the thought of social influence, user latent choices should be just like his friends supported in the probabilistic matrix resolving model [20–26, 29, 30]. Phelan et al. [31] gives news recommendation that utilizes time period. From all this, we can conclude that users having additional friends tend to get additional benefit. Chen et al. [32] discovered additional three measurements to be used in future recommender. Users topic interest, content sources and social preference are these three measurements. These dimensions proved that social preference method were useful in providing recommendations.

Around six recommenders were considered for examination of its quality of recommendations and usefulness [33]. It is observed that users friends were systematically providing higher recommendations. Outsized social network was analyzed by Java et al. [34]. Friends were divided into teams and this was helpful the acceptance of micro-blogging sites to examine user goals. Yuan et al. studied the impact of friendly relationship due to a sort of relation, membership, and its collective impact [35]. Two sorts of dissimilar social associations are united into CF primarily based on recommender via a resolving method. "Moves mutually needs, decides as you wish." similar to the emblem states, customer's alternative is often thoroughly relating to his/her private interest [36] (Table 1).

Table 1. Symbols and its descriptions

Symbol	Descriptions	Symbol	Descriptions
m	Total users	R	Rating matrix
X	Users set	n	Number of item
r	Rating by user	S	Similarity matrix
Fu	Set of users friend	Y	Set of items
Cir_u	Total number of users direct and indirect friends	Mx	Number of X's rating
I	Interpersonal rating function	H_u	Items rated set
$Z_{u,v}$	Factor Interpersonal similarity	$P_{u,j}$	Factor of personal interest
λ, β, η	Parameters to handle over-fitting	D_u, D_i	Latent feature matrices
A	Rating behavior matrix	$B_{u,v}$	Rating behavior similarity

3   Rating Prediction Modelling

In this paper, so as to forecast user-service rankings, we have a tendency to specialize in users' rating behaviors. We have a tendency to combine five factors, user's personal interest, similarity of relational interest, and similarity of relational rating behavior, and relational rating action diffusion, item similarity into latent feature model that is matrix factorization. From the listed factors, item similarity is the most contribution of our approach. Below we describe each of the factor in detail.

Personal Interest. Users with several rating records, typically select items all by themselves with very small effect by their friends because of their individuality. Though, to solve the cold start begin drawback, several existing works [24, 29] took the friends circle in social networks. It unheeded the uniqueness for skilled users but worked for cold start begin user with some records. In different words, the applicability of hidden feature vector of user and product relies on applicability of user interest D_u and item topic D_i to a precise point. Formally we can show the importance of personal interest of user u to the subject of item i in the recommender model.

$$P_{u,j} = Sim(D_u, D_i) \tag{1}$$

Relational Interest Similarity. Vector of user latent feature should be same as latent feature vector of his/her friends supported by their interest similarity. Price of similarity interest between user v and u is given by $Z_{u,v}$. All of the rows are normalized to unity $\sum_v Z_{u,v}^* = 1$

$$Z_{u,v} = \sqrt{\sum_i^n \left(A_r^u - A_r^v\right)^2} \tag{2}$$

Relational Rating Behavior Connection. Rating behavior pattern is crucial. This behavior or pattern couldn't be detached from temporal data. Therefore, we tend to outline this pattern or behavior as user activities with the system. This type of behavior or pattern demonstration stirs us the course of study plan. The plan assembles that course would we tend to take and after we should move to category. We control a rating plan or schedule for statistic of rating pattern or behavior using users historical ratings records. This can be illustrated with a state in which a user has given one star to an item one star and another three stars on Friday. It is understood that user has very tiny risk to require rating actions. We tend to control this sort of rating calendar to denote users' rating actions. This actions similarity might express user latent options similarity to certain level. To illustrate, course study of a student schedule might denote his study activities to a particular degree. If the student's course of study timetable is comparable with other student, we tend to might conclude that both need related study actions, and moreover, they will be colleagues. We tend to extend it with the rating timetable to compute conduct similarity of social rating.

Rating actions matrix is $A^u = \left[A^u_{r,d}\right]_{P \times Q}$. This shows that user rating behavior of user u. When user u in a day d has rated r stars, this count given by $A^u_{r,d}$. This is behavior count. Interpersonal rating behavior similarity is given by

$$B_{u,v} = \sqrt{\sum_{r=1}^{P} \sum_{d=1}^{Q} \left(A^u_{r,d} - A^v_{r,d}\right)^2} \qquad (3)$$

Rating behavior similarity between user's friends v and user u is denoted by $B_{u,v}$. Rating schedule of user u and his/her friend v should be same to some point. This is the fundamental notion of likeness of interpersonal rating conduct. Each row of B is normalized to unity $\sum_v B^{c^*}_{u,v} = 1$ so that similarity degree can be measured unbiased.

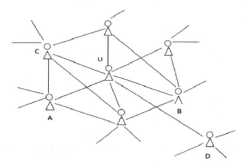

Fig. 1. Example of a user's social network. Friend having many mutual friends can be close and distant.

Relational Rating Behavior Diffusion. This paper is about the issue of rating pattern distribution of social users. We explore this distribution of rating pattern by relating span of users social linkage and temporal data of rating patterns. For user u, divide his linkage into parts like indirect, mutual and direct friends. It is displayed in the Fig. 1. User U have mutual friends like A, B, C can be treated as direct friends whereas D is indirect friend.

(a) It is quite clear that if an addict has number of common friends with user u, then these friends can be treated as close friends. For additional mutual friends they need closer or nearer friend unlike D. The weight $|F_{u \cap v}|/|Cir_u|$ is controlled as a coefficient of relational rating behavior distributions Number of mutual friends between user u and users friends is given by $|F_{u \cap v}|$. Total number of users direct and indirect friends is given by $|Cir_u|$.

(b) Another point is, when user's friend and user has rated the items, this gets diffused smoothly. So the social rating behavior is smoother. Additionally, we have a tendency to regard temporal rating actions as a very significant info to differentiate whether or not the dispersions are smooth. so we have a tendency to design this coefficient as $\sum_{j \in (R^c_u \cap R^c_v)} \exp\left(-|R_{u,j} - R_{v,j}| \times |D_{u,j} - D_{v,j}|\right)$.

Date difference between user u rating and friend v rating for item i is given by $|D_{u,j} - D_{v,j}|$. Rating difference that shows as positive or negative diffusion is given by $|R_{u,j} - R_{v,j}|$. It is quite clear that if these differences are small then diffusion is smooth.

(c) Last point, rating count is utilized to live skilled level users as in [29] due to the idea that rating behaviors could also be a lot of simply dispersed from skilled users. If v's followers have rated in class c and all of these trust v it is quite clear that v is knowledgeable in class c.

This factor is denoted as $M_v^c \times \sum_{x \in F_v^c} \frac{M_x^c}{M_x}$ where M_v^c represents number of v's ratings in class c, $x \in F_v^c$ indicates x and v are friends in class c, M_x implies the overall ratings given by x. Interpersonal rating actions diffusion to the users friend v from user u is measured. So, the smooth degree is given by

$$I_{u,v}^c = M_v^c \times \sum_{x \in F_v^c} \frac{M_x^c}{M_x} \times \frac{F_{u \cap v}^c + 1}{cir_u^c + 1} \times \left(1 + \sum_{j \in (R_u^c \cap R_v^c)} \exp(-|R_{u,j} - R_{v,j}| \times |D_{u,j} - D_{v,j}|)\right)$$

$$(4)$$

Each row of I is normalized to $\sum_v I_{u,v}^{c^*} = 1$. This is required for fair smooth degree measurement. We leverage $I_{u,v}^{c^*}$ to limit user v and u's hidden features. The elementary notion is that the simpler relational rating actions disseminations are, the more equivalent relational hidden features are.

For matrix factorization, personal interest of user, relational rating actions match and rating conduct diffusion are combined. Here is the proposed objective function

$$\psi(R^c, X^c, Y^c) = \tfrac{1}{2} \sum_u \sum_j E_{u,j}^{R^c}(R_{u,j}^c - \widehat{R_{v,j}^c})^2 + \tfrac{\lambda}{2}\left(\|X^c\|^2 + \|Y^c\|^2\right)$$
$$+ \tfrac{3\beta}{2} \sum_u ((X_u^c - KX_v^c)(X_u^c - KX_v^c)^T) + \tfrac{\gamma}{2}\sum_u \sum_j |H_u^{c^*}|(P_{u,j}^{c^*} - X_u^c Y_j^{cT})^2 \qquad (5)$$

Where, $K = \sum_{v \in F_u^c} \tfrac{1}{3}\left(I_{u,v}^{c^*} + B_{u,v}^{c^*} + Z_{u,v}^{c^*}\right)$

$Z_{u,v}^{c^*}$ is factor interpersonal interest similarity. $P_{u,j}^{c^*}$ is factor of personal interest. Matrix X and Y are user and item rating matrix with Frobenius norm to avoid over fitting. $E_{u,j}^{R^c}$ is indicator having value 1 if item i in class c is rated by user u, and have value 0 for other cases. $\widehat{R_{v,j}^c}$ is prediction of rating in category c for user u to item i. $|H_u^{c^*}|$ is absolute number of items which are normalized that an item from class c is rated by user u. It could measure the user understanding in c. Relational factors combined together to limit users' hidden features, that willower time complexity equated with previous work. Hidden feature X_u must match to the common of friends hidden features with the notion of relational factors imposed by second term with the weight of $\tfrac{1}{3}\left(I_{u,v}^{c^*} + B_{u,v}^{c^*} + Z_{u,v}^{c^*}\right)$ in c. In previous model, authors did work only on the user feature matrix by adding interpersonal factors like similarity of relational rating behavior, similarity of relational interest, and dissemination of interpersonal rating behavior into matrix factorization. Now, in our proposed model, we additionally work on the item feature matrix by fusing item popularity similarity matrix into matrix factorization.

Item Similarity. Similar things will help us to predict ratings in the item based CF methods [4], from typical item-based CF algorithms in [4]. So, it's necessary to search out things that have alike options. In this work, we have tendency to assume item's popularity will indirectly mirror its real ratings. We have a tendency to leverage users' ratings to infer item's popularity. Supported users' real ratings, we have a tendency to believe that similar ratings between two items have similar reputation. Based on the idea, we have a user set is $X = \{X_1, X...X_m\}$, there are m number of users. $R_{u,i}$ is the rating score given by user u to item i. So item i's reputation is $Q_i = \{R_{u1,i}, R_{u2,i...}, R_{um,i}\}$.

$Q_j = \{R_{u1,j}, R_{u2,j...}, R_{um,j}\}$ is the set of some items rated by same users which are referred as "virtual friends". Here, we take term virtual friends of item to denote that these two items rated by the same user. To find the relevance between the items virtual friends can be used. One important step in item-based collaborative filtering is to calculate similarity between two items but condition is that both of these items must be rated by the same user. The basic idea for similarity computation is that first separate the users who have rated items i and j and then apply cosine similarity computation technique to find out similarity matrix $S_{i,j}$. We measure similarity between two items i and j, we use cosine similarity measure as follows:

$$S_{i,j} = cosine(Q_i, Q_j)$$

We fuse item similarity matrix into objective function with the help of matrix factorization:

$$\psi(R^c, X^c, Y^c) = \frac{1}{2}\sum_u \sum_j E^{R^c}_{u,j} (R^c_{u,j} - \widehat{R^c_{v,j}})^2 + \frac{\lambda}{2} (\|X^c\|^2 + \|Y^c\|^2) + \frac{3\beta}{2}\sum_u((X^c_u - KX^c_v)(X^c_u - KX^c_v)^T) + \frac{\gamma}{2}\sum_u \sum_j |H^{c*}_u| (P^{c*}_{u,j} - X^c_u Y^{cT}_j)^2 + \quad (6)$$
$$\frac{\eta}{2}\sum_i((Y_i - \sum_j S_{i,j}Y_j)(Y_i - \sum_j S_{i,j}Y_j)^T)$$

Where, $K = \sum_{v\in F^c_u} \frac{1}{3}\left(I^{c*}_{u,v} + B^{c*}_{u,v} + Z^{c*}_{u,v}\right)$

The last term of the objective function is our contribution where ŋ is the constant parameter that is used to balance item similarity factor. Y_i and Y_j are the item i and item j's latent feature vectors respectively. When hidden feature of item Y_i is analogous to its friends hidden feature Y_j having weight of $S_{i,j}$ then it means that item reputation similarity matrix is imposed.

4   Experiments and Results

We have implemented experiments on Yelp dataset to evaluate performance of BaseMF (Model without social consideration), ContextMF (Model considering inter-personal influence and personal choice), Exploring Users Rating Behavior (EURB) model (Considers social users rating behavior) and our proposed model. Initially dataset details is given, next is various performance measures and parameter settings. At last we will discuss the performance graphs.

Dataset. One of the popular directory service provider with all social network data including user reviews is Yelp. It's one of the widely used review web site in America. Users can write their review about and rate the companies, communicate expertise, etc. Social networking and native reviews are combined to form an area online community. Yelp dataset contains eight classes, together with restaurants, home services, hotels, travel, active life, beauty &spas, night life, pets, and looking. We manage to work on restaurant, travel, shopping and pet class. We experiment with eightieth of every rating of knowledge indiscriminately because the training set and therefore the rest two hundredth of every rating knowledge because the test set in each class, to make sure all users' latent options are learned.

Measure of Performance. Once we get U_u and P_i which are the latent feature of user and item respectively, the performance of our algorithmic program are going to be embodied by the errors. Root Mean sq. Error (RMSE) and Mean Absolute Error (MAE) are the foremost standard accuracy measurements that are outlined as follows:

$$RMSE = \frac{\sqrt{\sum_{(u,i) \in R_{test}} \left(R_{u,i} - \widehat{R_{u,i}} \right)^2}}{|R_{test}|} \tag{7}$$

$$MAE = \sum_{(u,i) \in R_{test}} \frac{|R_{u,i} - \widehat{R_{u,i}}|}{|R_{test}|} \tag{8}$$

Where $R_{u,i}$ the actual rating price of user u for item i. $\widehat{R_{u,i}}$ is resultant expected rating price. R_test is set of pairs of user - item within test set. $|R_{test}|$ is the amount of user-item pairs within test set.

Parameter Setting. (a) k: It is hidden vector dimension. When value of k is too large, it'll be tough to compute similarity among users. However when value of k is just too little, it'll be troublesome to express item and user options. Regardless of k is, it's truthful for entire algorithms that are used for comparison.

Here in performance comparison, we have a tendency to set $k = 10$.

(b) λ_1 and λ_2: The parameters of trading-off over-fitting factor about (6) and (7).

(c) β: burden imposed by social factors within second term of the objective perform (6) and (7). We set $\beta = 30$

(d) η: it is factor of personal interest weight in the final term of (6) and (7). Consider value of $\eta = 30$

Performance comparison of various existing models like BaseMF, ContextMF, EURB and our proposed model on Yelp dataset is shown in Figs. 2 and 3. It is found that proposed model performs better than existing EURB model on performance.

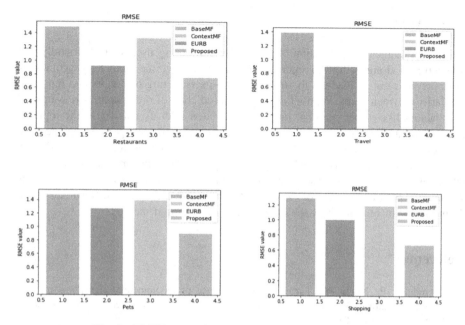

Fig. 2. RMSE comparison in various category of datasets

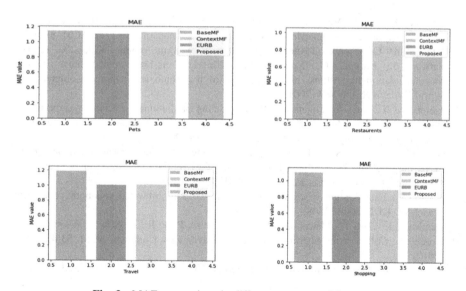

Fig. 3. MAE comparison in different category of datasets

Cold start users were considered in our proposed model. BaseMF, ContextMF, EURB and the proposed model gives performance as per the Restaurants, shopping, pets and travel in Yelp.

5 Conclusion

This paper suggests a prediction of ratings by combining user interest, friendly relationship info along with item reputation factor to improve the prediction accuracy. So as to predict ratings, we tend to concentrate on users' rating behaviors and reputation similarity between items. Within the proposed prediction of rating approach, we tend to fuse factors like—user personal interest (associated with user and also the topics of item), similarity of social interest (user interest related), rating of social behavior similarity (users behavior habits related), social rating behavior diffusion (users' rating behavior diffusions related), and item similarity. All these factors can be deduced by rating distributions of a user set into a framework that implements matrix factorization. We tend to perform a series of experimentations using Yelp dataset. Results obtained during experimentation show the convenience of proposed method.

References

1. Cremonesi, P., Koren, Y., Turrin, R.: Performance of recommender algorithms on top-n recommendation tasks. In: Proceedings of the Fourth ACM Conference on Recommender Systems - RecSys 2010 (2010)
2. Adomavicius, G., Tuzhilin, A.: Toward the next generation of recommender systems: a survey of the state-of-the-art and possible extensions. IEEE Trans. Knowl. Data Eng. **17**(6), 734–749 (2005)
3. Bell, R., Koren, Y., Volinsky, C.: Modeling relationships at multiple scales to improve accuracy of large recommender systems. In: Proceedings of the 13th ACM SIGKDD International Conference on Knowledge Discovery and Data Mining - KDD 2007 (2007)
4. Sarwar, B., Karypis, G., Konstan, J., Reidl, J.: Item-based collaborative filtering recommendation algorithms. In: Proceedings of the Tenth International Conference on World Wide Web - WWW 2001 (2001)
5. Jahrer, M., Töscher, A., Legenstein, R.: Combining predictions for accurate recommender systems. In: Proceedings of the 16th ACM SIGKDD International Conference on Knowledge Discovery and Data Mining - KDD 2010 (2010)
6. Xue, G.-R., et al.: Scalable collaborative filtering using cluster-based smoothing. In: Proceedings of the 28th Annual International ACM SIGIR Conference on Research and Development in Information Retrieval - SIGIR 2005 (2005)
7. Koren, Y.: Factorization meets the neighbourhood: a multifaceted collaborative filtering model. In: KDD 2008 (2008)
8. Candes, E.J., Plan, Y.: Matrix completion with noise. Proc. IEEE **98**(6), 925–936 (2010)
9. Koren, Y.: Collaborative filtering with temporal dynamics. Commun. ACM. **53**(4), 89–97 (2010)
10. Koren, Y.: Collaborative filtering with temporal dynamics. Commun. ACM. **53**(4), 89–97 (2010)
11. Wang, J., de Vries, A.P., Reinders, M.J.T.: Unifying user-based and item-based collaborative filtering approaches by similarity fusion. In: Proceedings of the 29th Annual International ACM SIGIR Conference on Research and Development in Information Retrieval - SIGIR 2006 (2006)
12. Herlocker, J.L., Konstan, J.A., Terveen, L.G., Riedl, J.T.: Evaluating collaborative filtering recommender systems. ACM Trans. Inf. Syst. **22**(1), 5–53 (2004)

13. Liu, N.N., Zhao, M., Yang, Q.: Probabilistic latent preference analysis for collaborative filtering. In: Proceeding 18th ACM Conference on Information and Knowledge Management - CIKM 2009 (2009)

14. Liu, Q., Chen, E., Xiong, H., Ding, C.H.Q., Chen, J.: Enhancing collaborative filtering by user interest expansion via personalized ranking. IEEE Trans. Syst. Man Cybern. Part B Cybern. **42**(1), 218–233 (2012)

15. Feng, H., Qian, X.: Mining user-contributed photos for personalized product recommendation. Neurocomputing **129**, 409–420 (2014)

16. Qian, X., Liu, X., Zheng, C., Du, Y., Hou, X.: Tagging photos using users' vocabularies. Neurocomputing **111**, 144–153 (2013)

17. Li, J., Qian, X., Tang, Y.Y., Yang, L., Mei, T.: GPS estimation for places of interest from social users' uploaded photos. IEEE Trans. Multimed. **15**(8), 2058–2071 (2013)

18. Jamali, M., Ester, M.: A matrix factorization technique with trust propagation for recommendation in social networks. In: Proceedings of the Fourth ACM Conference on Recommender Systems (2010)

19. Jamali, M., Ester, M.: *TrustWalker*: a random walk model for combining trust-based and item-based recommendation. In: KDD 2009: Proceedings of the 15th ACM SIGKDD International Conference on Knowledge Discovery and Data Mining (2009)

20. Huang, J., Cheng, X.Q., Guo, J., Shen, H.W., Yang, K.: Social recommendation with interpersonal influence. In: Frontiers in Artificial Intelligence and Applications (2010)

21. Ma, H., Yang, H., Lyu, M.R., King, I.: SoRec: social recommendation using probabilistic matrix factorization. In: Proceeding 17th ACM Conference on Information and Knowledge Management - CIKM 2008 (2008)

22. Ma, H., Zhou, D., Liu, C., Lyu, M.R., King, I.: Recommender systems with social regularization. In: Proceedings of the Fourth ACM International Conference on Web Search and Data Mining - WSDM 2011 (2011)

23. Yu, L., Pan, R., Li, Z.: Adaptive social similarities for recommender systems. In: [RecSys2011] Proceedings of the 5th ACM Conference on Recommender Systems (2011)

24. Jiang, M., et al.: Social contextual recommendation. In: Proceedings of the 21st ACM International Conference on Information and Knowledge Management - CIKM 2012 (2012)

25. Salakhutdinov, R., Mnih, A.: Probabilistic matrix factorization. In: Advances in Neural Information Processing Systems (NIPS 2008) (2008)

26. Zhao, G., Qian, X., Xie, X.: User-service rating prediction by exploring social users' rating behaviors. IEEE Trans. Multimed. **18**(3), 496–506 (2016)

27. Ma, H., King, I., Lyu, M.R.: Learning to recommend with social trust ensemble. In: SIGIR (2009)

28. Yang, X., Guo, Y., Liu, Y.: Bayesian-inference-based recommendation in online social networks. IEEE Trans. Parallel Distrib. Syst. **24**(4), 642–651 (2013)

29. Yang, X., Hill, M.: Circle-based recommendation in online social networks. In: KDD (2012)

30. Feng, H., Qian, X.: Recommendation via user's personality and social contextual. In: Proceedings of the 22nd ACM International Conference on Conference on Information & Knowledge Management - CIKM 2013 (2013)

31. Cui, P., Wang, F., Liu, S., Ou, M., Yang, S., Sun, L.: Who should share what? Item-level social influence prediction for users and posts ranking. In: Proceedings of the 34th International ACM SIGIR Conference on Research and Development in Information Retrieval - SIGIR 2011 (2011)

32. Chen, J., Nairn, R., Les, N., Nardi, B., Schiano, D., Gumbrecht, M.: Short and tweet: experiments on recommending content from information streams. In: Proceedings of the SIGCHI Conference on Human Factors in Computing System (2010)

33. Sinha, R., Swearingen, K.: Comparing recommendations made by online systems and friends. In: Proceedings of the DELOS-NSF Working Group on Personalisation and Recommender Systems in Digital Libraries (2001)
34. Java, A., Song, X., Finin, T., Tseng, B.: Why we Twitter: understanding microblogging usage and communities. In: Proceedings of the 9th WebKDD and 1st SNA-KDD 2007 Workshop on Web Mining and Social Network Analysis (2007)
35. Yuan, Q., Chen, L., Zhao, S.: Factorization vs. regularization: fusing heterogeneous social relationships in top-n recommendation. In: [RecSys2011] Proceedings of the 5th ACM conference on Recommender systems (2011)
36. Kim, H.R., Chan, P.K.: Learning implicit user interest hierarchy for context in personalization. Appl. Intell. **28**(2), 153–166 (2008)
37. Jain, G., Sharma, M., Agarwal, B.: Spam detection in social media using convolutional and long short term memory neural network. Ann. Math. Artif. Intel. **85**(1), 21–44 (2019)
38. Verma, S.K., Mittal, N., Agarwal, B.: Hybrid recommender system based on fuzzy clustering and collaborative filtering. In 4th International Conference on Computer and Communication Technology (ICCCT-2013), pp. 116–120 (2013)

A Comparative Study of Methods for Hiding Large Size Audio File in Smaller Image Carriers

Abraham Ayegba Alfa[1], Kharimah Bimbola Ahmed[1],
Sanjay Misra[2](✉), Adewole Adewumi[2], Ravin Ahuja[3],
Foluso Ayeni[4], and Robertas Damasevicius[5]

[1] Kogi State College of Education, Ankpa, Nigeria
`abrahamsalfa@gmail.com`, `kharimahahmed@ymail.com`
[2] Covenant University, Ota, Nigeria
`{sanjay.misra,wole.adewumi}@covenantuniversity.edu.ng`
[3] Shri Vishwakarma Skill University, Gurgaon, Haryana, India
[4] Southern University, Baton Rouge, USA
`foluso.ayeni@icitd.com`
[5] Kanus University of Technology, Kaunas, Lithuania
`robertas.damasevicius@ktu.lt`

Abstract. The last 13 years or more has been characterized by unmatched progress in multimedia content and undercover communication. Sound file is chief among mode of transmitting classified and confidential information across the cyberspace and channels. Steganography is a traditional solution to address the problems of data security on public channels and transmitters. In general, the practice of image steganography favored smaller size secret message transmission in larger size carrier approach. However, there is a major of challenge of concealing classified large size audio data inside a smaller image cover with lesser distortions and quality retention of the cover medium and original message respectively. To achieve these, the audio file is compressed to appropriate size before embedding it into two distinct image files (grayscale and RGB color) at binary level using both Most Significant Bits (MSBs) and Least Significant Bits (LSBs) methods. This paper established that the most suitable of MSBs and LSBs in terms of robustness, payload size, and retention of perception transparency of media understudied.

Keywords: Steganography · LSBs · MSBs · Grayscale · RGB colour · Binary level · Hiding · Carrier · Imperceptibility · Audio · Image

1 Introduction

The concept of Steganography is derived from the two Greek words *stegos* and *grafia* meaning *cover* and *writing*. It is simply the art and science of invisible exchange of information. Steganography is termed cover writing, which hiding information in other information to conceal the presence of the information [1]. One method of Steganography is called image steganography which involves information hiding

© Springer Nature Singapore Pte Ltd. 2019
A. K. Somani et al. (Eds.): ICETCE 2019, CCIS 985, pp. 179–191, 2019.
https://doi.org/10.1007/978-981-13-8300-7_15

exclusively in images or pictures [2]. The seeming popularity and availability of digital audio signal have informed the choice of scholars in transmitting secret information [3]. The advancements of the Internet technology are a major motivation in networking and several communications security solutions. There is massive data generation; processing, transmission and sharing in more recent time [4].

Presently, digital communication offers several benefits such as high speed, compression and better quality. Security is a major concern of digital data/information because of fear of theft during transmission. In effect, Steganography offers protection to the information before transmission. The primary structure of Steganography consists of three parts including the secret message, the stego key and the cover medium [5]. In most situations, the cover medium is one of a painting, an MP3, a digital image, or a TCP/IP packet. Cover medium are carrier of the secret message. A key enables intended recipient to decode/decipher/detect the secret message. Keys are a password, a blacklight, or a pattern [6].

Steganography field is concerned with ways of concealing secret message in a cover message. Cover message includes text, video, audio and image [7]. Cryptography is another field of information security that encrypt secret message prior their transmission in order to evade unauthorized users from intercepting sensitive/valuable information in communication channel [7, 8]. This paper explores the most suitable of both Steganography methods in transmitting large size audio within smaller size image cover.

2   Literature Review

In general, information hiding comprises several sub-disciplines such as watermarking and steganography. Steganography is the art of secret exchange of messages [21, 22]. The goal of steganography is to embed a message in a manner unknown except the sender and envisioned recipient [9]. In general, concealed information never attract focus and less susceptible to attacks [10]. Again, cryptography takes aim at making information inconceivable. The quest for intellectual property rights issues brought about copyright sub-field known as digital watermarking. Watermarks can be considered as imperceptible or visible depending on steganographic or non-steganographic scheme implemented. The renew focus of scholar to steganography came about after the extremist attacks of 11th September, 2001, when it became evident information hiding could potentially have been used to plot and execute attacks by criminals [11].

Still Imagery: It is popularly implemented for embedding secret messages into a digital image [12]. This scheme utilises human visual system (HVS) weakness. HVS is incapable of identifying the small alterations in colour vectors luminance at high frequency portion of the visual spectrum. Colour pixels setare used to stand for a picture. The individual pixels are depicted using their optical characteristics, that is, chroma, contrast and brightness which are mapped digitally as set of 0s and 1s.

Modification of Significant Bits of a Cover Image in Bitmap Format: It uses binary form that is dispersed throughout the MSBs or LSBs of each message's pixel before/after encoding. For a 24-bit bitmap, 8 bits match each of three colour values of red, green, and blue in each pixel. When, blue component of colour image is selected, 28 different values exist. In particular, the disparity between 11111111 and 11111110 in blue intensity value is highly unrealizable to the human visual system (HVS). Certain methods are capable of operating on compressed (JPEG) and uncompressed (bitmap) digital image. But, it is incapable of tolerating compression and formatting processes of image in steganographic schemes [13].

Image Steganography: An image is a picture copied or created and stored in electronic form, which can be described with regards to raster or vector graphics [14]. An image stored in raster form is referred to as a bitmap. An image map is a file encompassing information describing diverse positions on a particular image using hypertext links. An image can be described as a group of numbers corresponding to unique light intensities in diverse areas of the image. Picture elements (pixels) are numeric denotation of a grid and the individual points in an image. Grey scale images make use of 8 bits for each pixel capable of displaying 256 diverse colours or shades of grey. Usually, digital colour images are stored in 24-bit files for the RGB colour model called true colour. In practice, all colour variations for the pixels of a 24-bit image are derived from three primary colours of red, green and blue. Primary colours are denoted by 8 bits each. Therefore, there are 256 different components of red, green and blue for a specific pixel.

Image Compression: There are two kinds of compression schemes for images include: lossy and lossless compressions [14]. In lossless compression, every distinct bit of data and information in original file is retained in uncompressed file. GIF (Graphical Interchange Format) and BMP (Bitmap File) are most suitable image formats for lossless compression procedure. On the other hand, lossy compression decreases image file through permanent elimination of redundant information. The outcome of lossy uncompressed file offers a portion of the original information encompassed in the original image file, which is dissimilar to the original image. It is common for image format such as JPEG (Joint Photographic Experts Group) [15].

Spatial Domain Transformation: Spatial steganography hides data in image directly by altering certain bits in the pixel values of the image [16]. Least Significant Bit (LSB) is one of the modest techniques for concealing a secret message in the LSBs of pixel values with imperceptible distortions to human eye. Embedding of message bits embedding can be achieved through simply and random approaches. LSB replacement technique and Matrix embedding fall in to the category of the spatial domain transform steganography. The benefits of spatial domain LSB method include: the original image degradation is not easy; and increased hiding capacity of image information. However, LSB technique offer low robustness; and hidden data is susceptible to unassuming attacks [5; 15].

Transform Domain: In frequency domain, the secret message is place into transformed coefficients of image which provide increased capacity for information hiding and robustness to withstand attacks. This steganographic system is considered one of the strongest when compared those of LSB techniques because, information is hidden in portions of the image less exposed to destruction during image processing tasks such as compression and cropping. Recently, certain transform domain approaches are independent of underlying image format; this makes it beat lossy and lossless format conversions. The major transform domain approaches include: Discrete Fourier transformation (DFT), Discrete Wavelet Transformation (DWT), and Discrete Cosine Transformation (DCT) [5, 14].

Distortion: The information hiding approach makes use of original cover signal distortion to ascertain the degree of deviation in the decoding process. This distortion process requires the previous idea of the original cover image in order to recover the secret message during decoding process in which the decoder functions to seek out variances between the original cover image and the distorted cover image. Stego images are generated through the application of sequence of variations to the cover images, which is use to map the secret message to be transmitted. The pseudo-randomly is the most suitable encoding algorithm for message hiding in selected pixels [18]. The message bit of 1 denotes presence of difference between stego image and cover image for a particular message pixel; otherwise the message bit is a 0. There is possibility of an encoder altering the pixels value of 1 in way as to retain the image statistical properties. However, the receiver is capable of identifying modifications such as cropping, scaling or rotating to the stego-image by attackers [17, 18].

A steganography algorithm for concurrent encoding of classified message and small-size image into a large-size image was developed by [15]. LSB substitution method was utilized to encode secret data inside small grey-level image. The Discrete Wavelet Transformation (DWT) performed preprocessing before the eventually concealing the image in the cover image. The outcomes offered certain improvement in secret message protection and capacity.

The idea of echo hiding of binary message bits in the carrier signal. 2D-Discrete Haar Wavelet Transform coefficients are performed on the cover signal in [18]. This algorithm applies pseudorandom sequence to encode the data, which offered better efficiency and evade unintended decoding at any point in time. The echo hiding offered more immunity to stego signal against disturbances and noise during broadcast on susceptible channels.

Another method of embedding audio data into an image with LSB and wavelet transform was attempted by [19]. The speech was compressed to 1 kb size due to large size initially. The result revealed high imperceptibility and quality of recovered audio.

A study to deal with the shortfall of cryptography using steganography was carried out by [11]. Thereafter, the audio file is compressed with Discrete Cosine Transform (DCT). Again, Advance Encryption Standard (AES) is used to encrypt audio file for security. Then, secret message is embedded in the LSB position on cover image. But, LSB is inappropriate for concealing data in RGB colour image.

An enhanced LSB based Steganography by concealing data in the binary form of RGB colour image was proposed by [5]. The LSB algorithm modified resolution of the image to evade attacks by searching about the identical bits of the pixel values between the secret messages and cover image. The outcomes provided quality retention of image (secret data). Nevertheless, there is evidence of quality changes after encoding image. The quantity of data to be concealed in carrier relatively impacts on the nature of alteration of the resultant medium. Consequently, larger secret messages cause higher alterations and high susceptibility during transmission. According to [20], the bits-per-pixel of 0.4 or lower was appropriate for safely sharing information between parties in case of larger secret message with corresponding smaller cover media. This revealed that many techniques favour smaller size secret message in larger size carriers. However, the reverse concept image steganography is experimented in this paper.

3 Methodology

In this paper, the stages for demonstrating the concept of transmitting of large size audio file (secret message) using a smaller size image cover is illustrated in Fig. 1.

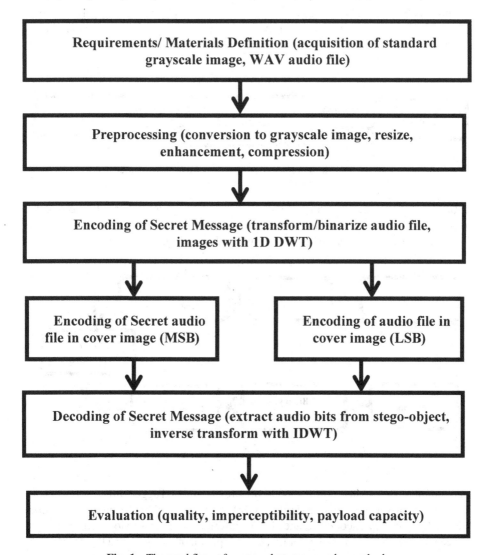

Fig. 1. The workflow of proposed steganography method.

3.1   Structure of Image Steganography

This paper understudies the concept of secure data communication that involves embedding large audio size in two distinct image covers types such as grayscale and RGB colour images, which is deviation from common practice. MSBs and LSBs substitution methods are used to conceal the presence of audio data (secret message) encoded into digital images (cover medium) to safeguard across highly suspicious communications channels.

The selection of cover images is benefit from bits replacements attributes distinct pixel locations which are available to many image formats used in present-day communication applications. Similarly, audio is selected due to widespread usages in communication channels and data. It possesses attributes that allow conveyance of bulky secret messages as against other media. The entire structure of the steganography methods experimentation is divided into frontend, intermediate and backend stage as illustrated in Fig. 2.

Fig. 2. The structure of steganography method experiment.

In Fig. 2, the frontend is initiate the process of hiding audio data into the selected cover images to generate an transitional form known as stego-object for the next stage. The entire procedure at the frontend is illustrated in in Fig. 3.

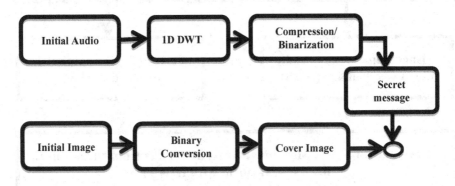

Fig. 3. The frontend of the steganography method

Intermediate: The lossless compression pictures (such as Tagged Image File Format (TIFF) and Graphic Interchange Format (GIF)) were selected as the appropriate intermediate form of the encoded image during information transmission and exchanges between sender and receiver in order to evade suspicion of attackers and third parties. At this stage, audio is transformed are embedded into the cover images along with stego key after preprocessing with MSB and LSB. This is broadcast across channels to receiver from sender with negligible alterations to the HSV as depicted in Fig. 4.

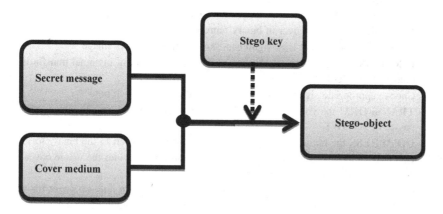

Fig. 4. The intermediate of the steganography.

Backend evaluates the effectiveness of steganography methods by reversing the entire process to recover the audio data hidden in the cover image using the stego key as illustrated in Fig. 5. This determines the quality audio file and cover image before and after encoding procedures.

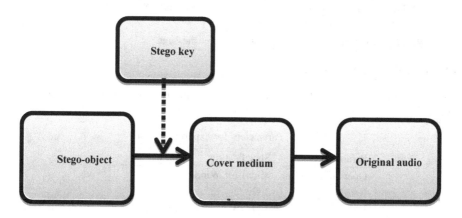

Fig. 5. The backend of steganography

Cover Image Embedding Bits Estimation: The cover image is converted into the binary form. Thereafter, the MSB and LSB pixel values for grayscale image (or cover image) are calculated which correspond to the sound notes signal (secret message) bits of the length of the sound notes wavelets to be embedded.

The entire processes of the proposed steganography concepts are described in Table 1.

Table 1. The proposed steganography method algorithm

INPUT: Voice notes/pictures	
OUTPUT: Stego-object and stego key	
1.	**INPUT** standard dimension of images for cover carrier: 512 × 512
2.	**REPRODUCE** image cover in binary representation
3.	Determine the most and least significant bits position on cover image
4.	**DECOMPOSE** the voice/speech note into smaller component in signal transform domain using ID DWT
5.	Choose appropriate signal vector representing the voice/speech note
6.	**TRANSFORM** the voice note into high and low frequency signals by means of filter passes of ID DWT
7.	**COMPRESS** the high frequency filter pass vector and convert to the binary format
8.	**ENCODE** the two portions of binarized **sound note vector** into the image cover at MSB and LSB positions respectively
9.	Create stego-object by MSB and LSB displacements of voice notes in cover image
10.	Work out the MSE and PNSR values of stego-object and initial image cover
11.	If PNSR > MSE or PSNR > 40 dB, then, quality of embedded data is appropriate
12.	**END** Do and transmit the embedded message across medium alongside stego key
13.	**IF NOT** secret message quality is inappropriate (low), continue at step 2
14.	The secret message is delivered to receiver
15.	Retrieve the stego-object and stego key from the transmission medium
16.	**OUTPUT** stego-object and stego key
17.	**REVERSE** the bits of stego-object to MSB and LSB arrays using stego key
18.	**EXTRACT** the encoded audio bits from the MSB and LSB positions i image cover
19.	**DECODE** the audio file encoded using IDWT
20.	**RETRIEVE INITIAL** or voice note or secret message

3.2 Experimental Settings

The minimum constraints for implementing steganography methods are presented Table 2.

Table 2. Parameters for experimentation

Parameters	Specification
Operating System	Microsoft Windows 8
RAM	2 GB
HDD	120 GB
GUI	MatLab R2013a, MLSB, 1D DWT
CPU Processor	671 GB
HD Graphics	2.4 GHz
System type	64-bit OS, x64-based processor
Compiler Machine	MatLab R2013a, C##
Image format	GIF, TIFF
Image size	512 × 512
Audio format	Wave (.wav)
Evaluation	PSNR, MSE, CR

4   Results and Discussion

4.1   RGB Colour Encoding

The outcomes of the audio encoding inside RGB colour cover image using pixel bits substitution method are shown in Fig. 6(a) and (b).

Fig. 6. (a) MSB encoding of audio bits (b) LSB encoding of audio bits

In Fig. 6(a) and (b), the audio bits substituted are fairly recognizable to the HVS because of white borders to the right and left of resulting images. The MSB and LSB are unsuitable for hiding audio signals reason being that it can be easily detected whenever large scale noise is embedded. The histogram plot representation of the initial RGB image and stego RGB image is shown in Fig. 7.

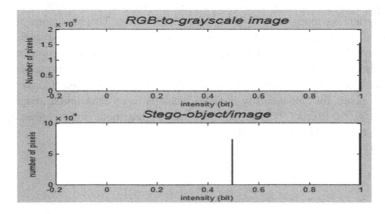

Fig. 7. Bits distribution before and after embedding for RGB colour image (Color figure online)

In Fig. 7, there is evidence of negligible variations in the qualities of the images compared because the amount of bits corresponding to full intensity in initial RGB colour image is twice as the stego image (that is, 160000 to 80000 bits). This implies the possibility of detection by skilled HVS.

4.2 Grayscale Encoding

The results of the audio encoding inside Grayscale cover image by means of pixel bits substitution method is shown in Fig. 8(a) and (b).

Fig. 8. (a) MSB encoding of audio bits (b) LSB encoding of audio bits

In Fig. 8(a) and (b), the audio bits substituted are fairly recognizable to the HVS because of white borders to the right and left of resulting images, which is analogous to RGB colour cover image. Unlike the MSB, LSB encoding method is less susceptible to third party attacks though with visibly large scale noise. The histogram plot representation of the initial grayscale image and stego images is shown in Fig. 9.

In Fig. 9, there is no disparity between the qualities of images compared to the HVS. Reason being that the amount of full intensity bits in initial grayscale image is larger than those of stego-image (that is, 150000–850,000 pixels). Conversely, the amount of pixels for half-intensity in stego image is larger than initial image (that is, 720,000–100,000 pixels). This supports the quality retention, higher imperceptibility and safety of embedded audio data.

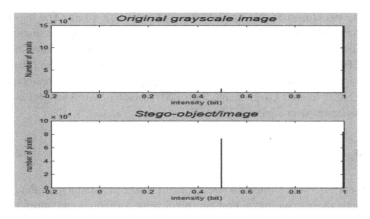

Fig. 9. Bits distribution before and after encoding for grayscale image.

4.3   Performance Measured

The parameters and values of metrics calculated for grayscale cover image and stego-object realized are contained in Table 3.

Table 3. Grayscale images calculated

	Mean	PSNR (db)	CR
MSB	7.45	39.06	11.12
LSB	9.81	38.25	10.55

In Table 3, there are little changes between the chosen images. Again, steganography methods provided good quality and common attributes to HSV of individuals because, the mean and PNSR for in both methods. But, the imperceptibility of MSB method is better than LSB because it is closest to 40.00 db (standards of quality image).

Similarly, the parameters and values of metrics calculated for RGB colour cover image and stego-object realized are contained in Table 4.

Table 4. RGB colour images calculated

	Mean	PSNR (db)	CR
MSB	16.90	35.88	11.80
LSB	7.97	39.15	12.00

In Table 4, there are negligible changes between the experimented images. The steganography methods provided quality and similar characteristics to an individual's HSV because, the mean and PNSR values. Again, the imperceptibility attributes of LSB and MSB are relatively the similar with LSB possessing the biggest PNSR values of 39.15 db, and closest to the standard image benchmark 40 db. The reason is that, values of audio bits are similar to those of cover image pixels at the unique location after encoding. In general, the stego-object and the initials images have high resemblances to HVS of individuals. In either scenario, the quality of media after decoding secret message is unaffected but high-grade to grayscale based steganography. More importantly, RGB colour cover image with LSB is better than when put side-by-side with the grayscale cover image with MSB.

5 Conclusion

This paper corroborated numerous researches in image steganography that utilised at least two methods such as spatial domain (pixel value variation) and transform domain (DWT). The undercover exchanges between parties for image encompass pixel MSBs and LSBs as usual practice in order to minimize risks on channels. The effectiveness of image steganography depends on the exploitation of HVS weaknesses in spotting insignificant alterations of the cover image pixels by advertent receivers.

This paper established that colour images with LSB have greater propensity of enhancing undercover broadcast of audio file with lesser likelihoods of being discovered by HVS/unwanted clients during transmission. However, this is attained upon preprocessing of standard grayscale and RGB colour images including noise removal, smoothing, enhancement, and resizing. Similarly, the audio sound file is contiguously reduced with DWT compression to simplify embedding process in cover image. The process of hiding the preprocessed voice notes involved bits substitution method, which replaced the MSBs and LSBs of image cover medium with those of voice notes. This paper recommends for further study a method for secret broadcast of large size audio data inside image covers by combining MSBs and LSBs in hiding slices of audio file into the cover image.

Acknowledgments. We acknowledge the support and sponsorship provided by Covenant University through the Centre for Research, Innovation, and Discovery (CUCRID).

References

1. Shelke, F.N., Dongre, A.A., Soni, P.N.: Comparison of different techniques for steganography in images. Int. J. Appl. Innov. Eng. Manag. **3**(2), 171–176 (2014)
2. Al-Hasainy, M.A.F., Al-Sewadi, H.A.A.: Full capacity image steganography using seven-segment display pattern as secret key. J. Comput. Sci. **14**(6), 753–763 (2018)
3. Singh, P.: A comparative study of audio steganography techniques. Int. Res. J. Eng. Technol. **03**(04), 580–585 (2016)
4. Qui, H.: An efficient data protection architecture based on fragmentation and encryption. Unpublished Ph.D. thesis, Institute of Sciences and Technologies, Paris Institute of Technology, Paris, France, pp. 1–136 (2017)
5. Krati, V., Pal, B.L.: A proposed method in image steganography to improve image quality with LSB technique. Int. J. Adv. Res. Comput. Commun. Eng. **3**(1), 5246–5251 (2014)

6. Jeevan, K.M., Krishnakumar, S.: An image steganography method using pseudo hexagonal image. Int. J. Pure Appl. Math. **118**(18), 2729–2735 (2018)
7. Kaul, N., Bajaj, N.: Audio in image steganography based on wavelet transform. Int. J. Comput. Appl. **79**(3), 7–10 (2013)
8. Jung, K.: Comparative histogram analysis of LSB-based image steganography. WSEAS Trans. Syst. Control **13**, 103–112 (2018)
9. Hanzlik, P.: Steganograhy in reed-solomon codes. Unpublished M.Sc. thesis, Department of Business Administration, Lulea University of Technology, Scandinavia, Sweden, pp. 1–68 (2011)
10. Arya, A., Soni, S.: A literature review on various recent steganography techniques. Int. J. Future Revolut. Comput. Sci. Commun. Eng. **4**(1), 143–149 (2018)
11. Dilawar, A.: Image steganography: for hiding audio messages within grayscale image using LSB, DCT and AES algorithm. Unpublished M.Tech thesis, Department of Computer Science and Software Engineering, International Islamic University, Islamabad, Pakistan pp. 1–80 (2014)
12. Kumar, N.M., Kumar, M.P., Rao, M.S.: Data hiding using image steganography. Int. J. Adv. Res. Dev. **3**(2), 53–56 (2018)
13. Ashita, K., Smitha, V.P.: Randomized steganography in skin tone images. Int. J. Comput. Sci. Eng. Inf. Technol. **8**(2–3), 1–8 (2018)
14. Devi, K.J.: A secure image steganography using LSB technique and pseudo-random encoding technique. Unpublished B.Tech thesis, Department of Computer Science and Engineering, National Institute of Technology, Rourkela, Odisha, pp. 1–40 (2013)
15. Badescu, I., Dumitrescu, C.: Steganography in image using discrete wavelet transformation. In: Advanced in Mathematical Models and Production Systems in Engineering, vol. 1, no. 1, pp. 69–72 (2013)
16. Al-Asadi, S.A.: Image steganography based on variable sized segments. J. Eng. Appl. Sci. **13**(1), 2282–2287 (2018)
17. Jayaram, P., Ranganatha, H.R., Anupama, H.S.: Information hiding using audio steganography – a survey. Int. J. Multimed. Its Appl. **3**(3), 86–96 (2011)
18. Goel, S., Rana, A., Kaur, M.: Comparison of image steganography techniques. Glob. J. Comput. Sci. Technol. **13**(4), 7–14 (2013)
19. Lahiri, S.: Audio steganography using echo hiding in wavelet domain with pseudorandom sequence. Int. J. Comput. Appl. **140**(2), 16–19 (2016)
20. Baluja, D.: Hiding images in plain sight: deep steganography. In: 31st Conference on Neural Information Processing Systems, Long Beach, USA, pp. 1–11 (2017)
21. Temiatse, O.S., Misra, S., Dhawale, C., Ahuja, R., Matthews, V.: Image enhancement of lemon grasses using image processing techniques (histogram equalization). In: Panda, B., Sharma, S., Roy, N.R. (eds.) REDSET 2017. CCIS, vol. 799, pp. 298–308. Springer, Singapore (2018). https://doi.org/10.1007/978-981-10-8527-7_24
22. Dhawale, C.A., Misra, S., Thakur, S., Jambhekar, N.D.: Analysis of nutritional deficiency in citrus species tree leaf using image processing. In: 2016 International Conference on Advances in Computing, Communications and Informatics (ICACCI), pp. 2248–2252. IEEE, September 2016

An Improved SMS Based Metering System

Modupe Odusami[1], Olusola Abayomi-Alli[1], Sanjay Misra[1(✉)],
Daniel Ugbabe[2], Ravin Ahuja[2], and Rytis Maskeliunas[3]

[1] Department of Electrical and Information Engineering, Covenant University,
Ota, Nigeria
{modupe.odusami, olusola.abayomi-alli,
misra.sanjay}@covenantuniversity.edu.ng
[2] University of Delhi, New Delhi, India
[3] Kaunas University of Technology, Kaunas, Lithuania
rytis.maskeliunas@gmail.com

Abstract. Electricity Metering has experienced significant changes through the years and the technology of e-metering continually advanced. This paper develops an improved SMS based metering system using GSM technology. An Arduino microcontroller is connected to a GSM-based wireless communication module. It monitors electrical pulses and the units consumed and the costs are calculated. This data is displayed on an LCD screen, and the data is also sent to the user via SMS. The user will be able to get SMS alerts of low and expired credit. The maximum and minimum units are set to 20 units and 0 units respectively. The alert on the LCD does not show until the units are getting close to the minimum. The experimental test was taken at different times at an interval of 5 min during the late hours of the day and the result shows that the proposed system is very effective.

Keywords: SMS · GSM · Wireless communication

1 Introduction

Electricity has made certain processes easier and faster thereby easing our way of life. All home appliances and devices, even some vehicles now, run on electrical power, thus, Electricity has turned out to be a basic need for human survival and advancement [1, 2]. Governments around the world have made the availability of electricity a priority as electricity is very necessary for developmental progress. The advantages of Electricity outweigh the disadvantages among which are: Electricity reduces the cost of production, costs, which are due to generator fuel expenses, machinery repairs due to electricity fluctuation etc.

Electricity invites investors and businesses, thereby increasing exportation and generating revenue, it gives room for productive innovation, inviting new and improved production methods and procedures. Electricity reduces the cost of living; goods and service charge less with available electricity. Therefore, Electricity is an important commodity which has proven to be a global necessity. To monitor consumption, service providers engage a metering system which decides the expended power per unit time and executes its calculation based on the rate of sale of energy per

© Springer Nature Singapore Pte Ltd. 2019
A. K. Somani et al. (Eds.): ICETCE 2019, CCIS 985, pp. 192–204, 2019.
https://doi.org/10.1007/978-981-13-8300-7_16

unit and other parameters this is done with the aid of an energy meter. An energy meter is a device used to measure accurately the number of units of electricity consumed by any home or modern foundation [3].

Often times consumers are reluctant to make effort to walk down to where the meter is mounted in order to check for the credit level. Here a new method of real-time monitoring of credit usage is introduced in this paper which will automatically sense the used energy, processed the reading, then send SMS to users via the GSM network. The proposed system enables the user to know credit level especially when it is low without making contact with the meter; it is beneficial to the customers as the system is made very user-friendly. This GSM based information gathering framework can be exceptionally quick, exact and productive.

The rest of the paper is sectioned as follows: Sect. 2 discussed the related work, the proposed system is detailed in Sect. 3, Results and Discussion is explained in Sect. 4, and Sect. 5 concludes the paper.

2 Related Work

The literature review presented below is a concise explanation of the technology used regarding previous methodologies used in metering system. Some of the metering and technologies considered under this review include ZIGBEE, General Packet Radio Services (GPRS), Ad-Hoc wireless communication, Radio Frequency, Power line Communication, Worldwide Interoperability for Microwave Access (WiMAX).

2.1 Zigbee

Authors in [4] developed Embedded Based Digital Energy Measurement for Improved Metering and Billing. The system consists of a digital energy meter with voltage and current sensors that measures the amount of energy consumed, and these measurements are sent to microcontroller IC; at both the customer and supplier end, which computes the power consumed and other values such as late bills and daily updates. The flaw in this system is that it is costly. Ashna and George [2] designed a Remote automatic meter reading system based on ZigBee and GPRS. The system sends the customer's energy consumption information to the billing company, where the bill is compiled and sent back to the consumer via SMS.

Kistler et al. [5] presented a system that used the Zigbee technology to construct a series of home area networks of connected devices was developed. The system has the ability to transmit to Zigbee points about 10 m apart. The system was designed to send a picture of the meter through a Zigbee network and waits on standby until the reader at the billing company connects. When the reader connects, the reader takes necessary consumption readings and generates the billing information. The system reduces the requirement for manpower; the consumer is still required to take pictures of the energy meter in his possession. It also requires Zigbee networks to be installed at different locations in the country, due to its short range.

2.2   General Packet Radio Services (GPRS)

Authors in [6] designed Wireless Energy Meter and Billing via 'SMS'. A special id number is assigned to every meter; this id number is interlinked with the SIM card number service number which is also unique. The microcontroller computes energy reading periodically, it could be hourly, weekly or monthly and it sends this information to the electricity providers. The bill SMS is sent to both the electricity company and also the consumer. This system provides flexibility for paying bills. In this system the electricity company can also cut the power to a consumer does not require it by sending an SMS. Wasi-ur-Rahman et al. [7] developed an intelligent SMS based remote metering system. The system makes use of a central server which exchanges information with the meters connected to it. The meter exchanges information with the central server through SMS. A GSM modem is connected to the meter via the serial port.

Dange et al. [8] designed Prepaid Energy Meter based on GSM Module using recharge protocol which involves a prepaid card similar to a mobile SIM card, when the prepaid card runs out of units, power is disconnected by the power utility and this power can be recharged by SMS from a remote location by the customer at his convenience. The billing information is also sent to the user's mobile phone. Authors in [9, 10] used the Internet and GSM modules to monitor power consumption. This system made it possible for the electricity provider to access customer billing information remotely through the internet. The system supports wide coverage area for communication, ease of maintenance and it is able to obtain real-time data from the energy meters. Authors in [11] developed a visual basic application for the utility center to receive data periodically.

2.3   Radio Frequency

Authors in [12] developed an Automatic Meter Reading (AMR) Using Radio Frequency. The system consists of a transmitter and receiver unit. A PIC16F877A microcontroller is used at the transmitter computes the energy consumed and transmits to the receiver unit through an RF Module the pulses from the meter are given to the microcontroller via an opt coupler and the LCD displays the unit, the time and date. The microcontroller at the receiver end sends its data through a MAX232 converter to a laptop or PC. The use of the PC does not make this system convenient and easily accessible.

2.4   Ad- Hoc Wireless Communication

Vijayaraj et al. [13] developed an Automated EB Billing System Using GSM and Ad-Hoc Wireless Routing. The system involves the wireless transfer of data between electricity company and the consumers the details of energy consumption by the customer is refreshed monthly to the electricity company using Radio Frequency. The use of unique id is for ease of identification for the electricity company to know the exact customer transmission is coming from or going to.

2.5   Power Line Carrier Communication

Newbury and Miller [14] developed a Multiprotocol routing for automatic remote meter reading using power line carrier systems. The system Information is obtained from component parts in the full duplex way from links connecting to a device. The data is being transmitted over existing power lines. A control circuit is provided at the customer end for monitoring the system.

2.6   Worldwide Interoperability for Microwave Access (WiMAX)

Authors in [15] developed an Automatic electric meter reading system consisting of four units: the sensing unit, a communication unit, a processing unit, and the billing unit. Readings from the energy meter are sent to the microcontroller for the necessary calculations. There are 2 modules, one at the service provider end and one at the consumer end. These data from the microcontroller are then sent over a WiMAX transceiver and then received by the module at the service provider end by another microcontroller; the collection of data is done with the aid of an application. In [16] a Prepaid Smart Metering Scheme Based on WiMAX Prepaid Accounting Model was designed. The system was based on a centralized verification and charging using WIMAX prepaid model and also its application to roaming of electrical vehicles was discussed. WiMAX technology is known for high performance and large coverage area but this system is costly to implement and complex.

2.7   Bluetooth

Authors in [17] presented a Bluetooth energy meter in which the energy meter computes consumption readings and sends the data to a remote reader which could be a PC. This reader, polls every Bluetooth meter reading to get the corresponding energy consumption readings of each resident. Singh et al. [18] designed the Ethernet Shield based Electricity Billing Framework that makes use of an Arduino microcontroller with an Ethernet shield and a pulse detection unit. Energy consumption is continuously recorded and sent to the electric billing company through the internet via an Ethernet shield. Then the billing information is compiled and sent to the consumer by GSM network.

2.8   Advanced RISC Machine (ARM)

Li and Zhang [19] presented an Automatic meter reading system using WIFI technology and an ARM-based Power meter with WIFI communication module. The ARM acts as the main controller; it compiles the billing information and communicates this information to the billing company and the consumers via WIFI. Authors in [20] designed an automatic meter reading system based on an ARM-based microprocessor which acts as the main process that computes and compiles the energy consumption and billing information. The meter provides two-way data communication between Electricity Company and the energy meter via GPRS. In this proposed system we make use of GPRS technology for data communication, this is because GPRS/GSM has high data rates and covers a wide area and it is not expensive to implement. Table 1 below shows the summary of related works.

Table 1. Summary of related work

Authors	Methodology	Strength	Limitations
Ashna et al. [2]	GPRS/GSM technology	Wide coverage area, high data rates	Post paid billing system
Kistler et al. [5]	GPRS/GSM technology	Wide coverage area, high data rates	Post paid billing system
Wasi-ur-Rahman et al. [7]	Use of Gem network and a temper detection unit	Cost-efficient, high data speed	Expensive, complex
Kehe et al. [9]	GPRS and Web services	Wide area coverage	Network Location dependent
Adegboyega et al. [10]	Intranet and Internet-based system	Wide coverage area, high data rate	Complex
Ali et al. [12]	Radio frequency	Efficient communication	Subject to noise interference
Vijayaraj and Saravanan [13]	Ad-hoc Wireless routing	Consumer identification	Complex
Ahmed et al. [15]	WiMAX technology	Large coverage area, efficient and reliable	Expensive, complex
Khan et al. [16]	WIMAX	Reliable, Secure	Expensive, complex
Singh et al. [18]	Ethernet shield and Arduino	Wide coverage and security	Expensive
Li et al. [19]	Advanced RISC Machine processor	High efficiency and reliability	Complex and expensive

In addition to above works and technology - a prepaid electricity system using LUHN algorithms [21] is also proposed. In the same environment - energy consumption forecast using demographic data is also presented [22].

3   SMS Based Metering Architecture

The developed SMS based system consists of four main parts: a power supply unit, pulse detection and switching, processing, and output. Overview and functional block details are shown in Fig. 1.

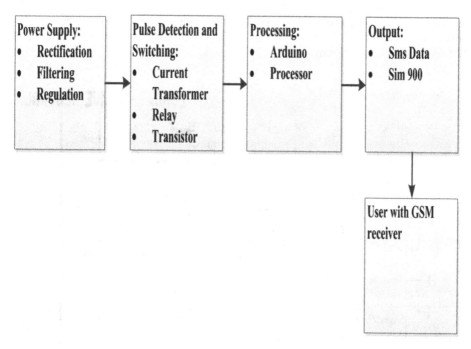

Fig. 1. Bock diagram of the design

From Fig. 1, it can be deduced that power supply stage is a regulated supply powering the system. The power which comes in as AC voltage is stepped down, rectified to DC voltage, filtered and regulated before reaching the main circuit, this is to ensure proper performance of the system. This stage involves the following processes; Transformation, rectification, filtering and voltage regulation. The pulse detection and switching stage is responsible for sensing current pulses and feeds this information to the microcontroller. The processing stage is where all the computation and processing are done. The output is the final stage of the project, it is where the billing information obtained from the above processes is sent to the consumer via a SIM 900 GSM module.

3.1 Detailed Design of Proposed System

The SMS based Metering System is constructed using a current transformer SC701 100–5 mA and a burden resistor of 33 ohms, Arduino Uno R3 Microcontroller with 14 digital input/output, GSM module SIM900, and 20 × 4 LCD display. The current transformer is responsible for sensing current pulses and feeds this information to the microcontroller. The pulses received by the sensing are used to compute energy consumption and billing information and also the power cut function. These calculations are done by the microcontroller (Arduino). GSM module is interfaced with the Arduino to send billing information to the user via a mobile GSM network.

The detailed design circuit showing Configuration of the different section for performing functionality is shown in Fig. 2. The design specification of the developed SMS based Metering system is shown in Table 2.

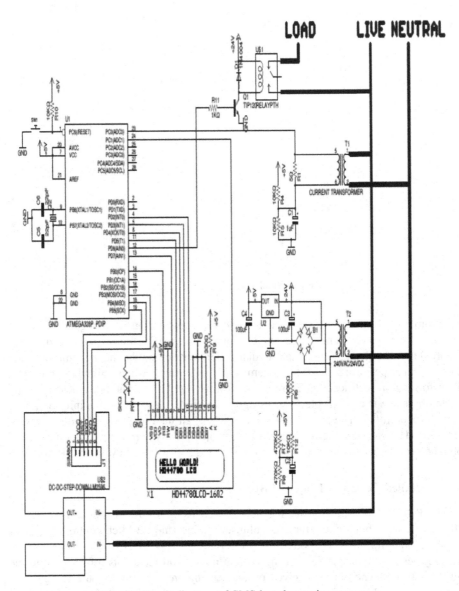

Fig. 2. Circuit diagram of SMS based metering system

The schematic diagram in Fig. 2 describes how all components were connected and interfaced to fulfill the intended project objectives. Power from the 230 V AC supply line is stepped down to 24 V by the step-down transformer then rectified to 24 V DC by the bridge rectifier this is further stepped down to 5 V by voltage regulator the voltage regulator also ensures steady voltage flow. This 5 V powers the Arduino microcontroller, the LCD and the GSM module. The reading energy consumption is obtained from the current transformer, the Arduino computes this reading and sends them to the GSM module for SMS transmission with user's mobile. The relay performs the switching operation, turning the power off when units run out. The embedded software design has the C code loaded into the Arduino Uno.

Table 2. Design specifications Of SMS based metering system

Parameter	Specification
Transformer	Input voltage: 240 V AC
	Output voltage: 24 V AC
	Operating frequency: 50 Hz
	Current rating: 500 mA
	Turns ratio: 20
Rectifier	1n4004
Voltage Regulator	L7805
Microcontroller	Arduino Uno
	Input voltage: 5 V-DC
GSM module	SIM900
Test load	Power rating: 60–100 W
Current transformer/Current sensing	SC701 100–5 mA

3.2 Processing Unit

This is the part of the system where all the computation and processing are done. The pulses received by the sensing are used to compute energy consumption and billing information and also the power cut function. These calculations are done by the microcontroller (Arduino) and it is an integrated circuit that has a microcontroller at its heart. A microcontroller is basically a tiny computer. It has kilobytes of Random access memory (RAM) to store data and a few kilobytes of electronic erasable programmable memory (EEPROM) for storing programs and commands. The processing unit is the core aspect of this study. The system flowchart is shown in Fig. 3.

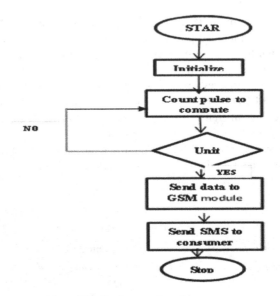

Fig. 3. System flowchart

3.3 Algorithm for the Processing Unit

The following is the algorithm that explains the processing unit of the system:
Step 1: Start
Step 2: Initialize system to start counting units
Step 3: check if the units remaining is less than 5 units
Step 4: if no continue counting else
Step 5: proceed to send a message to the GSM module.
Step 5: send a low unit alert to the user's mobile.
Step 6: if units left is 0 cutoff power.

3.4 Implementation

The unit level from the meter is measured inform of pulses. The signal from the meter through Optocoupler is normally high (5 V) and the high to low transition of this voltage wave indicates the occurrence of a pulse. The Arduino Microcontroller (Atmega 328) uses C language programming, and the microcontroller is programmed to read data from the metering IC every second. When microcontroller reads the power consumption, it is stored and current reading is incremented in its software. In this design, meter is calibrated such that for 1 unit of energy (kWh) consumption, it generates 3200 pulses in LED.

$$\text{Energy per count}, \text{Enpc} = \frac{I_{max} \times V_{rms}}{3200} \tag{1}$$

Where: I_{max} is the maximum load current
V_{rms} is the RMS voltage.

$$\text{Energy per LED pulse, Enpp} = \frac{1000 \times 3600}{\text{Mpr}} \qquad (2)$$

Where:

Mpr is the pulse rate of the meter in impulse/kWh.

Practical set up for the system prototype developed using the designed detailed circuit shown in Fig. 2 is presented in Fig. 4.

Fig. 4. System prototype

A green LED is situated at the front side of the 20 × 4 LCD display indicating the status of the system and also to display the reading processed from the Arduino Microcontroller.

Fig. 5. Sample results on display. (Color figure online)

Figure 5 indicates the final testing of the entire system to ensure proper functionality of the collective functionality of all the units together.

4   Results and Discussion

The maximum and minimum units are set to 20 units and 0 units respectively. The alert on the LCD does not show until the units are getting close to the minimum, it starts to display the alert message when units rundown from 5 units and then displays the power cut message at the minimum limit set which is 0 units. At the user's mobile, the alert pops up at 5 units and 2 units then display the power cut message at 0 units. This test was taken at different times at an interval of 5 min during the late hours of the day. Table 3 represents the effective functionality of the system.

Table 3. Message on Users mobile showing effective functionality of the system.

Time (minutes)	LCD display	Alert on LCD	SMS (user's mobile)
0	20 units	–	–
5	17 units	–	–
10	14 units	–	–
15	11 units	–	–
20	8 units	–	–
25	5 units	Low units! Recharge 5	Low units! Recharge 5 units left
27	2 units	Low units! Recharge 2	Low units! Recharge 2 units left
30	0 units	Cut off power	Power cut

5   Conclusion

This research describes the design and implementation of an advanced metering system with GSM/GPRS communication. Different technologies which have been used for metering system are very costly and the designs are very complex. The research gives a significant advancement to current energy metering. It makes use of two-way communication between user and energy meter. The system alerts the user when energy units are low so the user will either reduce the load or recharge as soon as possible. This research designs a model with improved features. An Arduino microcontroller was used to compute and communicate data from the meter to the GSM module which acts as a transmitter to send the data from the Arduino to the user's mobile. This eliminates the stress of always going to check your energy meter for units left.

The development of an SMS based metering system is a viable means of automatically viewing the energy level from the electricity meter by consumers. Consumers can get SMS from distances longer than the coverage distance of Bluetooth or ZIGBEE based metering system. The results that are generated over a period of time show that the SMS based metering system is an efficient system. For future work, this research can be upgraded in such a way that the purchase and the checking of unit balance through SMS communication between the meter and the user's mobile will be possible.

Acknowledgments. We acknowledge the support and sponsorship provided by Covenant University through the Centre for Research, Innovation, and Discovery (CUCRID).

References

1. Robinson, N.H.: Electricity. https://www.britannica.com/science/electricity
2. Ashna, K., George, S.N.: GSM based automatic energy meter reading system with instant billing. In: 2013 International Multi-Conference on Automation, Computing, Communication, Control and Compressed Sensing (iMac4s), pp. 65–72 (2013)
3. Jain, A., Kumar, D., Kedia, J.: Design and development of GSM based energy meter. Int. J. Comput. Appl. **47**, 41–45 (2012)
4. Gopinath, S., Suresh, R., Devika, T., Divya, N., Vanitha, N.S.: Embedded based digital energy measurement for improved metering and billing system (2013)
5. Kistler, R., Knauth, S., Klapproth, A.: EnerBee-example of an advanced metering infrastructure based on ZigBee. In: Proceedings of 2nd European ZigBee Developers Conference, EuZDC, pp. 1–11 (2008)
6. Roja, P.S., Babu, B.K., Kumar, V.S.D.: Wireless energy meter and billing via 'SMS'. IJSR **2**, 474–478 (2013). India Online ISSN 2319-7064
7. Wasi-ur-Rahman, M., Rahman, M.T., Khan, T.H., Kabir, S.L.: Design of an intelligent SMS based remote metering system. In: 2009 International Conference on Information and Automation, ICIA 2009, pp. 1040–1043 (2009)
8. Dange, K.M., Patil, S.S., Patil, S.P.: Prepaid energy meter using GSM module. Int. J. Eng. Sci. Inven. **6**, 80–85 (2017)
9. Kehe, W., Xiaoliang, Z., Yuanhong, W., Yuhan, X.: Design and implementation of web services based GPRS automatic meter reading system. In: 2010 3rd International Conference on Advanced Computer Theory and Engineering (ICACTE), pp. V4-360–V4-363 (2010)
10. Adegboyega, A., Gabriel, A.A., Ademola, A.J., Victor, A.I., Nigeria, K.: Design and implementation of an enhanced power billing system for electricity consumers in Nigeria. Afr. J. Comput. ICT **6**, 49–58 (2013)
11. Tan, H.R., Lee, C., Mok, V.: Automatic power meter reading system using GSM network. In: 2007 International Power Engineering Conference, IPEC 2007, pp. 465–469 (2007)
12. Ali, A., Saad, N.H., Razali, N., Vitee, N.: Implementation of automatic meter reading (AMR) using radio frequency (RF) module. In: 2012 IEEE International Conference on Power and Energy (PECon), pp. 876–879 (2012)
13. Vijayaraj, A., Saravanan, R.: Automated EB billing system using GSM and ad-hoc wireless routing. Int. J. Eng. Technol. **2**, 343–347 (2010)
14. Newbury, J., Miller, W.: Multiprotocol routing for automatic remote meter reading using power line carrier systems. IEEE Trans. Power Deliv. **16**, 1–5 (2001)
15. Ahmed, T., Miah, M., Islam, M., Uddin, M.: Automatic electric meter reading system: a cost-feasible alternative approach in meter reading for Bangladesh perspective using low-cost digital wattmeter and WiMAX technology. arXiv preprint arXiv:1209.5431 (2012)

16. Khan, R.H., Aditi, T.F., Sreeram, V., Iu, H.H.: A prepaid smart metering scheme based on WiMAX prepaid accounting model. Smart Grid Renew. Energy **1**, 63 (2010)
17. Singh, P., Sharma, D., Agrawal, S.: A modern study of Bluetooth wireless technology. Int. J. Comput. Sci. Eng. Inf. Technol. (IJCSEIT) **1**, 55–63 (2011)
18. Koay, B., Cheah, S., Sng, Y., Chong, P., Shum, P., Tong, Y., et al.: Design and implementation of Bluetooth energy meter. In: Proceedings of the 2003 Joint Conference of the Fourth International Conference on Information, Communications and Signal Processing, 2003 and Fourth Pacific Rim Conference on Multimedia, pp. 1474–1477 (2003)
19. Li, L., Hu, X., Zhang, W.: Design of an ARM-based power meter having WIFI wireless communication module. In: 2009 4th IEEE Conference on Industrial Electronics and Applications, ICIEA 2009, pp. 403–407 (2009)
20. Wu, C.-H., Chang, S.-C., Huang, Y.-W.: Design of a wireless ARM-based automatic meter reading and control system. In: 2004 Power Engineering Society General Meeting, pp. 957–962. IEEE (2004)
21. Jonathan, O., Azeta, A., Misra, S.: Development of prepaid electricity payment system for a university community using the LUHN algorithm. In: M. F. Kebe, C., Gueye, A., Ndiaye, A. (eds.) InterSol/CNRIA -2017. LNICST, vol. 204, pp. 107–114. Springer, Cham (2018). https://doi.org/10.1007/978-3-319-72965-7_9
22. Aderemi, O., Misra, S., Ahuja, R.: Energy consumption forecast using demographic data approach with Canaanland as case study. In: Bhattacharyya, P., Sastry, H., Marriboyina, V., Sharma, R. (eds.) NGCT 2017. CCIS, vol. 827, pp. 641–652. Springer, Singapore (2017). https://doi.org/10.1007/978-981-10-8657-1_49

Cervical Cancer Detection Using Single Cell and Multiple Cell Histopathology Images
Do You Have a Subtitle? If So, Write It Here

Mithlesh Arya$^{(\boxtimes)}$, Namita Mittal, and Girdhari Singh

Malaviya National Institute of Technology, Jaipur, India
mithlesharya@gmail.com, nmittal.cse@gmail.com,
girdharisingh@rediffmail.com

Abstract. Cervical cancer is the second most common cancer in females in India. A Pap smear screening is most efficient and prominent to detect the abnormality in cells. Pap smear test is time-consuming and sometimes gives the wrong result by human experts. In India, a shortage of pathologist is there in rural areas. Automated systems using image processing and machine learning techniques help the pathologist to take correct decisions. In this paper, two data sets are generated from one pathologist center. The first data set contains 300 single cells and the second contains 50 multiple cell images for the validation of work. In a single cell, nucleus and cytoplasm both are extracted from the cell, but in multiple cells, only the nuclei are extracted due to overlapping of cells. Edges have been enhanced by sharpening function, and the multi-threshold values and morphological operations have been used for the segmentation of cell. Shape-based features have extracted from a multiple cell and single cell images. Support Vector Machine (SVM) and Artificial Neural Network (ANN) is applied to improve the performance of classification using 10 fold cross-validation.

Keywords: Cervical cancer · Multi-threshold ·
Morphological operation · Shape-based features · ANN · SVM

1 Introduction

Cancer is an uncontrolled division of cells in any tissue of the body. According to a survey by the Indian Council of Medical Research (ICMS) [1]. About 70–75% population of India lives in rural area. In a rural area, females are generally not aware of cancer. There are many types of cancer in females like breast, lung, cervical, mouth, uterus, etc. After breast cancer, cervical cancer is the second largest cause of cancer-related mortality in India [2]. Cervical cancer develops in the lower part of the uterus called cervix. In the early stages, there are no accurate sign and symptoms of cervical cancer. Prognosis of cervical cancer

© Springer Nature Singapore Pte Ltd. 2019
A. K. Somani et al. (Eds.): ICETCE 2019, CCIS 985, pp. 205–215, 2019.
https://doi.org/10.1007/978-981-13-8300-7_17

depends on the stage it is detected. In various studies, it has been proved that early detection of cancer shows a better outcome. In the line of early detection of cancer, routine screening of prone age group can be beneficial. Pap smear is one of the most commonly used screening methods for screening of cervical cancer [3]. In Pap smear screening sample is taken by the scrapping of the cervical lining, and then the sample is placed on a glass slide and dyed. The sample obtained is then studied under a microscope by the pathologist. Availability of pathologists in developing countries like India especially in rural parts is very limited. To overcome this limitation we develop an automated system. According to the international classification system, cervical cancer is divided into three different class's viz., Normal, LSIL (low grade squamous intraepithelial lesion) and HSIL (high grade squamous intraepithelial lesion). But in this study, we are comparing shape-based features of single and multiple cells. On that basis, cells are categorized into normal and abnormal classes.

Many semi-automated and automated methods are proposed in the last ten years. Some have worked on multiple cells, and some have worked on single cells. According to Pap smear images, literature is mainly categorized into two factors:

1. Single-cell images and
2. Multiple cell images

In some papers, for single cell image only the features of the nucleus are gathered, and in some other papers, the features of nucleus and cytoplasm both are extracted. In most of the papers, for multiple cells, only nuclei features have been obtained because the cells are highly overlapped in real images. As we can see in Fig. 1, the cells are highly overlapped and unable to identify the boundaries of the cell. The dark area in the image shows the nucleus and light blue area is overlapped cytoplasm of the cells. There are three steps for the designing of an automated system. The first step is pre-processing and segmentation method to get the cell from an image. The second step is the features observation and extraction based on color, texture, and shape of the cell. The final step is classification of the cells according to calculated features, into normal and dysplasia classes.

The segmentation methods have been adopted for the identification of the region of interest (ROI) according to the requirement. Distinct segmentation methods have been suggested by the researchers on the segmentation of nucleus and cytoplasm. The segmentation has been implement using Watershed technique [4,5], K-means clustering [6–8], and Fuzzy C-mean [9], etc. In paper [4] morphological operations are applied to find the region of centroid using regional minima and boundaries of nuclei are detected using watershed transform method. In paper [5] multi-scale morphological watershed segmentation method applised on single cell images to find the features of the cells. In paper [6] ellipse detection method is used to find the nuclei of cells and cluster method is used to remove the duplicate from the original image. In paper [7] color clustering is adopted to slice the nucleus and cytoplasm from the background of a image. In paper [8] RGB images are converted into CIELAB format, then k-mean clustering is used to generate two clusters: background and cytoplasm. Threshold and morphological operation are used to segment the nuclei from the second cluster. Then shape-based features of nuclei are extracted from the second cluster. Fuzzy C-Mean is used as a classifier to make clusters of cells according to features in two clusters. In paper [9] fuzzy C-mean is used to generate the three cluster using the centroid values to find the region of the nucleus, cytoplasm, and background.

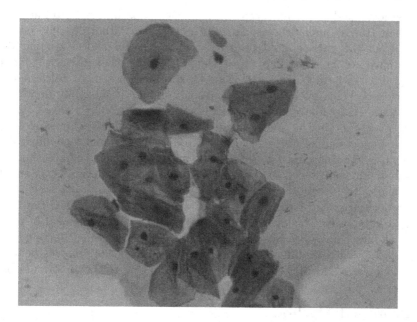

Fig. 1. Overlapped cells in smear image (Color figure online)

Literature can also be identified based on different features like shape based [10,11], texture based [12–14] and color based features. In shape-based features compactness, area, homogeneity, perimeter, major axis, minor axis, eccentricity, and N/C (Nucleus/Cytoplasm) ratio of nucleus and cytoplasm are calculated. Shape-based features represent the pattern of the nucleus and cytoplasm for normal cell and abnormal cell. If the area of the nucleus is high, it means the cell is abnormal, but if the area of the nucleus is small so, the cell is normal. Shape-based features are calculated using the number of pixels presents in the ROI(Cell). In textural based features, the texture of the cell and intensity distribution of pixel is mainly used for the classification of the cell. Texture-based features are calculated using the intensity value of the pixel and its neighbor pixel values. In the paper [14] histogram, GLCM, LBP, DWT and laws energy map methods are adopted to find the texture features of the cell on single cell image data-set. Using texture based pattern total twenty four feature values are calculated.

Finally, classification is the last but most important category to design an automated system. The classification has been performed using basic machine learning classifiers [15–18]. According to literature survey ANN, SVM, Ensemble, and KNN is used for the classification of the cells into 2 classes, 4 classes, and 7 classes.

2   Data-Set Collection

In our study, the data-set has been generated at the pathology center at Jaipur, Rajasthan (India) under the guidance of cytologist. Pathologist collected the sample from the patients and prepared the Pap smear slides. For our data-set, we have collected 40 Pap smear slides from a pathologist with different patients. Images have captured by using high-resolution camera Leica mounted on microscope Leica IGG50 using 40× resolution and 24 bits color depth using smear slides. Images have stored in JPEG format with 2560 × 1920 resolutions. Our data-set contains 50 smear images. Multiple cell images are also known as smear image. Out of these 20 images are normal, and the remaining 30 are abnormal. From this data-set, single cells are cropped to generate single cell data-set. In this single cell data-set; 150 are dysplasia/abnormal cells and 150 are normal cells.

Another data-set for the validation of work is DTU/HERLEV Pap smear benchmark data-set. This data-set was developed by the department of pathology at Herlev University Hospital, Denmark [19]. It contains 917 single cell images belonging to 7 different classes. In this data-set, 241 images belong to the normal class, and the remaining 676 images belong to the abnormal class.

3   Proposed Method

Segmentation is the very first and difficult step to get the ROI. According to the literature survey, a variety of segmentation methods have been recommended to determine an appropriate region. Segmentation is classified into three categories namely, edge detection/boundary detection, threshold and region detection. In this paper threshold method is used to measure rigorous region. In the first step, the nucleus of a single cell is computed. Then nuclei of multiple cells are figured. In the second step, the cytoplasm of a single cell is determined. Below two subsections describe the segmentation of single cell and multiple cells.

3.1   Segmentation : Nucleus Identification

Recognition of nucleus boundary from single cell image and multiple cell image is the first step. In the acquisition of the nucleus boundary first, the RGB image is transformed into a gray-scale image because working with gray image is simple. To simplify the segmentation process, sharpening and contrast enhancement are an important step. To sharpen the boundary of the nucleus and cytoplasm sharpening function used, then a median filter is used to remove the noise. Contrast enhancement function makes the nucleus darker so, the contrast enhancement function is applied on both data-sets. These steps are applied to a single cell as well as on multiple cells. In single cell nucleus and cytoplasm, both are extracted. In multiple cells, only nuclei are extracted because few images have more than 30 cells in a single image and cell overlapping is very high. In this paper, multiple threshold values are applied for the detection of the boundary of the nucleus and cytoplasm. In this paper first, a loop from 0 to 255 with 10% increment is used, to find out the correct threshold value for cell and smear image. After applying these iterations on multiple images, the appropriate values of the threshold comes out. This threshold value is compared with the prime image. The threshold value which gives the accurate result as prime image is selected as the final threshold value. For the single cell threshold value is 60 and for smear image threshold value are 80. By using these threshold values, the foreground and background space are captured.

3.2 Cytoplasm Detection

After nucleus segmentation, the next step is to extract the boundary of cytoplasm from the single cell image. For the separation of cytoplasm from single cell image threshold value is 150. This threshold value has been done by the previous method. Then, we removed all the objects which are less than 500 pixels using connected component region. These small objects are noise (RBC or inflammatory cells) and removing of noise is necessary because it may affect the classification phase. Fill hole is the next step to fill the space in the cytoplasm region. Finally, we performed morphological operation dilation and erosion with disk-shaped structuring element (SE) with radius value 5. Dilation is used to expand the boundaries of cytoplasm and erosion is used to remove the unwanted pixel values from the boundary to find the proposed shape of the nucleus and cytoplasm. Below equation shows the erosion function on the dilated image using a structuring element.

$$I_{erosion} = I_D \ominus SE \tag{1}$$

Finally, the outcome is a binary image with extracted nucleus and cytoplasm for single cell images. Figure 2 shows the input images, filtered images, segmented nucleus and segmented cytoplasm of the single cell. Then we have derived the features of nucleus and cytoplasm. But in multiple cells, many images have 30 cells and the area of nuclei of normal cell various from 20 to 500 pixels, so no objects are removed. Figure 3 shows input, filtered and segmented nuclei in multiple cell images. Then we have obtained the number of nuclei using the connected component with 8 neighbors. Then we determine the features for all the extracted nuclei. In multiple cells image, the numbers of nuclei are more, so the mean values of all the features are calculated for the better performance of the classifier.

RGB Image   Preprocessed Image   Segmented Nucleus   Segmented Cytoplasm

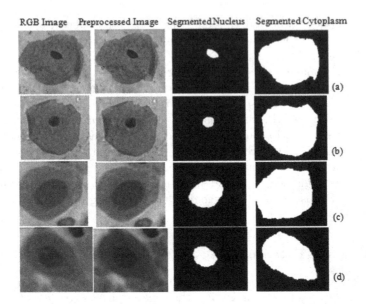

Fig. 2. Single cell: input image, filtered image, segmented nucleus and segmented cytoplasm. Normal cell (a–b), abnormal cell (c–d)

3.3   Feature Derivation

Feature extraction is the second and important step after segmentation. A set of features are derived on the bases of pattern of cell. For single cell images, features of both nucleus and cytoplasm are extracted. But for multiple cells, we

Input Image   Filtered Image   Extracted Nucleus

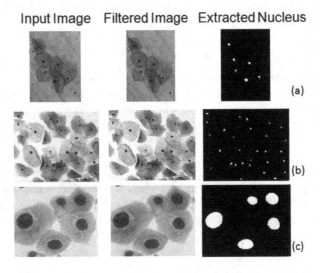

Fig. 3. Multiple cell: input image, filtered image and segmented nuclei. Normal cells (a–b), abnormal cells (c)

have extracted features of nuclei only because cells are highly overlapped. Hence, the list extruded features on shape-based for nucleus and cytoplasm are:

1. Area (A_n, A_c): A_n represents the no. of pixels in the nucleus and A_c represents the number of pixels in the cytoplasm.
2. Perimeter (P_n, P_c): P_n represent the no. of pixels in the boundary of the nucleus and P_c represents the no. of pixels in the boundary of cytoplasm.
3. Major Axis (H): H represents the major axis of the nucleus.
4. Minor Axis (L): L represents the minor axis of the nucleus.
5. Compactness (C_n, C_c): $C_n, = (P_n * P_n / A_n)$ and $C_c = (P_c * P_c / A_c)$
6. Eccentricity (E): $E = H/L$ show the shape of the nucleus.
7. Nucleus Cytoplasm ratio (N/C): $N/C = A_n / A_c$

For single cell image, 10 features of nucleus and cytoplasm have been calculated. In the case of multiple cells image, only 6 features of nuclei are calculated. If the normal cell turn into an abnormal cell, the size and shape of the nucleus changed. In the case of the normal cell nucleus is circular in shape. In an abnormal cell the shape of nucleus is turned into an elliptical shape. The maturity of the cell is expressed by N/C ratio of a cell, if the cell is normal its nucleus is small in size and ratio is low but in case of a dyspalsia cell its nucleus size generally increased and the ratio is high. Eccentricity value shows the roundness of the nucleus. If the value is near zero means the cell is normal; otherwise, it is abnormal.

3.4 Classifier

The descriptions of the classifier are widely available in the literature, so only a brief introduction of the classifier is provided. Basic classifiers like decision-tree, k- nearest neighborhood and Bays classifier are not working well with our data-set. The performances of ANN and SVM classifier have evaluated on our two data-sets. Artificial Neural Network (ANN) work as a nonlinear statistical classifier. Its classify the data well when relation between input and output is complex. In ANN Multiple hidden layers are there with different neurons in each layers. The number of neurons value in hidden layer depend on the derived features of the cell. In this study, 4 hidden layers with 30 neurons are used. Support Vector Machine (SVM) constructs the hyper plane between the clusters of the data-set by extracted features. The data has been mapped into a different cluster with the help of kernel function. In this paper, the radial basis function (RBF) kernel is used with 10 cross fold.

4 Experimental

In this paper two self-generated and one standard data-sets is used for the assessment of our proposed work, and the data-sets are classified into two classes only: normal and abnormal. In a single cell data-set, features of nucleus and cytoplasm both are obtained. But in multiple cell data-set, only features of nuclei are obtained. In multiple cell data-set, numerous nuclei are acquired, so the mean values of all the obtained features of nuclei are calculated. By comparing the features of single cell and multiple cell, we want to grab the best feature out of ten proposed features. The Table 1 shows the extracted features of the normal and abnormal cell for the single cell data-set. First three entries are for normal cell, and the remaining three are for abnormal cells, these six values are randomly picked from the list of extracted features. As shown in the table the area and perimeter of the nucleus for the normal cell are comparatively less than the area of the nucleus for the abnormal cell. Area of nucleus varies from 2000 to 5000, but for abnormal cells, it varies from 5000–9500. Area of cytoplasm is comparatively high for the normal cell than an abnormal cell because in the case of abnormality the cell is divided and equivalent to the nucleus. As we can see that the N/C ratio of a normal cell is less because the size of nucleus is low in the normal cell comparatively to an abnormal cell. The Table 2 shows the extracted features of the second data-set that is multiple cells. First three entries are for normal cell, and remaining three are for an abnormal cell. Area and perimeter of the nucleus for normal cells are comparatively less than the area and perimeter of an abnormal cell. The values of the nucleus area for normal cell varies from 10 to 500 and for abnormal cells, it varies from 500 to 6000. Compactness and eccentricity value of the normal cell is comparatively less than an abnormal cell.

The performance of the classifier is evaluated using accuracy, specificity, and sensitivity. Below three equation shows the formula for evaluation parameters. The accuracy calculates the perfectly classified data into normal and dysplasia classes.

$$Accuracy = \frac{TP + TN}{TP + FP + TN + FN} \tag{2}$$

$$Specificity = \frac{TN}{TN + FP} \tag{3}$$

$$Sensitivity = \frac{TP}{TP + FN} \tag{4}$$

According to the equations, TP is true positive value, means abnormal cell determine as abnormal cell and TN is true negative value, means the normal cell is determine as a normal cell. FP is false positive value, means abnormal cell is determined a normal cell and FN is false negative value, means normal cell is determined as an abnormal cell. The specificity shows the percentage of correct classification of a normal cell. The sensitivity shows the percentage of correct classification of the abnormal cell.

Table 3 shows the performance evaluation of ANN and SVM. Classification accuracy of single cell using both nucleus and cytoplasm is best, i.e., 99.0%, but when the classification is done by nucleus only for a single cell, accuracy decreased. Sensitivity and specification for single cell data-set using nucleus and cytoplasm are better than the nucleus only. Accuracy calculation using multiple data-sets is low when ANN classifier is used. By using SVM classifier, the accuracy of a single cell with a nucleus and cytoplasm features is 95.1% but when features of the nucleus are used it decreases. For multiple cells, accuracy is 92%. According to these results, ANN is working better than SVM. In Fig. 4 the performance of individual feature for a single cell is shown in the form or graph. In this graph N/C ratio is giving the best accuracy that is 99.2% for ANN classifier. In this paper for multiple cell images data-set features of nuclei are extracted, but according to individual feature performance, N/C is giving the best results. So in future work, the challenge will be feature extraction of nuclei and cytoplasm for us from the multiple cell images due to overlapping of cells.

Table 1. Features of a normal and an abnormal single cell

Cell	Area	Perimeter	Eccentricity	Area of cell	N/C ratio
Normal	4431	276.35	1.59	128526	0.034
Normal	2012	173.12	1.31	162441	0.012
Normal	3909	254.45	1.75	147887	0.026
Abnormal	9064	387.06	1.22	31848	0.284
Abnormal	7184	340.39	1.37	28480	0.252
Abnormal	9270	376.43	1.01	23943	0.387

Table 2. Features of a normal and an abnormal multiple cells

Cell	Area	Perimeter	Compactness	Eccentricity
Normal	59.11	26.53	11.91	1.68
Normal	21.25	15.59	11.44	1.62
Normal	34.28	19.02	10.56	1.44
Abnormal	590.1	101.35	17.41	1.64
Abnormal	3408.1	255.9	19.22	1.79
Abnormal	4000.3	258.5	16.71	1.79

Table 3. Performance evaluation of ANN and SVM

Cell	Accuracy	Sensitivity	Specificity
Single cell (ANN)	99.0	100	98.03
Single cell nucleus only	90.1	90	90.19
Multiple cell nuclei only	96.0	93.8	100
Single cell (SVM)	95.1	100	91.07
Single cell nucleus only	87.1	89.4	85.18
Multiple cell nuclei only	92.0	93.3	90.0

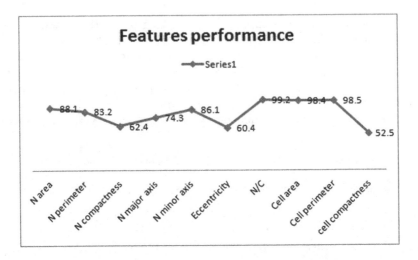

Fig. 4. Performance of individual features on single cell data-set

5   Conclusion

The suggested technique separates the nuclei from multiple cell data-set and nucleus and cytoplasm from single cell data-set using user defined threshold value and morphological operations. Using the features of nucleus and cytoplasm of the single cell, ANN is giving the best result. For a normal cell, the N/C ratio varies from 3% to 5% and for the abnormal cells, it varies from 13.4% to 41.9%. But when we are just using the features of the nucleus from a single cell, the accuracy decreased. We can see that only the features of nucleus do not give good results. By using both features, we can also improve the accuracy of multiple cells. Therefore, our next work will be to separate out the cytoplasm from multiple overlapped cells and to increase the data-set size.

Acknowledgements. We should like to thank Dr. Archana Pareek and Dr. Mukesh Rathore for providing us Pap smear slides from her pathology lab and for helping us to capture images with the help of microscope.

References

1. Cancer Cases in India Likely to Soar 25% By 2020: ICMR - Times of India. www.timesofindia.indiatimes.com/india/Cancer-cases-in-India-likely-to-soar-25-by-2020-ICMR/articleshow/52334632.cms. Accessed 16 June 2017
2. Sreedevi, A., Javed, R., Dinesh, A.: Epidemiology of cervical cancer with special focus on India. Int. J. Women's Health **7**, 405–14 (2015)
3. Tan, S.Y., Tatsumura, Y.: George Papanicolaou (1883–1962): discoverer of the pap smear. Singap. Med. J. **56**(10), 586 (2015)

4. Plissiti, M.E., Nikou, C., Charachanti, A.: Watershed-based segmentation of cell nuclei boundaries in Pap smear images. In: Proceeding of the 10th IEEE International Conference on Information Technology and Application in Biomedicine, pp. 1–4 (2010)
5. Kenny, S.P.K., Victor, S.P.: A comparative analysis of single and combination feature extraction techniques for detecting cervical cancer lesions. ICTACT J. Image Video Process. **6**(3), 1167–1173 (2016)
6. Peng, Y., et al.: Clustering nuclei using machine learning techniques. In: 2010 IEEE/ICME International Conference on Complex Medical Engineering (CME), pp. 52–57 (2010)
7. Cheng, F.-H., Hsu, N.-R.: Automated cell nuclei segmentation from microscopic images of cervical smear. In: International Conference on Applied System Innovation (ICASI), pp. 1–4 (2016)
8. Arya, M., Mittal, N., Singh, G.: Fuzzy-based classification for cervical dysplasia using smear images. In: Sa, P.K., Bakshi, S., Hatzilygeroudis, I.K., Sahoo, M.N. (eds.) Recent Findings in Intelligent Computing Techniques. AISC, vol. 708, pp. 441–449. Springer, Singapore (2018). https://doi.org/10.1007/978-981-10-8636-6_46
9. Kaaviya, S., Saranyadevi, V., Nirmala, M.: PAP smear image analysis for cervical cancer detection. In: 2015 IEEE International Conference on Engineering and Technology (ICETECH), pp. 1–4 (2015)
10. Mahanta, L.B., Nath, D.C., Nath, C.K.: Cervix cancer diagnosis from pap smear Images using structure based segmentation and shape analysis. J. Emerg. Trends Comput. Inf. Sci. **3**(2), 245–249 (2012)
11. Lakshmi, G.K., Krishnaveni, K.: Multiple feature extraction from cervical cytology images by Gaussian mixture model. In: World Congress on Computing and communication Technologies (WCCCT), pp. 309–311 (2014)
12. Athinarayanan, S., Srinath, M.V.: Classification of cervical cancer cells in PAP smear the screening test. ICTACT J. Image Video Process. **6**(4), 1234–1238 (2016)
13. Sukumar, P., Gnanamurthy, R.K.: Computer aided detection of cervical cancer using pap smear images based on hybrid classifier. Int. J. Appl. Eng. Res. **10**(8), 21021–21032 (2015). Research India Publications
14. Arya, M., Mittal, N., Singh, G.: Texture-based feature extraction of smear images for the detection of cervical cancer. IET Comput. Vis. **12**(8), 1049–1059 (2018)
15. Kale, A., Aksoy, S.: Segmentation of cervical cell images. In: 20th International Conference on Pattern Recognition (ICPR) Istanbul, pp. 2399–2402 (2010)
16. Poonam, S.N., Vivek, M., Sharan, P.: Automated cervical cancer detection using photonic crystal based bio-sensor. In: IEEE International Advance Computing Conference (IACC), pp. 1174–1178 (2015)
17. Sharma, M., Singh, S.K., Agrawal, P., Madaan, V.: Classification of clinical dataset of cervical cancer using KNN. Indian J. Sci. Technol. **9**(28), 1–5 (2016)
18. Sajeena, T.A., Jereesh, A.S.: Automated cervical cancer detection through RGVF segmentation and SVM classification. In: 2015 International Conference on Computing and Network Communications (CoCoNet), pp. 663–669 (2015)
19. Martin, L., Exbrayat, M.: Pap-smear classification (2003)

IoT Based Alert System for Visually Impaired Persons

Vaibhav Bhatnagar[1(✉)], Ramesh Chandra[2], and Vikram Jain[1]

[1] S.S. Jain Subodh PG College, MCA Institute, Rambhag Circle, Jaipur, India
vaibhav.bhatnagar15@gmail.com
[2] Amity Institute of Information Technology,
Amity University Rajasthan, Jaipur, India

Abstract. Internet of Things (IoT) is now becoming the emerging issue in the field of research. IoT is used in home appliances, health monitoring, Industry Internet and many more. Sensors and Actuators play important role in the Internet of Things. In this paper, a prototype is proposed that will warn the visually impaired persons for nearest obstacles and hurdles. Ultrasonic sensor is used to design the proposed system. The visually impaired person will receive warning through its smart-phone and the location will be saved. Whenever visually impaired person will come near to that place it will warn again to the person. This will be a low-cost system and will replace the stick that is generally used by every visually impaired person.

Keywords: IoT · Alert system · Blind person · Joystick · Ultra-sonic sensor

1 Introduction

Internet of Things [1] is now entering in the daily life of human being. Internet of Things (IoT) is system that connects objects (things) that can be accessible via internet. The keyword thing can be any object in which an IP address can be assigned and have capability of sensing and transfer a data over a network. These things have built in sensors that connected to an Internet platform (server) which integrates data from the things from different places and applies an analytics so that useful information can be generated. These high configured platforms are able to identify which information is useful and which information can be ignored. Overview of Internet of Things is depicted in (Fig. 1).

As an example [2], a pulse rate monitor is attached with the hands of a patient that regularly sense the pulse rate. When the pulse rate becomes down or high, it will send the current location of that patient to his relative. It is a low-cost system in which pulse rate sensor is of only Rs. 215 (EC-0567) and GSM module (SmartElex) is of only Rs. 1500 in which a heart patient is safe to move everywhere.

© Springer Nature Singapore Pte Ltd. 2019
A. K. Somani et al. (Eds.): ICETCE 2019, CCIS 985, pp. 216–223, 2019.
https://doi.org/10.1007/978-981-13-8300-7_18

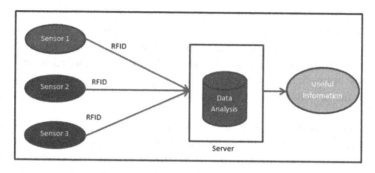

Fig. 1. Overview of Internet of Things

Visual impairment is the limitation of eyes for vision system. A person is called a Visual Impaired if he lost of visual inability and acuity of the person to see objects as clearly like other healthy person or he has inability see as wide an area like other healthy person or he cannot look at light or he has problem of double vision. There are large numbers of problems faced by these visually impaired persons such as identification of currency during local business transaction, study of a book that does not have Braille script, purchasing small commodities like vegetables & fruits and most hazardous problem is chances of accident due to walking.

Large numbers of accidents are faced by visually impaired persons due to unplanned cities and undisciplined traffic, so taken this problem into serious consideration, this paper proposed a system for visually impaired persons that warn for upcoming obstacle and hurdle. The location will be saved in the mobile of person and warn again whenever that person goes near to that location. This will replace the dependency of the visually impaired person on the stick. Followed by Introduction this paper is divided into 3 parts in which II part is Literature Review, III part is Design of Proposed, IV part is conclusion and future work.

2   Related Work

Researchers have proposed some IoT based solutions for physically challenged persons [12]. In this section some endeavors are highlighted:

Domingo [3] depicted the used of IoT for physically challenged persons. He explained that the main components of these systems are the sensors and actuators that sense the nearby object that are hazardous for these persons. IoT offers the assistance and support of physically challenged people for achieve a safe and secure life and facilitate them to enjoy the economic and social life. In this paper, an overview of the Internet of Things for people with disabilities is provided. He also described the challenges to introduce such systems. Customization for people with their disabilities is the main challenge described by author. Since these persons have their own needs and requirement, the IoT based system must be adapted for their customized needs.

Turcu and Turcu [4] shown how radio frequency identification, multi-agent and Internet of Things technologies can be used to improve people's access to quality and affordable healthcare services, to reduce medical errors, to improve patient safety, and to optimize the healthcare processes. Some agents will be developed in order to integrate various things into an Internet of Things platform, to assist users in various activities on the IoT platform, etc. All of them must be developed based on personal preferences.

Lopes [5] proposed IoT architecture for disabled people and intends to identify and describe the most relevant IoT technologies and international standards for the stack of the proposed architecture. In particular, the paper discusses the enabling IoT technologies and its feasibility for people with disabilities. At the end, it presents two use cases that are currently being deployed for this population. This work proposes an IoT architecture specific for people with disabilities that addresses the technological challenges of the current Internet and identifies and describes the most relevant technologies and standards for the first four layers of the proposed architecture. This architecture was specifically designed to be suitable for disabled people however we believe that it could be also suitable for other IoT application domains with similar requirements.

Vanishree et al. [9] addressed the challenges of Visually Impaired persons using IoT. Authors described micro and nano-sensors that can be attached to the body of person such as etinal prosthesis for storing vision affected by retinitis pigmentosa. Another device Vibratory Belt along with the Virtual White Cane can also be used that enhanced mobility of the user.

3   Design of Proposed System

The proposed system is specifically designed for visually impaired person that will sense the object coming in front while the walking and alerts the visually impaired person through his smart-phone. The main feature of this proposed system is that it will make the visually impaired person free from the stick which is now become the symbol of visually impaired persons. The proposed system is a low cost system that works on the 10000 mAh battery which can be easily kept in the pocket of the visually impaired person. The proposed system has two parts one part is sensing the nearby object and warns the user and second part is retrieve the location again warns the user whenever visually impaired person go nearby that location. Layout of proposed system is shown in Fig. 2.

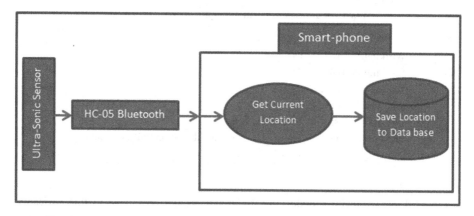

Fig. 2. Layout of proposed system

The following methodology is adopted for design of proposed model:

Step 1: Detecting the Obstacle: The first step is to detect the obstacle coming towards the visually impaired system. To identify the nearby object Arduino based Ultra-sonic sensor [6] is used and fixed into the spectacles. These ultrasonic sensors use Sound Navigation and Ranging (SONAR) technique to identify the distance from an object. These sensors give non-contact range detection with stable readings. The operation is not affected by sunlight or black material, although acoustically, soft materials like cloth can be difficult to detect. It comes complete with ultrasonic transmitter and receiver module. This Ultra-sonic sensor will be attached in the spectacles of blind person. Wherever the person will move (Right, Left and down direction) his neck the sensor will sense the obstacle. HC-SR04 and SEN-13959 are some commonly used ultra-sonic sensors. Figure 3 and Table 1 [7] depicts the connection and specification of sensor for Arduino connectivity.

Fig. 3. Ultra-sonic sensor with Arduino connection

Table 1. Specification of ultra-sound sensor (HC-SR04).

Particular	Specification
Working voltage	DC 5 V
Working current	15 mA
Working frequency	40 Hz
Measuring angle	15°
Trigger input signal	10uS TTL pulse
Echo output signal	Input TTL lever signal & range in proportion
Dimensions	45*20*15 mm

Figure 4 illustrate the pseudo code for detecting the obstacles. In this code, the micro-seconds taken by the sensor are converted into the inches, and the measured inches are compared with distance_value. distance_value is a variable that can be decided by the visually impaired person that how much distance is to be required for giving the waning. Whenever the inches value is greater than the distance_value, the Bluetooth chip will become active and will start sending the data to the smart-phone. This pseudo code can be directly implemented Sketch IDE.

```
void loop()
{
    int dur, inches;
    pinMode(trigger, OUTPUT);
    digitalWrite(trigger, LOW);
    delayMicroseconds(2);
    digitalWrite(trigger, HIGH);
    delayMicroseconds(10);
    digitalWrite(trigger, LOW);
    pinMode(echoPin, INPUT);
    dur = pulseIn(echoPin, HIGH);
    inches = microsecondsToInches(dur); // // convert the time into a distance
    delay(100);
    if(inches > distance_value)
    {
        Bluetooth.print("GO STOP"); /// Send data to bluetooth
    }
}
int microsecondsToInches(int microseconds)
{
    return microseconds / 74 / 2;
}
```

Fig. 4. Pseudo code for detecting obstacle using ultra-sound sensor

Step 2: Recording the Location: The second step is to save the current location when an obstacle is sensed. This smart-phone is connected with ultra-sonic sensor using the Bluetooth chip- HC-05. When the sensor will sense the object, a notification will be received by the connected smart-phone through the Bluetooth connection and the current location will be added into the database.

```
protected void onCreate(Bundle savedInstanceState)
{
locationManager = (LocationManager) getSystemService(Context.LOCATION_SERVICE);
locationManager.requestLocationUpdates(LocationManager.GPS_PROVIDER, 0, 0, this);
}
@Override
public void onLocationChanged(Location location)
{
txtLat = (TextView) findViewById(R.id.textview1);
txtLat.setText("Latitude:" + location.getLatitude() + ", Longitude:" + location.getLongitude());
String location=txtLat.getText();
SQLiteQuery = "INSERT INTO Iot_Location VALUES('"Location"');";
SQLITEDATABASE.execSQL(SQLiteQuery);
}
```

Fig. 5. Pseudo code for recording location into the database

In the Fig. 5, the pseudo code for adding the current location is shown, this code will activated only when the Bluetooth of smart-phone will receive the signal from the sensor. Using the location Manager, current latitude will be saved into the table named as Iot_Location. The location is retrieved from the GPS (Global Position System) into the variable location. The variable location will contain information about the current location (Latitude, Longitude) for the visually impaired person.

Step 3: Warning the Person: The last but important and critical step is to warn the user, whenever he is moving towards the same direction where he faced the obstacle. This warning system will be carried out with the help of GPS of the smart-phone. In this proposed the system, user will be warned four precision before the already saved location. So, in the database, not only the current location but also four near values of latitude and longitude will also be saved in the database. These four values are the adjacent coordinates (East, West, South and North) of that particular location as shown in Table 2. Whenever user will visits in these positions, the smart-phone will start giving a warning message through push notification. The push notification is enabled with sound, so that visually impaired person can be alerted.

Table 2. Adjacent coordinates of a location

S.N.	Position	Neighbor locations	Moving nearby from the current position
1	26.81978, 75.85726	26.81978, 75.85826	East Side from current position
2		26.81978, 75.85626	West Side towards current position
3		26.81878, 75.85726	South Side from current location
4		26.82078, 75.85726	North Side from current location

These fives values (current location with four adjacent coordinates) will be saved in the database. These fives values will continuously matched from the current position of the user. This step will avoid any miss happening or accident for the visually impaired person. This warning can not only be used in case of avoiding collision but can also be used in hazardous place such as heavy traffic, pit on a road etc.

4 Discussion and Result

In this research paper, IoT based solution is proposed for visually impaired persons. The proposed system is compared [10, 11] with Smartcane i.e. smart stick used by blind people in Table 3.

Table 3. Comparison of proposed system with smart cane

Parameters	Proposed system	Smart-stick used by blind persons
Cost	Rs. 219	Rs. 3675
Size	Wearable with Spectacles	Around 3 Feet
Alert user by	Buzzer that trigger from the smart-phone	Only Vibrate
Detection distance	Max 450 cm	Max 300 cm
Requirements	Smart-phone with ear-phone	No requirements

It is perspicuous clear that proposed is better than the smart-stick in terms of cost, detection distance and size. Moreover, this white stick is becoming the identity as a result some peoples do not prefer to use it, but this proposed system is wearable at spectacles and goggles. Another significant parameter is that proposed system warns the user whenever he is moving towards an obstacle which was faced earlier. This warning will be given by push up notification using smart phone caring by him.

5 Conclusion and Future Work

Internet of Things (IoT) is now becoming the emerging issue of research. Nowadays IoT is being used in daily life. In this paper, IoT is also becoming a boon to visually impaired person. In this paper, a prototype model of alert system is proposed to avoid collusion and accident for visually impaired person. Ultra-sound sensor is used to sense the obstacle in front of person. This ultra-sound sensor can easily be fixed in the spectacles of the visually impaired person, so whenever and wherever he will move his neck, the obstacle would be sensed. This sensor is connected to the smart-phone of user through Bluetooth chip attached to the sensor. When an obstacle is sensed, Bluetooth chip will send a message to the smart-phone and smart-phone will warn the person using a push-notification. Further, whenever the person will go near to that position where obstacle was send, another push notification will be sent to the user to not to move that side. This work will replace the white stick which has now becoming the identification of visually impaired person. As a future, the four adjacent coordinates will be enhanced with more coordinates so that user can be alert from all directions.

References

1. Lee, I., Lee, K.: The Internet of Things (IoT): applications, investments, and challenges for enterprises. Bus. Horiz. **58**(4), 431–440 (2015)
2. Kumar, R., Rajasekaran, M.P.: An IoT based patient monitoring system using raspberry Pi. In: International Conference on Computing Technologies and Intelligent Data Engineering (ICCTIDE). IEEE (2016)
3. Domingo, M.C.: An overview of the Internet of Things for people with disabilities. J. Netw. Comput. Appl. **35**(2), 584–596 (2012)
4. Turcu, C.E., Turcu, C.O.: Internet of things as key enabler for sustainable healthcare delivery. Procedia - Soc. Behav. Sci. **73**, 251–256 (2013)
5. Lopes, N.V., et al.: IoT architecture proposal for disabled people. In: 2014 IEEE 10th International Conference on Wireless and Mobile Computing, Networking and Communications (WiMob). IEEE (2014)
6. Swan, M.: Sensor mania! the internet of things, wearable computing, objective metrics, and the quantified self-2.0. J. Sens. Actuator Netw. **1**(3), 217–253 (2012)
7. Topic: Specification of HC-SR04. https://cdn.sparkfun.com/datasheets/Sensors/Proximity/HCSR04.pdf. Accessed 1 Dec 2018
8. Misra, P., Enge, P.: Global Positioning System: Signals, Measurements and Performance, 2nd edn. Ganga-Jamuna Press, Massachusetts (2006)
9. Vanishree, M.L., Sushmitha, S., Roopa, B.K.: Addressing the challenges of visually impaired using IoT. Int. J. Recent. Innov. Trends Comput. Commun. **5**(1), 182–186 (2017)
10. Topic: Smart Cane. https://www.snapdeal.com/product/smart-cane-white-electronic-travel/62628493122. Accessed 12 Jan 2019
11. Topic: Ultra-Sonic sensor. https://www.amazon.in/Adraxx-HC-SR04-Ultrasonic-Distance-Measuring/dp/B01LXFUAFV. Accessed 12 Jan 2019
12. Durugkar, S.R., Poonia, R.C.: An era of micro irrigation - a priority driven approach to enhance the optimum utilization of water and a way towards "Intelligent Farming". In: Recent Patents on Computer Science, vol. 11, no. 4, pp. 247–254(8), December 2018

Join Query Optimization Using Genetic Ant Colony Optimization Algorithm for Distributed Databases

Preeti Tiwari[✉] and Swati V. Chande

International School of Informatics and Management, Jaipur, India
preeti.tiwari@icfia.org, swatichande@gmail.com

Abstract. With the increase in geographical spread of data both in terms of quality and quantity, attention on the storage, retrieval and modification of this distributed data has become a prime area of research. The focus is on efficient, accurate and timely availability of information extracted from various underlying data centers. Processing of queries from these distributed database environments has become a challenging task for the database researchers because as the number of relations increases in the database, the join order complexity also increases. There are N! ways of solving a particular query where N represents the number of Relations in the join query. The success of query processed in the Distributed Database Environment depends largely on the search strategy implemented by the query optimizer whose task is to search an optimal Query Evaluation Plan in minimum time amongst the various query plans that can minimize the consumption of computer resources. Various search strategies beginning from Deterministic Algorithms to the most recent and modern Evolutionary Algorithms have contributed incalculably towards query optimization but they bear their own set of limitations and drawbacks. This research paper focuses on the implementation of a hybrid strategy of Evolutionary Algorithms for the optimization of join queries in DDBMS. The hybrid strategy is an integration of Ant Colony Optimization Algorithm and Genetic Algorithm and has been coined as GACO-D (Genetic Ant Colony Optimization Algorithm for Distributed Database). This paper focuses on the search of an optimal Join Order in minimum response time using GACO-D and also compares its performance with existing strategies.

Keywords: Distributed database · Join query · Query optimization · GA · ACO

1 Introduction

The role of Information Management System and its availability has made it unmanageable for Information Providers to concentrate the data at one large mainframe site. The enhanced network traffic and reduced efficiency have enforced partitioning of data at multiple locations with each location having their own storage and local processing capabilities. This led to the development of Distributed Databases that play a significant role in providing reliable and accurate information to the end user. The advancements

© Springer Nature Singapore Pte Ltd. 2019
A. K. Somani et al. (Eds.): ICETCE 2019, CCIS 985, pp. 224–239, 2019.
https://doi.org/10.1007/978-981-13-8300-7_19

in hardware, software, protocols, storage and networks have changed the outlook of the data requirements and the exponential growth of data has presented new challenges to database researchers. The fast processing of complex queries and quick access to underlying information with minimum utilization of resources has become a necessity. With data spread over geographically separated machines that are interconnected via communication network, the processing of queries is a complex task.

As compared to centralized database, Distributed Database Management System (DDBMS) processes queries in multiple processing phases with many internal procedures. The most important component of query processing is query optimization whose task is to generate multiple equivalent Query Evaluation Plans (QEP) of the input query, evaluate each one of them on the basis of the predefined cost function and select the BEST QEP for query execution. Large number of factors affect the query optimizer like use of data statistics, choice of decision site, selection of join sites, network topology, replicated fragments, semi-joins, order of query operators, order of join operators, Site Order and Search Strategy (Aljanaby et al. 2005).

Among these factors, the Search Strategy implemented for query optimization problem is one of the most projected and challenging problems in today's scenario because the measure of performance of the input query is time based analysis and is continuously being minimized from seconds to milliseconds and from milliseconds to nanoseconds. Strategies are being developed that bear high processing speed with high computational solving abilities so as to generate end results with minimum execution times. Strategies like Deterministic Strategies (including Dynamic Programming, Iterative Dynamic Programming and Greedy Algorithms), Randomized Strategies (including Simulated Annealing, 2-Phase Optimization and Iterative Improvement techniques) and Evolutionary Strategies (including Genetic Algorithm and Ant Colony Algorithm) have already been researched upon and implemented to find optimal solutions to query optimization problems in Distributed Database Environment (Sevinc and Cosar 2009; Doshi and Raisinghani 2011; Zhou et al. 2009).

The main aim of this research paper is to propose a powerful hybrid search strategy for distributed database queries to handle time based optimization of large join queries and to validate its effectiveness against other existing similar solutions in terms of Query Response Time. The proposed search strategy is a hybrid of two Evolutionary Algorithms i.e. integration of Ant Colony Optimization Algorithm and Genetic Algorithm whose task is to generate an optimal Join Order of the input query in minimum Response Time. The paper is organized as follows: Sect. 2 discusses the join query optimization problem in Distributed Database. It also provides a succinct description of the search strategies implemented so far by the researchers for effective query optimization in distributed database and the need of proposing new strategy for the same. Section 3 focuses on the proposed Search Strategy. Section 4 focuses on the experimental setup followed by results analysis and discussion in Sect. 5. The research paper concludes in Sect. 6.

2 Join Query Optimization Problem in Distributed Database

The Join Operation is the most domineering factor of distributed query optimization. The order of join operation directly impacts the performance of the system i.e. higher the number of relations, higher is the complexity and vice versa (Apers, *et al.* 1983; Ceri and Negri 1984; Dong and Liang 2007). When a query enters a distributed environment, it is first parsed, checked and validated semantically and syntactically by Semantic Data Controller and then processed in four phases.

The Query Decomposition phase decomposes the validated query into a sequence of relational operations expressed in relational algebra form on the Global Conceptual Schema. The Data Localization Phase uses the Materialization and Access Planning Program to map the Global Conceptual Schema to the Fragmentation Schema and Allocation Schema of the distributed data. In the Global Query Optimization phase, equivalent Query Evaluation Plans (QEP) are constructed on the basis of permutations on the order of relational operators. These QEPs represent the Search Space of the optimizer and have different data transmission speed, execution cost, data statistics and execution time (Ghaemi *et al.* 2008). If there are N relations in a query exhibiting a join operation, then there is O (N!) possible QEPs because the Join Operation exhibits commutative and associative properties. The Global Query Optimizer evaluates these plans according to the predefined cost model with the fundamental mission to generate an Optimal Join Order with minimum cost that represents the "Best" Query Execution Plan (QEP). This QEP is passed onto the Local Query Optimization Phase where the Global Execution Monitor ensures the execution of the query as per the best QEP generated by the Global Query Optimization Phase. The Query Optimization at the Local Query Optimization Phase is supported by the Local Conceptual Schema.

The efficiency of the distributed database query optimizer is always a trade-off between the total execution and the quality of the QEPs generated. The cost based optimizers gauge upon the system resources and run time costs utilized for generating and processing of QEPs. They assign a cost to each QEP and embrace the one with least possible cost. These cost estimations are described in terms of I/O Operations, amount of disk buffer space, disk storage service time and relevant factors described in data dictionary. Most of the factors defined above are no longer considered as constraints for optimization problems because of the development of fast speed processors and availability of large primary and secondary storage spaces (Ban et al. 2015; Hameulain and Morvan 2009).

The concern and focus of the optimizers have shifted to query execution time and cost based optimizers are now generally referred as time based optimizers where the evaluation of best QEP is done on the basis of speed of processing of CPU and is generally attributed as Query Response Time which can be measured in terms of milliseconds and seconds. In a distributed database, join query optimization problem is defined as a Combinatorial Optimization Problem (Hameurlain and Morvan 2009) wherein feasible solution sets are generated from the finite set of objects satisfying a given condition and the candidate solutions are reduced to a discrete one. These problems are also defined as NP-Hard (Non-Deterministic Polynomial Time) Problems because the set of feasible solutions are built only at run time on the basis of theoretical construction procedures that cannot be forecasted by anybody (Kumar et al. 2014).

The optimization of distributed queries of primitive distributed databases (like SDD-1, Distributed INGRES, System R*) started in 1976 by developing optimization procedures on techniques like Hill Climbing Technique (Bernstein *et al.* 1981; Wong and Youssefi 1976; Stonebraker *et al.* 1976), Fragment and Replicate Techniques (Stonebraker and Neuhold 1977; Stamos and Young 1993) and nested loop and merge scan techniques (Selinger and Adiba 1980; Lohman *et al.* 1985). These optimizers mainly focused on the Processing Cost of the query and could generate optimal Join Order of the queries with extremely small number of relations (4–7 joins).

The Dynamic Programming Approach along with Greedy Algorithms and Iterative Dynamic Programming of Deterministic Strategies pioneered query optimization by implementing Bottom-Up Technique that deterministically search the solution space either by complete traversal or by applying heuristics for pruning the space (Selinger *et al.* 1979; Selinger and Adiba 1980; Ono and Lohman 1990; Vance and Maier 1996; Steinbrunn *et al.* 1997; Palermo 1974; Swami 1988; Shekita and Young 1997; Kossmann and Stocker 2000). The Deterministic Algorithms posed reasonable polynomial complexity and in order to generate good plans, these algorithms often lost tracks of pruning of alternative paths at replicated sites. They generated the problems of space complexity and tractability and could not optimize queries with more than 12–15 joins.

Randomized Strategies (Ioannidis and Wong 1987) work on Transformational Optimization Technique where randomness is a part of its logical structure in a hope of achieving good performance from "average solution cases" (Lanzelotte *et al.* 1993, Galindo-Legaria *et al.* 1994). Iterative Improvement, Simulated Annealing and 2-Phase Optimization Algorithms are cost based query optimizers that generated better results than Deterministic Algorithms but they do not guarantee optimal paths (highly unpredictable) as they are in-deterministic in nature (Swami and Gupta 1988; Martin *et al.* 1990). These algorithms bear a constant space overhead constraint and it is difficult to predict the execution time of randomized algorithms (Olken and Rotem 1986).

The need of high performance systems with large databases that could provide flexibility, high computation speed and customizability according to the dimension of the input problems led to the development of Evolutionary Algorithms. These algorithms mimic the mechanism of biological evolution of artificial breeding programmatically. Among these algorithms, Genetic Algorithms (GA) are the most researched upon optimization algorithms that fits well to all types of unpredictable problems. They are advantageous for query optimization problems as they require restricted parameter settings and can self-initialize from potential solutions sets rather than from a solitary solution (Ab Wahab et al. 2015). One of the main drawback of GA is its inability to converge quickly towards the optimal values (Uzel and Koc 2012; Devooght 2010) and it does not give any guarantee of finding global maxima but there are possibilities of finding local maxima at an early stage.

The Ant Colony Optimization Algorithm (ACO) also became increasingly popular over other metaheuristics for query optimization problems because of its positive feedback mechanism that promotes rapid solution finding and distributed computing ability to avoid premature convergence of solution sets (Abreu et al. 2011; Selvi and Umarani 2010). ACO lacks an organized initialization mechanism during its initial construction phase and hence displays slower convergence towards optimal solution (Dorigo and Stützle 2002; Kumar et al. 2015; Hlaing and Khine 2011; Golshanara et al. 2014; Wagh and Nemade 2017). Particle Swarm Optimization Algorithm (PSO) is an effective global search algorithm that is very simple to implement and its biggest strength is parallelization for concurrent processing (Gong et al. 2009; Bai 2010). However, PSO shows quick convergences which often leads to local optimum problem with no guarantee to generate optimal and constant results.

Looking at the limitations and weaknesses of aforesaid strategies, it was observed that their drawbacks can be overcome if one algorithm can complement the incapability of another. The integration of two or more algorithms creates immense scope to explore the capabilities, computational speed, accuracy and effectiveness in various types of combinatorial optimization problems. This paper proposed the integrating two Evolutionary Algorithms i.e. ACO and GA for optimization of join queries in DDBMS. GA is chosen because it is extremely adaptable to optimization problems and various hybrid strategies like GA with Learning Automata (Nasiraghdam et al. 2010), Immune Vaccine Construction (Yao 2017), Max-Min Ant System (Ban et al. 2015), Teacher Learner Based Optimization Algorithm (Lakshmi And Vatsavayi 2016), Fuzzy C-Mean Clustering Algorithm (Liu and Xu 2016) have been applied for query optimization problems in relational and distributed databases. ACO is the next widely and successfully implemented effective search strategies for optimization problems in databases and by integrating GA with ACO, the limitation of slow convergence towards optimal solutions of ACO can be eliminated by building a systematic and organized initialization phase of ACO. The next section discusses the conceptualization of the proposed algorithm which is coined as GACO-D (Genetic Ant Colony Optimization Algorithm for Distributed Database).

3   GACO-D for Distributed Database Query Optimization

GACO-D is an integration of two widely implemented search metaheuristics i.e. ACO and GA wherein both complement each other by taking advantage of their strengths. The GA component of GACO-D instantiates the optimization of the queries by initializing chromosomes and evolving them with genetic operators over a predefined number of generations. The semi-optimized results of the GA Component are then used to generate ants and initialize pheromone as an initial stage of ACO component and the final optimization takes place with the ACO Component of the GACO-D (Zhu et al. 2014; Zang and Lu 2012; Yao et al. 2008; Zhang and Gao 2009). This gives an advantage of quick conversion along with parallelism.

GACO-D is the Search Strategy of the Global Query Optimizer of DDBMS Query Processing and contributes towards improvement in Query Response Time. When a fragmented query is received by the global query optimizer, the GA component of GACO-D, creates multiple Query Evaluation Plans (QEPs) and represents them in the form of chromosomes. The best fit parent chromosomes undergo transformations via crossover and mutation operator over a number of generations to create better QEPs with lower costs. The best QEPs are passed as input to the ACO component of GACO-D and are initialized as ants wherein multiple ants represent multiple QEPs. Using the probabilistic rule of ACO, each ant in parallel explores new query plans to generate optimal Join Order.

GACO-D contributes towards efficient query optimization in the following ways:

(a) It allows both metaheuristics to perform completely which adds advantage of processing and optimization capabilities of both metaheuristics.

(b) It overcomes the initial unorganized behavior of ants by initializing it's trail with semi-optimal QEP and pheromones. This helps in overcoming Slow Convergence drawback of ACO.

(c) Parallelism can be implemented by GACO-D as both ACO and GA Component of the hybrid can start their processing from multiple points. Search of multiple solutions in parallel allows the optimizer to converge quickly towards global optimum.

(d) The integration of ACO and GA is made precise with accurate control on the number of times each component of GACO-D would be executed. This helps in retrieval of optimal Join Order with accurate execution time.

The algorithm of GACO-D is presented below (Algorithm 1.1).

Algorithm 1.1: GACO-D Algorithm

Input: Number of Joins, Number of Ants (m), Pheromone Influence Factor (α, alpha),
Heuristics Factor (β, beta), Pheromone Influence Factor (ρ, Rho),
Pheromone Increase Factor (q), Maximum Iterations for GACO Termination
(MaxTime), Population Size (PopulationSize), Maximum_Generations
(Generation), Mutatin_Rate (MutationPercent)

Output: Optimal Join Order Sequence, Minimum Response Time (Seconds),
Query Execution Cost f(c)

1.	**Begin**
2.	**Start Time: Query Response Time (milliseconds / seconds)**
3.	**Initialization:** Create Initial Population of chromosomes Randomly using Permutation Representation
4.	**Evaluate:** Find Fitness using customized function f(c) of Population
5.	**Create New Generation:** Initialize Generation=0
6.	While (Generation $\neq$ m) do
7.	Calculate **Fitness** using the customized function f(c)
8.	Parent Selection: **Rank Based Parent-Selection:** Select two parents to generate two off-spring
9.	Crossover: **Davis Order1 Crossover :** Two-Point Random Selection
10.	Mutation: **Swap Mutation** off-springs with predefined Mutation Rate
11.	Mutation Rate: **Experimentally Determined**
12.	Replacement Strategy: **Elitism Ranking Mechanism**
13.	Calculate **Fitness** using the customized function f(c)
14.	**Sort Population** in Descending order of Fitness
15.	End While
16.	Initialize **Ants = Chromosomes,** Pheromones on Trail Paths
17.	Assign JoinQueryCost[ ][ ] to each TrailPath
18.	Calculate Query Cost of Each TrailPath
19.	Retrieve BestTrailPath Query Cost of BestTrailPath
20.	Retrieve Query Cost of BestTrailPath as BestQueryCost
21.	While (MaxTime) do
22.	For each Ant[ ] ϵ Ants [ ][ ]
23.	Construct New Paths using **State Transition Rule**
24.	Apply **Pheromone Update**
25.	Calculate NewBestTrailPath(Ants[ ][ ], JoinQueryCost[ ][ ])
26.	Calculate NewBestQueryCost (NewBestTrailPath[ ], JoinQueryCost[ ][ ])
27.	End For
28.	If (NewBestQueryCost < BestQueryCost)
29.	BestTrailPath[ ] = NewBestTrailPath[ ]
30.	BestQueryCost = NewBestQueryCost
31.	End If
32.	End While
33.	**Stop Time: Query response Time (milliseconds/seconds)**
34.	Display Total Response Time in milliseconds/seconds
35.	Display Optimal Join Order as BestTrailPath[ ]
36.	Display Query Execution Cost as BestQueryCost
37.	**End**

For a given Distributed Query Optimization Problem, the GACO-D begins by creating a random pool of candidate solutions representing population. This population has chromosomes consisting of genes and the size of the chromosomes depends on the number of relations in the input query. Each solution set is assigned a fitness value based on the objective function associated with the underlying optimization problem. Based on the non-negative fitness function the fit individuals gain high probability of getting selected for reproduction and the less fit individuals are likely to get discarded. The candidate solutions with higher fitness value undergo crossover (recombination) and mutation to create fitter off-springs often defined as new generation. This procedure of selecting fit candidate solutions for evolving new generation is in line with the Darwinian Theory of "Survival of the Fittest". By restricting the selection of weak offsprings, GACO-D eliminates not only that candidate solution but also all of its descendants. This makes the algorithm converge towards high quality solutions within a few generations. The hybrid repeats its process of evolution to create better individuals or solution sets until termination criterion is met. The better individuals evolved by crossover and mutation are used to initialize ants and pheromones.

Ants are considered to be blind social agents that are capable of completing complex task and have the capability of finding the shortest path from nest to food. The hybrid algorithm further iterates over three segments. The first stage is Constructing Ant Solution wherein set of m simulated ants constructs solutions from elements of a finite set of available solution components. It includes the routines for ants to construct solutions incrementally. The first node is selected randomly as the starting point and one node at a time is added to the ants' traversal path. The decision of the next node selection is based on pheromone trail and heuristic information. The second stage consists of applying Local Search where the solutions obtained by ants are improved before updating pheromones and the final stage is Pheromone Updation. In this phase Evaporation of Pheromones takes place on the least visited paths, to avoid an unlimited increase of pheromone values and to allow the ant colony to forget poor choices done previously. Also Pheromone Deposition is done by increasing the value of the pheromone on promising solutions. The major objective of this phase is to increase the concentration of pheromones on good solutions and decrease the value of pheromones on bad solutions (Dorigo et al. 2006; Dorigo and Stützle 2002; Dorigo et al. 2006; Dorigo et al. 1999).

4 Experimental Setup

To access the performance of the GACO-D in terms of Query Response Time, simulation experiments are conducted for different number of joins. The significance of conducting experiments is to validate the strong knowledge base and theoretical predictions stated for query optimization in DDBMS. The validity of the results generated by GACO-D cannot be justified unless they are compared with other existing algorithms. The response time taken by GACO-D to generate an optimal Join Order is measured in milliseconds (converted to seconds). The effectiveness and performance of GACO-D is conducted by comparing the time taken by other existing strategies to derive the results from the same set of studies i.e. when the same query is executed by them. The response time results are compared with existing similar search algorithms applied for query optimization in distributed databases i.e. ACO and GA. The comparison of GACO-D with only these search strategies is done because both these evolutionary algorithms work on time based optimization and are able to generate optimal join orders with measurable response time (expressed in milliseconds and seconds). Also because these algorithms are capable of handling large join complex queries and do not pose problems of space complexity or tractability unlike Deterministic and Randomized strategies.

The hardware platform used for the experiments is with the following specifications: Intel® Core ™ i5-4200U CPU @ 1.60 GHz 2.30 GHz, 6 GB RAM, 500 GB HDD and 64-Bit Operating System. The experiments are conducted on the .Net Framework 4.5 under the Microsoft Visual Studio Package 2012. The input to GACO-D is the count of the total number of relations involved in the join operation. The relations are represented as nodes in a connected graph G = (N, E) where N = Nodes and E = Edges that represents the connections between these nodes. Each relation is encoded as an integer with some initial weight assigned to it randomly. The weight represents the total number of records (cardinality) involved in the join-operation and contributes significantly towards the retrieval of an optimal join order based on the defined fitness function.

The performance of GACO-D is measured in terms of Total Response Time (Seconds). Many researchers have also evaluated the performance of the of their proposed optimizer in terms of execution time (Kadkhodaei and Mahmoudi 2011; Ban et al. 2015; Alamery et al. 2010; Zhou et al. 2009; Zhou 2007; Dökeroğlu and Coşar 2011). The system has been evaluated on the termination criterion and the results are recorded after the last termination criterion of GACO-D. For each input of number of joins, the algorithm is made to run independently twenty (20) times and the average of the query response time is calculated to estimate the average time taken by GACO-D to converge towards optimal solutions. The independent runs helped to generate unbiased results that helped to effectively test the performance of the GACO-D and to prove its efficiency and accuracy.

The most significant components of GACO-D are the parameters (Table 1) that are explained below. The values of these parameters greatly impact the performance and efficacy of the algorithm (Eiben et al. 1999; Fidanova 2006, Nowotniak and Kucharski 2012; Saremi 2007). The values of these parameters are provided before the processing of the algorithm begins and they stay static during the execution of GACO-D.

Table 1. Discrete parameter settings for GACO-D, ACO and GA

S. No.	Symbol	Parameter name	Acceptable range	GACO-D	ACO	GA
1	(m)	Number of ants (Tiwari and Chande 2019)	Empirical analysis	Ant Ratio = 0.2	Ants = 10	NA
2	(α)	Pheromone relativity importance	$0 < \alpha < 5$	1	1	NA
3	(β)	Relative heuristic factor	$0 < \beta < 15$	5	3	NA
4	(ρ)	Pheromone evaporation coefficient	$0 \leq \rho < 1$	0.1	0.9	NA
5	(q)	Pheromone increase factor	>1	2	2	NA
6	TC	Termination condition of GACO-D	Empirical analysis	50	100	NA
7	NOJ	Number of joins	Input	5 to 85	5 to 85	5 to 85
8	P_m	Mutation rate (swap mutation)	Empirical analysis	0.1	NA	0.015
9	P_c	Crossover rate	Davis order one	50%	NA	60%
10	PopSize	Population size	Empirical analysis	1500	NA	100

The table also displays the parametric setting of Ant Colony Optimization Algorithm (Li *et al.* 2008) and Genetic Algorithm (Sevinc and Cosar 2009) implemented by respective researchers in their studies. (NA: Parameter that is Not Applicable in the given Algorithm). The acceptable range of Pheromone Relativity Importance, Relative Heuristic Factor, Pheromone Evaporation Coefficient and Pheromone Increase Factor are adopted from the experimental results carried out by various researchers and have contributed effectively towards the performance of ACO (Hei and Du 2011; Li and Zhu 2016; Liu *et al.* 2008; Xiao and Tan 2009; Shweta and Singh 2013). The values of Termination Criterion, Mutation Rate, Population Size and number of ants have been determined experimentally in the complete research

5   Results and Analysis

The effectiveness of the GACO-D is evaluated in terms of time taken to generate the optimized Join Order. The search time here is expressed in seconds and is denoted as Query Response Time. Table 2 displays the comprehensive analysis of Query

Response Time (seconds) taken by respective algorithms for searching and retrieval of optimal join order from multiple equivalent Query Evaluation Plans. The Query Response Time is expressed in seconds.

Table 2. Comprehensive Analysis of Query Response Time of GACO-D, GA and ACO for the generation of Optimal Join Order (Seconds)

NOJ	GACO-D	GA	ACO
5	0.0235	0.1474	0.0168
10	0.0903	0.1614	0.0695
15	0.1118	0.4916	0.0894
20	0.1273	0.5642	0.1789
25	0.3269	1.0318	0.4891
30	0.3577	1.1479	0.6420
35	0.4573	2.0820	0.7750
40	0.5656	2.4098	0.9830
45	0.7474	2.9547	1.2690
50	1.0692	3.3325	1.8222
55	1.3307	4.1019	2.7095
60	1.8496	4.7924	3.5367
65	2.5008	5.1408	4.8093
70	3.1622	5.4783	6.3841
75	4.0266	6.2623	8.5594
80	5.1198	7.1075	10.5816
85	6.6078	8.1495	13.5308

The comparative analysis shows that the Query Response Time taken by GACO-D is lesser than the response time taken by other proposed solutions for extremely high number of joins. For example, if the number of joins in the input query is 35, GACO-D takes 0.4573 s to generate its search space by defining all Query Evaluation Plans, evaluating each QEP and retrieve an optimal Join order. GA performs the same task in 2.0820 s, whereas ACO accomplishes it in 0.7750 s. Whether the input queries possesses small number of joins (5 to 25 joins) or large number of joins (beyond 25 joins) GACO-D exhibits its efficiency in terms of query response time to search and retrieve an optimal join order. For 85 joins, GACO-D generates optimal join order in 6.6078 s where as the GA takes 8.1495 s and ACO utilizes 13.5308 s for the same number of joins. The comprehensive graphical analysis of the Query Response Time is represented in Fig. 1.

The figure presents XY scatter chart with horizontal axis representing the total number of joins in the input query where as the vertical axis represents query response time in seconds. It can be observed that GACO-D and ACO show linear increase of Query Response Time with the increase in the number of joins in the input query and the time taken by ACO to generate optimal join order is more than the time taken by GACO-D. Also with extremely complex queries, ACO shows slowest convergence. The performance of Genetic Algorithm with increasing complexity of queries is quite

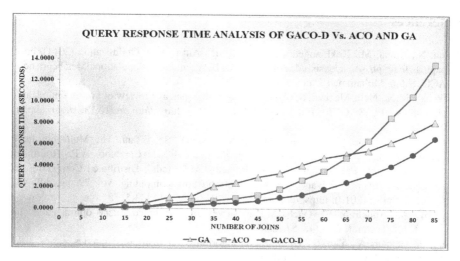

Fig. 1. Query response time analysis of GACO-D versus ACO and GA

unpredictable (Kadkhodaei and Mahmoudi 2011; Dong and Liang 2007). GA tends to show higher Query Response Time as compared to ACO and GACO-D when number of joins range between 25 and 65. After 70 joins, the convergence speed of ACO decreases and it shows higher Query Response Time. On the other hand, GACO-D exhibits slower convergence with joins lesser than fifteen when compared with ACO but as the number of joins goes beyond 20, GACO-D exhibits higher convergence. Hence it can be concluded that of the three algorithms studied, GACO-D exhibits best performance for optimization of large join queries by generating optimal join order in minimum Query Response Time.

6 Conclusion

GACO-D shows its competency in solving large-scale distributed database query optimization problems by performing extended exploration and exploitation mechanism which is supported by heuristics information and pheromone concentration values. It solves the query optimization problem by implementing multi-start approach of Combinatorial Optimization Problem. The social agents (ants) of GACO-D inherit the capability of constructing multiple QEP concurrently and then evaluating them for the best construct and then improving them repeatedly until the finest QEP is achieved. The construction and improvement procedure implemented repeatedly by the GACO-D for the quest for Optimal Join Order has contributed by not only by solving QO problem as Dynamic Query Optimization Problem but also has prevented the system from falling into Local Optimum. Hence the overall evaluation time of query processing was reduced. This gives GACO-D an edge towards the development of better QEPs within a short span of its execution. GACO-D proves to be an effective and viable solution for DDBMS Query optimization Problem when compared with existing strategies like ACO and GA.

References

Abreu, N., Ajmal, M., Kokkinogenis, Z., Bozorg, B.: Ant Colony Optimization (2011). http://paginas.fe.up.pt/~mac/ensino/docs/DS20102011/Presentations/PopulationalMetaheuristics/ACO_Nuno_Muhammad_Zafeiris_Behdad.pdf

AbWahab, M.N., Nefti-Meziani, S., Atyabi, A.: A comprehensive review of swarm optimization algorithms. PLoS ONE **10**(5), e0122827 (2015). https://doi.org/10.1371/journal.pone.0122827

Alamery, M., Faraahi, A., Javadi, H.H.S., Nourossana, S., Erfani, H.: Multi-join query optimization using the bees algorithm. In: de Leon, F., de Carvalho, A.P., Rodríguez-González, S., De Paz Santana, J.F., Rodríguez, J.M.C. (eds.) Distributed Computing and Artificial Intelligence. Advances in Intelligent and Soft Computing, vol. 79, pp. 449–457. Springer, Berlin (2010). https://doi.org/10.1007/978-3-642-14883-5_58

Aljanaby, A., Abuelrub, E., Odeh, M.: A survey of distributed query optimization. Int. Arab J. Inf. Technol. **2**(1), 48–57 (2005)

Apers, P., Hevner, A., Yao, S.: Optimization algorithms for distributed queries, series. IEEE Trans. Softw. Eng. **9**(1), 57–68 (1983)

Bai, Q.: Analysis of particle swarm optimization algorithm. Comput. Inf. Sci. **3**(1), 180–184 (2010)

Ban, W., Lin, J., Tong, J., Li, S.: Query optimization of distributed database based on parallel genetic algorithm and max-min ant system. In: 8th International Symposium IEEE Computational Intelligence and Design (ISCID), vol. 2, pp. 581–585 (2015)

Bernstein, P.A., Goodman, N., Wong, E., Reeve, C., Rothnie, J.B.: Query processing in a system for distributed databases (SDD-1). ACM Trans. Database Syst. **6**(4), 602–625 (1981)

Ceri, S., Negri Pelagatti, M.: Distributed Database Principles and System. McGraw-Hill, New York (1984)

Devooght, R.: Multi-Objective Genetic Algorithm, pp. 1–39 (2010). epb-physique.ulb.ac.be/IMG/pdf/devooght_2011.pdf

Dökeroğlu, T., Coşar, A.: Dynamic programming with ant colony optimization metaheuristic for optimization of distributed database queries. In: Gelenbe, E., Lent, R., Sakellari, G. (eds.) Computer and Information Sciences II, pp. 107–113. Springer, London (2011). https://doi.org/10.1007/978-1-4471-2155-8_13

Dong, H., Liang, Y.: Genetic algorithms for large join query optimization. In: Proceedings of the 9th Annual Conference on Genetic and Evolutionary Computation - GECCO 2007, pp. 1211–1218 (2007)

Dorigo, M., Birattari, M., Stützle, T.: Ant colony optimization – artificial ants as a computational intelligence technique. IEEE Comput. Intell. Mag. 28–39 (2006)

Dorigo, M., Caro, G.D., Gambardella, L.M.: Ant algorithms for discrete optimization. Artif. Life **5**(2), 137–172 (1999)

Dorigo, M., Stützle, T.: The ant colony optimization meta-heuristic: algorithms, applications, and advances. In: Glover, F., Kochenberger, G.A. (eds.) Handbook of Metaheuristics International Series in Operations Research & Management Science, vol. 57, pp. 251–285. Springer, Boston (2002). https://doi.org/10.1007/0-306-48056-5_9

Doshi, P., Raisinghani, V.: Review of dynamic query optimization strategies in distributed database. In: 3rd International Conference on Electronics Computer Technology (ICECT), vol. 6, pp. 145–149 IEEE (2011)

Eiben, A.E., Michalewicz, Z., Schoenauer, M., Smith, J.E.: Parameter control in evolutionary algorithms. In: Parameter Setting in Evolutionary Algorithms Studies in Computational Intelligence, pp. 19–46 (1999). IEEE Trans. Evol. Comput

Fidanova, S.: Simulated annealing: a Monte Carlo method for GPS surveying. In: Alexandrov, Vassil N., van Albada, G.D., Sloot, Peter M.A., Dongarra, J. (eds.) ICCS 2006. LNCS, vol. 3991, pp. 1009–1012. Springer, Heidelberg (2006). https://doi.org/10.1007/11758501_160

Galindo-Legaria, C.A., Pellenkoft, A., Kersten, M.L.: Fast, randomized join-order selection: why use (1994)

Ghaemi, R., Fard, A.M., Tabatabaee, H., Sadeghizadeh, M.: Evolutionary query optimization for heterogeneous distributed database systems. World Acad. Sci. **43**, 43–49 (2008)

Golshanara, L., Rouhani Rankoohi, S.M., Shah-Hosseini, H.: A multi-colony ant algorithm for optimizing join queries in distributed database systems. Knowl. Inf. Syst. **39**(1), 175–206 (2014). https://doi.org/10.1007/s10115-012-0608-4

Gong, D., Lu, L., Li, M.: Robot path planning in uncertain environments based on particle swarm optimization. In: Evolutionary Computation, CEC 2009, pp. 2127–2134. IEEE Congress (2009)

Hameurlain, A., Morvan, F.: Evolution of query optimization methods. In: Hameurlain, A., Küng, J., Wagner, R. (eds.) Transactions on Large-Scale Data- and Knowledge-Centered Systems I. LNCS, vol. 5740, pp. 211–242. Springer, Heidelberg (2009). https://doi.org/10.1007/978-3-642-03722-1_9

Hei, Y., Du, P.: Optimal choice of the parameters of ant colony algorithm. J. Converg. Inf. Technol. **6**(9), 96–104 (2011)

Hlaing, Z.C.S.S., Khine, M.A.: An ant colony optimization algorithm for solving traveling salesman problem. In: International Conference on Information Communication and Management, vol. 16, pp. 54–59 (2011)

Ioannidis, Y.E., Kang, Y.: Randomized algorithms for optimizing large join queries. In: Proceedings of ACM SIGMOD International Conference on Management of Data - SIGMOD 90, vol. 19, no. 2, pp. 312–321 (1990). https://doi.org/10.1145/93597.98740

Ioannidis, Y.E., Wong, E.: Query optimization by simulated annealing. ACM **16**(3), 9–22 (1987)

Kadkhodaei, H., Mahmoudi, F.: A combination method for join ordering problem in relational databases using genetic algorithm and ant colony. In: IEEE International Conference on Granular Computing, pp. 312–317 (2011). https://doi.org/10.1109/grc.2011.6122614

Kossmann, D., Stocker, K.: Iterative dynamic programming: a new class of query optimization algorithms. ACM Trans. Database Syst. (TODS) **25**(1), 43–82 (2000)

Kossmann, D.: The state of the art in distributed query processing. ACM Comput. Surv. (CSUR) **32**(4), 422–469 (2000). ISSN 0360-0300

Kumar, M.S., Srikanta, P., Dulu, P.: Implementation of query optimization techniques in distributed environment through genetic algorithm. Eur. J. Acad. Essays **1**(3), 89–93 (2014)

Kumar, T.V., Singh, R., Kumar, A.: Distributed query plan generation using ant colony optimization. Int. J. Appl. Metaheuristics Comput. **6**(1), 1–22 (2015). https://doi.org/10.4018/ijamc.2015010101

Lakshmi, S.V., Vatsavayi, V.K.: Query optimization using clustering and genetic algorithm for distributed databases. In: Proceedings of International Conference on Computer Communication and Informatics (ICCCI), pp. 1–8. IEEE (2016)

Lanzelotte, R.S., Valduriez, P., Zaït, M.: On the effectiveness of optimization search strategies for parallel execution spaces. In: VLDB, vol. 93, pp. 493–504 (1993)

Li, K., Kang, L., Zhang, W., Li, B.: Comparative analysis of genetic algorithm and ant colony algorithm on solving traveling salesman problem. In: IEEE International Workshop on Semantic Computing and Systems (2008). https://doi.org/10.1109/wscs.2008.11

Li, P., Zhu, H.: Parameter selection for ant colony algorithm based on Bacterial Foraging Algorithm. Math. Probl. Eng. **2016**, 1–12 (2016)

Liu, L.Q., Dai, Y.T., Wang, L.H.: Ant colony algorithm parameters optimization. Comput. Eng. **11**(34), 208–210 (2008)

Liu, S., Xu, X.: Distributed database query based on improved genetic algorithm. In: Proceedings of International Conference on Information Science and Control Engineering, pp. 348–351. IEEE Computer Society (2016)

Lohman, G.M., et al.: Query processing in R*. In: Kim, W., Reiner, D.S., Batory, D.S. (eds.) Query Processing in Database Systems. TINF. pp. 31–47, Springer, Heidelberg (1985). https://doi.org/10.1007/978-3-642-82375-6_2

Martin, T.P., Lam, K.H., Russell, J.I.: An evaluation of site selection algorithms for distributed query processing. Comput. J. **33**(1), 61–70 (1990)

Masrom, S., Siti, A.Z., Hashimah, P.N., Rahman, A.A.: Towards rapid development of User Defined (2011)

Nasiraghdam, M., Lotfi, S., Rashidy, R.: Query optimization in distributed database using hybrid evolutionary algorithm. In: International Conference on Information Retrieval and Knowledge Management, (CAMP), pp. 125–130. IEEE, March 2010

Nowotniak, R., Kucharski, J.: GPU-based tuning of quantum-inspired genetic algorithm for a combinatorial optimization problem. Bull. Pol. Acad. Sci.: Tech. Sci. **60**(2), 323–330 (2012)

Özsu, M.T., Valduriez, P.: Distributed Database Systems, 2nd edn. Prentice Hall (1999). ISBN 0-13-659707-6

Olken, F., Rotem, D.: Simple random sampling from relational databases. In: Proceedings of 12th International VLDB Conference, Kyoto, Japan, pp. 160–169 (1986)

Ono, K., Lohman, G.M.: Measuring the complexity of join enumeration in query optimization. In: Proceedings of the 16th International VLDB Conference, Brisbane, Australia, vol. 97, pp. 314–325 (1990)

Palermo, F.P.: A data base search problem. In: Tou, J.T. (ed.) Information Systems. Springer, Boston (1974). https://doi.org/10.1007/978-1-4684-2694-6_4

Pramanik, S., Vineyard, D.: Optimizing join queries in distributed databases. IEEE Trans. Softw. Eng. **14**(9), 1319–1326 (1988)

Saremi, A., Elmekkawy, T.Y., Wang, G.G.: Tuning the parameters of a memetic algorithm to solve vehicle routing problem with backhauls using design of experiments. Int. J. Oper. Res. **4**(4), 206–219 (2007)

Selinger, P.G., Astrahan, M.M., Chamberlin, D.D., Lorie, R.A., Price, T.G.: Access path selection in a relational database management system. In: Proceedings of the 1979 ACM SIGMOD International Conference on Management of Data – SIGMOD, vol. 79, pp. 23–34 (1979). https://doi.org/10.1145/582095.582099

Selinger, P.G., Adiba, M.E.: Access path selection in distributed database management systems. In: ICOD, pp. 204–215 (1980)

Selvi, V., Umarani, D.R.: Comparative analysis of ant colony and particle swarm optimization techniques. Int. J. Comput. Appl. (0975–8887) **5**(4), 1–6 (2010)

Sevinc, E., Cosar, A.: An evolutionary genetic algorithm for optimization of distributed database queries. In: 24th International Symposium on Computer and Information Sciences (2009). https://doi.org/10.1109/iscis.2009.5291839

Shekita, E.J., Young, H.C.: Iterative dynamic programming system for query optimization with bounded complexity. U.S. Patent 5, 671,403 (1997)

Shweta, K.M., Singh, A.: An effect and analysis of parameter on ant colony optimization for solving travelling salesman problem. Int. J. Comput. Sci. Mob. Comput. **2**(11), 222–229 (2013)

Stamos, J.W., Young, H.C.: A symmetric fragment and replicate algorithm for distributed joins. IEEE Trans. Parallel Distributed Syst. **4**(12), 1345–1354 (1993)

Steinbrunn, M., Moerkotte, G., Kemper, A.: Heuristic and randomized optimization for the join ordering problem. VLDB J. Int. J. Very Large Data Bases **6**(3), 191–208 (1997)

Stonebraker, M., Held, G., Wong, E., Kreps, P.: Design and implementation of INGRES. ACM Trans. Database Syst. (TODS) **1**(3), 189–222 (1976)

Stonebraker, M., Neuhold, E.: A distributed database version of INGRES. In: Proceedings of Second Berkeley Workshop Distributed Data Management and Computer Networks, pp. 19–36 (1977)

Sukheja, D., Singh, U.K.: A novel approach of query optimization for distributed database systems. Int. J. Comput. Sci. **8**(1), 307 (2011)

Swami, A.N., Gupta, A.: Optimization of large join queries in distributed database. In: Proceedings of ACM-SIGMOD Conference on Management of Data, pp. 8–17 (1988)

Tiwari, P., Chande, Swati V.: Optimal ant and join cardinality for distributed query optimization using ant colony optimization algorithm. In: Rathore, V.S., Worring, M., Mishra, D.K., Joshi, A., Maheshwari, S. (eds.) Emerging Trends in Expert Applications and Security. AISC, vol. 841, pp. 385–392. Springer, Singapore (2019). https://doi.org/10.1007/978-981-13-2285-3_45

Uzel, Ö., Koc, E.: Basics of Genetic Programming. Graduation Project I, pp. 1–25 (2012). http://mcs.cankaya.edu.tr/proje/2012/guz/omer_erdem/Rapor.pdf

Vance, B., Maier, D.: Rapid bushy join-order optimization with Cartesian products. In: ACM SIGMOD Record, vol. 25, no. 2, pp. 35–46 (1996)

Wagh, A., Nemade, V.: Query optimization using modified ant colony algorithm. Int. J. Comput. Appl. **167**(2), 29–33 (2017)

Wong, E., Youssefi, K.: Decomposition—a strategy for query processing. ACM Trans. Database Syst. (TODS) **1**(3), 223–241 (1976)

Xiao, H.F., Tan, G.Z.: Study improvement of the fusing genetic algorithm and ant colony algorithm in virtual enterprise partner selection problem on fusing genetic algorithm into ant colony algorithm. J. Chin. Comput. Syst. **30**(3), 512–517 (2009)

Yao, M.: A distributed database query optimization method based on genetic algorithm and immune theory. In: 2017 8th IEEE International Conference on Software Engineering and Service Science (ICSESS), pp. 762–765. IEEE, November 2017

Yao, Z., Liu, L., Wang, Y.: Fusing genetic algorithm and ant colony algorithm to optimize virtual enterprise partner selection problem. IEEE Congr. Evol. Comput. (IEEE World Congr. Comput. Intell.) 3614–3620 (2008). https://doi.org/10.1109/cec.2008.4631287

Zhang, X.R., Gao, S.: solving traveling salesman problem by ant colony optimization genetic hybrid algorithm. Microelectron. Comput. **4**, 024 (2009)

Zhang, W.G., Lu, T.Y.: The research of genetic ant colony algorithm and its application. Proc. Eng. **37**, 101–106 (2012)

Zhang, Y., Wu, L.: A novel genetic ant colony algorithm. J. Convergence Inf. Technol. **7**(1), 268–274 (2012)

Zhou, Y., Wan, W., Liu, J.: Multi-joint query optimization of database based on the integration of best-worst ant algorithm and genetic algorithm. In: IET International Communication Conference on Wireless Mobile & Computing (CCWMC 2009), pp. 543–550 (2009)

Zhou, Z.: Using heuristics and genetic algorithms for large-scale database query optimization. J. Inf. Comput. Sci. **2**(4), 261–280 (2007)

Zhu, S., Dong, W., Liu, W.: Logistics distribution route optimization based on genetic ant colony algorithm. J. Chem. Pharm. Res. **6**(6), 2264–2267 (2014)

Son, L.H., et al.: ARM–AMO: an efficient association rule mining algorithm based on animal migration optimization. Knowl.-Based Syst. **154**, 68–80 (2018). https://doi.org/10.1016/j.knosys.2018.04.038

Solving an Intractable Stochastic Partial Backordering Inventory Problem Using Machine Learning

Nidhi Srivastav[1(✉)] and Achin Srivastav[2]

[1] Department of Computer Science and Engineering, SKIT, Jaipur, India
nidhi03.srivastav@gmail.com
[2] Department of Mechanical Engineering, SKIT, Jaipur, India

Abstract. This paper addresses the intractability of order crossover in a partial backordering inventory problem. Here, the Artificial Neural Network (ANN), which is a machine leaning algorithm is used to solve a stochastic inventory problem. The results for examining order crossover with the back-propagation ANN shows notable reduction in inventory cost in comparison to linear regression method. A numerical study is taken to demonstrate the findings. This paper further draws insight on effectiveness of machine learning in comparison to regression.

Keywords: ANN · Order crossover · Machine learning

1 Introduction

Success and progress of any business organisation depends on the good management of inventories, as there has been a significant amount of money is tied with the inventories. Due to globalization, demanding customers and recent advances in technology, such as machine learning, artificial intelligence, internet of things (IOT), online retail chains; businesses are striving hard to sustain in the market. In recent scenario, customers are remained no more loyal to a particular organisation as they have got numerous options to place their orders. Nowadays, customers prefer to place small orders frequently, in comparison to earlier occasional large orders. With the large businesses mostly using Just-in-time (JIT), leads to an upsurge in small and frequent ordering. This frequent ordering leads to crossover of orders. The scenario when orders receive in different sequence as they were placed is referred as order crossover [1]. There could be numerous reasons for order crossover. Some of the possible reasons include the availability of global suppliers, frequent ordering from retailers (downstream member) to suppliers (upstream members), distant locations of suppliers and retailers and the presence of different modes of transportation. Different delivery times of the suppliers, results in items being stockout and in excess inventory problems. Due to presence of order crossover inventory systems often becomes intractable. Presence of stochastic lead times of suppliers, increases the complexity of inventory systems, that leads to erroneous computation of order quantity and safety factor.

© Springer Nature Singapore Pte Ltd. 2019
A. K. Somani et al. (Eds.): ICETCE 2019, CCIS 985, pp. 240–249, 2019.
https://doi.org/10.1007/978-981-13-8300-7_20

Recently, most of the literature on studying order crossover in inventory systems used linear regression to establish relationship between time between placing the orders and standard deviation of effective lead time (ELT). Prominent works on stochastic inventory systems used regression to examine order crossover [2–7]. In the past, most of the research studied order crossover for complete backorder inventory system [3, 4]. The works that examined order crossover in partial backorder inventory system and determined the optimal cost, service level and order size [5–7]. Few works examined order crossover for a partial backordering multi-objective inventory system [7].

From literature review on order crossover, it can be said all the papers either on complete backorders or partial backorder case have used regression under ELT approach. One of the key factors affecting the correct determination of stock level and computation of how much stock to carry are often unpredictable when merely depending on regression models. Surplus and idle stock basically signifies tied-up capital that could be put to better use. Declining stock levels requires exact calculations of upcoming demand, which is becoming more accessible due to machine learning technology. However, none of the earlier works has formulated the optimal inventory policy for the inventory systems considering machine learning.

This article attempts to consider the order crossover case for the partial backordering inventory model considering deterministic demand and stochastic lead time through machine learning using back propagation Artificial Neural Network (ANN). The problem statement is defined below:

Problem Statement

To optimize the inventory of a single retailer and multi supplier, single stage supply chain. It is assumed that the retailer is experiencing order crossover when orders are not received in sequence, to which they were issued to suppliers. Shortages in the inventory system are considered as a mixture of backorder and lost sales.

The noteworthy contributions of the paper are shown below:

- Optimal inventory policy is developed for stochastic inventory system experiencing order crossover using machine learning.
- Back propagation ANN is used to compute optimal order size and service factor by establishing the relationship between lead times and ordering interval.
- Proposed inventory system is applicable to situations when shortages occur as pure backorder, pure lost sales or the combination of backorder and lost sales.
- The proposed machine learning based inventory system is applicable to the inventory optimization of the fast-moving consumer goods (FMCG).
- Remarkable reduction inventory cost (46%) and order quantity (12%) is observed using proposed Machine learning based inventory system in comparison to existing regression-based inventory systems.

2   Methodology

Artificial Neural Network (ANN) works as a biological neuron and solve complex problems by learning and establishing relationships between inputs and outputs. ANN comprises of several neurons that are typically organized in layers through weights

[15, 16]. There is an activation function for each neuron. Neural Network functions can be easily fitted to available data, model it and predict on unseen data. Non linear sigmoid function is the common example of neural network transfer function. The fundamentals of ANN work found in early works on Neural Networks [8].

There are numerous applications of machine learning can be found in forecasting. A lot of research papers have used ANN in forecasting [9–12]. There are very few applications which uses ANN in inventory systems [13, 14]. In the present work, ANN is used for solving stochastic inventory problem. Multi-layer neural network is used in this paper. Back propagation learning algorithm is used for neural network and training set optimization and for computing the weight coefficients and the thresholds. Figure 1 shows the ANN structure.

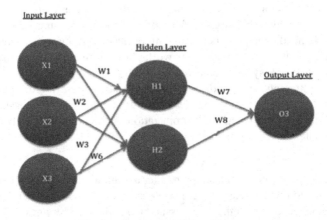

Fig. 1. A general ANN architecture

3 Inventory System

The notations are given below:

$C(Q, z_0)$ = total cost of inventory
A = ordering cost per order
D = demand per year
h = holding cost per unit per unit time
q = fraction of shortages backordered
$(1 - q)$ = fraction of lost sales
π = backorder cost per unit shortage
π_0 = lost sales cost per unit lost sale
Q = order size
σ_x = lead time demand standard deviation
σ_{LT} = lead time standard deviation
σ_{ELT} = effective lead time standard deviation
z_0 = safety stock factor, with $z \sim N(0, 1)$
$G(z_0) = \int_{z_0}^{\infty} (z - z_0) f(z) dz_0$ expected shortage per replenishment cycle

$f(z_0) = \int_{z_0}^{\infty} zf(z)d(z)$ [17]

s = reorder point

Assumptions:

1. A single item having a similar supplier in which lead time is independent and identically distributed [1].
2. Shortages are divided as combination of backorders and lost sales. $(1 - q)$ proportion of demand is lost sales and rest (q) proportion is backordered.
3. A fixed quantity of order Q will be placed at a fixed interval of time (T).
4. The rate of demand is constant and hence the period length is constant for given order size.
5. The cost functions are approximate.

The problem under consideration represents more practical situation since it is reasonable to assume that a proportion (q) of the demand during the stockout period is backordered, as some customers whose needs are not critical at that time can wait for the item to be satisfied. Thus, only a proportion $(1 - q)$ of total shortages is lost in inventory systems (results in cost saving). Total inventory cost can be written as sum of ordering, holding and shortage costs as shown in Eq. 1.

$$C(Q, z_0) = \frac{AD}{Q} + h\left(\frac{Q}{2} + z_0 \sigma_x\right) + \left[h(1 - q) + \frac{D}{Q}[\pi + \pi_0(1 - q)]\right]\sigma_x G(z_0) \quad (1)$$

Equation (1) shows the total inventory cost for the partial backorder inventory system.

4 Numerical Problem

Consider an inventory system with D = 1600, A = Rs. 4000, h = Rs. 10, π = Rs. 2000, π_0 = Rs. 5000, fraction of backorders, q and exponential lead time with mean 5. Determine the optimum total cost, order quantity and safety stock factor, which is operating at 80% backorders and remaining lost sales.

5 Proposed Machine Learning - Back Propagation ANN Approach to Solve the Intractable Problem of Inventory System

In this section, the proposed Back Propagation Feedforward ANN approach is used to solve the intractable inventory problem as explained in the below flow chart. Back propagation neural network is used to find the optimal inventory cost and order size. Initially the objective of inventory problem is identified. Then neural network has been trained using the dataset. For this purpose, Levenberg-Marquardt training algorithm is used. Finally, the trained network is tested and validated to obtain the coefficients which are further used to compute the effective lead times standard deviation. Lastly, the effective lead times standard deviation is used to calculate optimal cost of inventory, order size and safety stock factor. The above-mentioned methodology for determination of optimal parameters for the stochastic inventory problem is shown in Fig. 2.

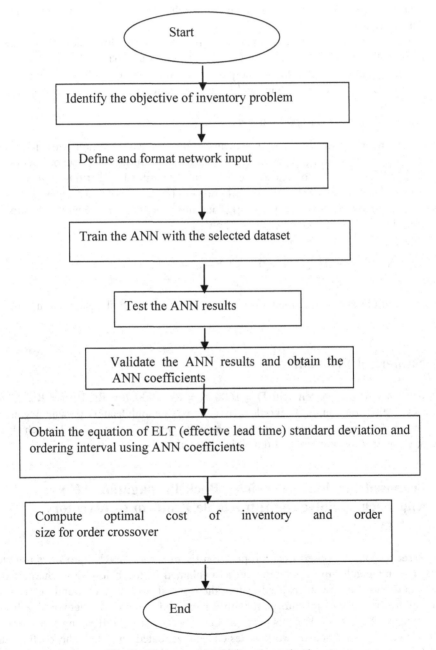

Fig. 2. Flow chart for finding the optimal cost

6   Results Obtained

The results obtained from (i) ignoring order crossover using classical optimization (ii) considering order crossover with Regression and (iii) examining order crossover in inventory system with ANN.

6.1   Results Obtained Using Classical Optimization

To find out the optimal value of order size, differentiating Eq. (1) w.r.t Q.

Equation (2) shows the optimal order quantity in absence of order crossover, when orders are presumed to reach in the identical sequence to which they were issued.

$$Q_{WOC}^* = \sqrt{\frac{2D(A + (\pi + \pi_0(1 - q))\sigma_x G(z_0))}{h}} \tag{2}$$

Or

Equation (2) can be rewritten as Eq. (3) in terms of economic order quantity (EOQ) in the absence of order crossover as below.

$$Q_{WOC}^* = EOQ\sqrt{\left(1 + \frac{(\pi + \pi_0(1 - q))\sigma_x G(z_0)}{A}\right)} \tag{3}$$

where $EOQ = \sqrt{\frac{2AD}{h}}$ is economic order quantity.

Similarly, differentiating Eq. (1) w.r.t z_0 results in the equation of tail probability.

$$P(z > z_0) = \frac{h}{(1 - q)h + \frac{D}{Q}[\pi + \pi_0(1 - q)]} \tag{4}$$

Solving the Eqs. (3) and (4) simultaneously, results in total inventory cost under without order crossover equals to Rs. 238208, order quantity equals to 5654 and safety stock factor equals to 2.7.

6.2   Results Obtained Using Regression

The effective lead time standard deviation, is estimated as below [3].

$$\sigma_{ELT} \approx a + bT = a + b\frac{Q}{D} \tag{5}$$

Table 1 shows the experimental data obtained during experiments conducted for studying order crossover.

Table 1. Experimental data of time between orders and standard deviation of ELT

Experiment no.	T	σ_{ELT}
1	0.0125	0.3843
2	0.0625	0.4197
3	0.1125	0.5318
...	...	...
39	1.9125	2.1004
40	1.9625	2.1257

The regression coefficients are obtained from Eq. (5) when time between orders (T) is considered as $0 > T > 2$. The regression statistics is shown in Table 1 shows $R^2 = 0.94$ which is a good fit of data.

The Table 2 shows the value of R square and regression coefficients.

Table 2. Regression statistics and coefficients

Regression statistics	
R Square	0.946792845
Adjusted R Square	0.945392657
Standard Error	0.124182098
Observations	40
Coefficients	
Intercept	0.585446932
T	0.884625891

Now using a = 0.5854 and b = 0.8846, total inventory cost under order crossover equals to Rs. 61,563.33 order quantity equals to 484 and safety stock factor equals to 3.09.

6.3 Results Obtained Using ANN

The artificial neural network is trained with the dataset of ordering periodicity and effective time lead time as shown in Table 1. This dataset is taken from the paper of Srivastav and Agrawal [5]. Hayya and Harrison [3] was first generated the dataset for examining order crossover in the inventory system. The dataset [5] for different values of time between orders and the corresponding effective lead times standard deviation has been generated with the following steps.

Step 1. Orders are placed at fixed interval and their stochastic lead times are randomly generated.

Step 2. Arrival time of the orders are computed with adding the lead time in ordering time.

Step 3. Afterwards, arrival time of orders are sorted on the basis of earliest arrival.

Step 4. Subsequently, effective lead time for each order is computed. Effective lead time of ith-order is the time lapses between the ith order and i[th] arrival regardless to the sequence to which it has been placed.

Step 5. Standard deviation of distribution effective lead times is computed.

Thus, for different values of time between orders, lead times are generated and their corresponding effective lead time are computed.

Feed Forward Back Propagation Neural Network has used Levenberg-Marquardt training algorithm for computing the results. The training stops when the validation error failed to decrease for six iterations. MSE (Mean Square error) and R values are used to detect performance of the network. The below regression plots show the network outputs relating to targets for the training, validation, and test sets. For a ideal fit, the data should be along a 45° line, where the network outputs are identical to the targets. For this particular problem, the fit is reasonably good enough for all data sets, with R values in each case of 0.99 or above (Fig. 3).

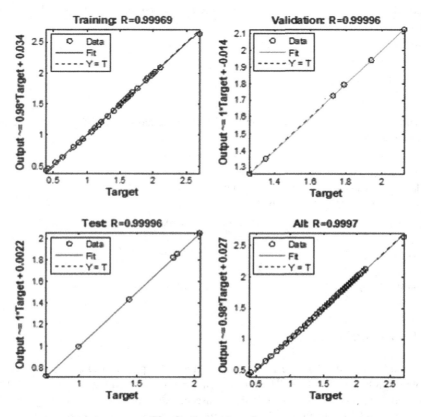

Fig. 3. Regression plots

Now total inventory cost using ANN with order crossover equals to Rs. 32,815.66 order quantity equals to 425 and safety stock factor equals to 3.13.

Table 3 shows the performance of ANN and Regression.

Table 3. Comparison of accuracy between Machine Learning (ANN) and Regression

Approach	MSE	R^2
ANN	0.0142	0.9999
Regression	0.0150	0.9467
% Change on using ANN w.r.t. regression	−5.33%	5.61%

Table 4 displays the comparison of Machine Learning (ANN) with Regression for order crossover.

Table 4. Comparison of Machine Learning (ANN) and Regression for order crossover case

Approach	Total cost (in Rs)	Order quantity	Safety stock factor
ANN	32815	425	3.13
Regression	61563	484	3.09
% Change on using ANN w.r.t. regression	−46.69%	−12.19%	1.29%

Table 5 shows the comparison of Machine Learning (ANN) with Classical Optimization.

Table 5. Comparison of Machine Learning (ANN) for order crossover with Classical Optimization ignoring order crossover

Approach	Total cost (in Rs)	Order quantity	Safety stock factor
ANN	32815	425	3.13
Classical optimization	238208	5654	2.7
% Change on using ANN w.r.t. classical optimization	−86.26%	−92.48%	15.95%

7   Conclusion

It is found that with Machine Learning, the stochastic partial backordering inventory problem can be easily solved. There has been a noteworthy reduction in total inventory cost considering order crossover with the use of ANN is observed in comparison to Regression and Classical Optimization approaches. A significant amount of savings in optimal inventory cost (86%) and increase in the safety factor is achieved with ANN with respect to Classical Optimization.

Furthermore, the machine learning is also found more appropriate in comparison to Regression to solve such stochastic inventory problems. There has been a noteworthy reduction in total cost (46%) and order quantity (12%) with ANN with respect to Regression. It can be said that most of the real-world stochastic problems can be easily modelled and solved using Machine Learning in comparison to the existing Regression and Classical Optimization approaches.

References

1. Riezebos, J.: Inventory order crossovers. Int. J. Prod. Econ. **104**, 666–675 (2006)
2. Hayya, J.C., Bagchi, U., Kim, J.G., Sun, D., et al.: On static stochastic order crossover. Int. J. Prod. Econ. **114**(1), 404–413 (2008)
3. Hayya, J.C., Harrison, T.P.: A mirror image inventory model. Int. J. Prod. Res. **48**, 4483–4499 (2010)
4. Wensing, T., Kuhn, H.: Analysis of production and inventory systems when orders may cross over. Ann. Oper. Res. **231**(1), 265–281 (2015)
5. Srivastav, A., Agrawal, S.: On single item time weighted mixture inventory models with independent stochastic lead times. Int. J. Serv. Oper. Manag. **22**, 101–121 (2015)
6. Srivastav, A., Agrawal, S.: On a single item single stage mixture inventory models with independent stochastic lead times. Oper. Res. Int. J. (2018). https://doi.org/10.1007/s12351-018-0408-z
7. Srivastav, A., Agrawal, S.: Multi-objective optimization of mixture inventory system experiencing order crossover. Ann. Oper. Res. (2018). https://doi.org/10.1007/s10479-017-2744-4
8. Haykin, S.: Neural Networks: A Comprehensive Foundation. Macmillan Publishing, New York (1994)
9. Zhang, G.P.: Time series forecasting using a hybrid ARIMA and neural network model. Neurocomputing **50**, 159–175 (2003)
10. Zhang, G.P., Qi, M.: Neural network forecasting for seasonal and trend time series. Eur. J. Oper. Res. **160**(2), 501–514 (2005)
11. Adhikari, R., Agrawal, R.K.: A combination of artificial neural network and random walk models for financial time series forecasting. Neural Comput. Appl. **24**(6), 1441–1449 (2014)
12. Tealab, A., Hefny, H., Badr, A.: Forecasting of nonlinear time series using ANN. Futur. Comput. Inform. J. **2**(1), 39–45 (2017)
13. Šustrová, T.: A suitable artificial intelligence model for inventory level optimization. Trends Econ. Manag. **25**(1), 48–55 (2016)
14. Šustrová, T.: An artificial neural network model for a wholesale company's order-cycle management. Int. J. Eng. Bus. Manag. (2016). https://doi.org/10.5772/63727
15. Agarwal, B., Ramampiaro, H., Langseth, H., Ruocco, M.: A deep network model for paraphrase detection in short text messages. Inf. Process. Manag. **54**(6), 922–937 (2018)
16. Jain, G., Sharma, M., Agarwal, B.: Spam detection in social media using convolutional and long short term memory neural network. Ann. Math. Artif. Intell. **85**, 21–44 (2019). https://doi.org/10.1007/s10472-018-9612-z
17. Silver, E.A., Pyke, D.F., Peterson, R.: Inventory Management and Production Planning and Scheduling. Wiley, New York (1998)

An Approach to Suggest Code Smell Order for Refactoring

Thirupathi Guggulothu[(✉)] and Salman Abdul Moiz[(✉)]

University of Hyderabad, Hyderabad, Telangana, India
thirupathi.gugguloth@gmail.com, salman@uohyd.ac.in

Abstract. Code smell is an indicator of issues in source code qualities that may hinder maintenance, and evolution. Source code metrics are used to measure the quality of the code. In the literature, there are many code smells, refactoring techniques, and refactoring tools. However, a software project often contains thousands of code smells and many of them have no relation with design quality. It is a challenge for developers to decide which kind of code smell should be refactored first. We have proposed an approach that suggests a code smell order based on two aspects: (1) finding relevant metrics for each code smell dataset with the help of feature selection technique (2) analyzing the internal relation among the code smells with those relevant metrics. With this analysis, we are suggesting code smell order for developers to save their effort in the refactoring stage. The suggested order is evaluated on simple java source code.

Keywords: Code smell · Refactoring · Maintenance · Design quality · Code smell order · Feature selection technique

1 Introduction

Code smells are inherent property of software that results in code or design problems i.e., if the software system contains code smell then it indicates that there are issues with code quality such as, understanding, maintenance, and evolution. Martin et al. [1] defined, 22 informal code smells. One way to remove them is refactoring techniques. Refactoring is a technique that makes better internal structure of the code without altering the behavior of the software shown by Opdyke et al. [2]. It is one of the most widely accepted phenomena, when it comes to the software quality enhancement.

In the literature, Rasool et al. [3] has given a clear evidence that they are many tools and techniques used to detect many different code smells in a software project. As the other big challenge for the developers is that which code smell has to be refactored first. Change in the ordering can lead to: (1) Loss of actual code leads to new errors/bugs which give raise to new code smells, (2) With different efforts require for different ordering and leads to different quality improvement, (3) Among all the code smells some of them are not so important for design

© Springer Nature Singapore Pte Ltd. 2019
A. K. Somani et al. (Eds.): ICETCE 2019, CCIS 985, pp. 250–260, 2019.
https://doi.org/10.1007/978-981-13-8300-7_21

quality of the system, (4) If all possibilities are tried to resolve the code smells it will lead to budget problem for developers to find order in refactoring. Hence, we can resolve this problem by prioritizing the code smells.

There are various techniques [4–6] proposed by several researcher to prioritize the code smells. Among them, Liu et al. [4] has saved the efforts of developers by scheduling the code smells detected and resolution to them. The reduced efforts for the developer are shown through two projects. Ouni et al. [5], has proposed automated refactoring approach to correct the code smells during correction process by giving highest priority to the riskiest code smells first. Vidal et al. [6] has developed a tool called SPIRIT to rank the code smells based on some three important criteria's such as type of code smell, based on past modification to the component and modification possibility for the system.

However, Liu et al. [4], schedule the code smell order to save the effort of developers. But, the authors schedule the code smell in abstract way. In this paper, we are shown the code smell order based on the impact of relevant metrics. For example, when X smell is having more impacted compared to other code smell relevant metrics then the considered X smell gets highest priority. Because, of this improvements, the developers efforts may reduced while refactoring other code smells. In this scenario, efforts means reducing some possible refactoring techniques or may remove other code smells also.

In this paper, we order the 4 code smell based on two factors: (1) finding relevant metrics for each code smell dataset with the help of feature selection technique. For 4 code smells, 4 datasets are considered. The dataset which we are considering is taken from the Fontana [7]. Fontana, proposed a machine learning (ML) technique that will able to detect four code smells. The authors experimented on 74 Java systems which are manually validated instances of trained dataset and used 16 different classification algorithms, and there combination with a boosting techniques. From these datasets, we are getting relevant metrics for each code smell with the help of machine learning technique called feature selection. (2) After that, we are analyzing the internal relation among the code smells with the help of those relevant metrics. With this analysis, we suggest code smell order for developers to save their efforts in refactoring phase. The suggested order is evaluated on simple java source code.

The paper is arranged as follows; The second section introduces a work related to order the code smells; The third section defines the four code smell to evaluate the proposed approach and the characteristics of dataset that considered in this work; The fourth section, approach to order the code smells; The fifth section presents experimental results of the approach which are considered; and the final section gives conclusion and future directions.

2   Related Work

One of the factor that makes the software to evolve and maintain becomes harder is due to code smells. But those code smells are inherent property of software which results in design or code problems. Martin et al. [1] has defined 22 informal code smells. Code smell is a type of code that violates the Booch et al. [8] basic design principles (such as abstraction, hierarchy, encapsulation, modularity and modifiability) and will have bad impact on the quality of design. In order to improve the quality of the design a technique called refactoring is addressed by Martin et al. [1] and Opdyke et al. [2].

Code smells can be detected and refactored by means of several automated tools. Even then large software applications often consist of thousands of code smells, refactoring them takes more time and effort for the developer. Moreover, not all code smell are important to the design problem of the system [9].

Liu et al. [4], has solved the efforts of the developer by scheduling which bad code smell detected can be refactored first. The author has reduced the efforts of the developer by showing the empirical results of two open source projects which has been reduced from a range of 17.64 to 20%.

Vidal et al. [6], proposed a tool called SPIRIT which is used to rank the code smells based on three criteria's. They are (1) To the extent the software can be modified, (2) Modifications done to the past entities, and (3) Selecting a code smell that is relevant in some way in its kind.

Pietrzak et al. [10], detected new code smells by analyzing the relation among 22 code smells of fowler. There are six different relation for the code smells which were proposed by the author. They are mutual support, rejection, inclusion, plain support, transitive support and aggregate support.

When we observed, the major difference of the previous work with respect to proposed approach is: We are ordering the code smell in a new way i.e., Smell which is first refactor when the impact of this smell is more on other code smells. Here, impact or improvement is on relevant metrics of each code smell.

3   Evaluated Code Smells

In this paper, out of 22 code smells defined by [1], four code smells were used in the datasets. These code smells cover the potential problems related to object oriented quality dimensions viz., coupling, cohesion, complexity, size, encapsulation and data abstraction. In Table 1, lists the four code smells, with their affected entities and quality dimensions. A brief introduction is presented here so that it is needful while analyzing the internal relation among the code smells.

3.1   Long Method

A code smell is said to be long method when the method has too many parameters, having high functional complexity, more number of code lines, and difficult to understand. Extract Method is one kind of refactoring techniques to solve this code smell.

3.2   Feature Envy

Feature Envy is defined as a method which uses other class data more when compared to it's own class data. It accesses more foreign data than local one. Extract Method is the one kind of refactoring techniques to solve this code smell.

Table 1. Selected code smells

Name of the code smell	Affected entities	Intra/Inter	Impacted on OO quality dimensions
Long method	Method	Intra	Coupling, cohesion, complex, Size
Feature envy	Method	Inter	Coupling
God class	Class	Intra	Coupling
Data class	Class	Intra	Data abstraction, encapsulation

3.3 God Class

Classes that have more responsibilities are said to be god class or large class. The increase in responsibility is due to many methods in the class, having much data, increased size and complexity. Hence, it's hard to understand and maintain. Extract Class is one kind of refactoring technique to solve this code smell.

3.4 Data Class

Data class consists of fields, get/set methods. The data of this class is used by other classes. Hence, its responsibility is to handle the data used by other classes or outsiders. Move Method is the one kind of refactoring techniques to solve this code smell.

4 Approach to Suggesting Code Smell Order for Refactoring

Flow chart of our approach can be shown as Fig. 1. Following sections will give a detailed explanation of our approach.

Fig. 1. An approach to suggest code smell ordering for refactoring

4.1 Characteristics of Datasets

In this study, four datasets of four code smells considered from Fontana et al. [7] are listed in the Table 2. The reason for choosing these datasets is because they are manually evaluated instances. Table shows that, out of total (420) evaluated instances, 1/3 (140) of them are positively detected and 2/3 (280) are negatively detected code smells. To balance the dataset authors [7], has considered these fractions. The ML algorithms tend to perform badly when the dataset is imbalanced. The datasets are taken on the web (http://essere.disco. unimib.it/reverse/MLCSD.html).

Table 2. Summary of datasets

DataSets	Evaluated instances	Number of detected instances
God class	420	140
Feature envy	420	140
Long method	420	140
Data class	420	140

4.2   Feature Selection Technique

Feature selection is the process of selecting relevant features or feature subsets according to the target variable. In this study, features are source code metrics and target variables are code smells. Total 61 and 82 metrics are used for class and method level code dataset respectively. Not all the metrics actually impact on the prediction of the target variable (code smell). For this reason, we first reduce the number of metrics by means of feature selection, i.e., we consider only those metrics which impact the most predictive code smells. To this aim, we have employed the widely-adopted Gain Ratio Feature Evaluation algorithm [11], identifying the features having more weight in the code smell detection. Decision tree algorithm given best performance on the evaluated code smells in the paper [7] and this algorithm uses entropy and gain ratio to construct the model. The output of the algorithm is a ranked list with metrics having higher GainRatio. The evaluated code smells gain ratio information is available in the (https:// drive.google.com/open?id=1v7Lj33MSI9BXv_Ppoikrtuz1TSWkmqM8).

4.3   Software Metrics Related to Code Smells

After applying feature selections techniques on the datasets the results are shown in the Table 3. From the table, we can observe that, the metrics which are obtained from the feature selection is more similar to the characteristics of the code smells that are defined in Sect. 3. The names of this metrics are shown in the appendix Fig. 3.

4.4   Analyze Internal Relationship Among the Code Smells

This is core section of our approach. In this section, we are analyzing the internal relationships among code smells with the help of relevant metrics shown in the Table 3.

Table 3. Metrics related to code smells

God class	Relevant metrics Related to code smell
God class	WMC/WMCNAMM, LOC/LOCNAMM, RFC, ATFD-class, CBO
Data class	NOAM, WMC/WMCNAMM, WOC, LOC/LOCNAMM
Long method	LOC-method, CYCLO, NOLV, MAXNESTING
Feature envy	ATFD-method, FDP, LOC-method LAA

Data Class Versus Other Code Smells. The data class code smell used by other code smells are defined in the Sect. 3.

DataClass → God Class: God class and data class are contradict one another i.e., they can't occur together by their given definitions. Table 3 shows that, WMC/WMCNAMM and LOC/LOCNAMM are relevant to both of the code smells. But, these metrics differ in their thresholds. For example, to detect the god class the threshold value is LOC $>=$ 100, but for data class it's LOC $<=$ 10. 25.62% of god class uses the features (ATFD) of data class confirmed by [12]. With this findings, refactor (using move method) data class first than god class, we can improves the metrics of god class that is shown in Table 4. Table reports that, ATFD metric improve by reducing the coupling between the god class and data class with the help of move method refactoring technique. CBO, RFC are more correlated to the ATFD metric. We found this correlation using spearman method with the values 0.68 and 0.60 respectively. WMC/WMCNAMM and LOC/LOCNAMM also improved because of the features which are accessed by god class are moved to the data class. Figure 2, shows that, the customer summary view class has affected god class, then lines 11–12 has features of data class code smell which are moved to the corresponding classes. With this move, the metrics ATFD, LOC/LOCNAMM, CBO, RFC are improved.

Instead of data class, if we refactor god class first, then line 11–12 are extracted as a new class. But, extracted class still has a coupling with data class i.e., LOC/LOCNAMM, WMC/WMCNAMM are improved but ATFD, CBO, RFC metrics increase in their values. With this analysis, we should refactor data class first than god class.

Data Class → Feature Envy: As discussed above, feature method class also uses the features (data) of data class using the ATFD metric shown in Table 3. With this, if we refactor data class first than feature envy, we can improves the ATFD, FDP and LOC that is shown in Table 4. FDP, LOC are much correlated to the ATFD metric with the values 0.99 and 0.49 respectively. Figure 2, shows that, the customer summary view class, lines 11-12 has features of data class code smell which are moved to the corresponding classes. With this move, the metric ATFD is improved and also improves the correlated metrics (FDP,LOC).

Data Class → Long Method: Long method class also uses the features (data) of data class using the ATFD metric. With this, if we refactor data class first than long method, we can improves the LOC that is shown in Table 4. LOC is correlated to the ATFD, NOLV metric with the values 0.495 and 0.64 respectively. Figure 2, shows that, the customer summary view class, lines 11–12 has features of data class code smell which are moved to the corresponding classes. With this move, the metric ATFD improved and also improves the correlated metrics (LOC, NOLV).

Feature Envy Versus Other Code Smells. The feature envy smell says that a method is more interested in other class than where it is actually placed. God class and Long method uses the features of envy method class. Refactoring the feature envy before the god class and long method will results in decomposition of classes. Figure 2 shows that, the customersummaryview class has a method getSummary which is affected by feature envy and long method. If we refactor feature envy, the lines 11, 12 are moved to corresponding classes. As a results, the metrics which improves with this refactoring is shown in Table 4.

Instead of feature envy, (1) if we refactor long method, some of the metrics WMC/WMCNAMM and LOC/LOCNAMM will increases in their sizes (2) If we refactor god class, coupling related metrics increases in their values i.e., affected with high coupling. These can be shown in Table 5.

Long Method Versus Other God Class. It seems very natural that long method and god class are correlate with one other because of the containment relation (method contain in a class). By this, we mean that if some class has one or more methods which are long methods then same class would turn into a god class smell. Refactoring long method first then, metrics which are improved shown in the Table 4. CYCLO, LOC, and MAXNESTING of long methods improves the metrics of WMC/WMCNAMM. That is, if we make long methods into smaller ones, it may remove the god class automatically.

Instead of long method, if we refactor god class first complexity metrics (WMC/WMCNAMM) improves but, it increases the coupling metrics.

Table 4. Suggested code smell order

Code smell order	Improved metrics			
	Data class	God class	Feature envy	Long method
Data class Vs other code smells	WMC/WMCNAMM, LOC/LOCNAMM, WOC	WMC/WMCNAMM, LOC/LOCNAMM, ATFD, CBO, RFC	ATFD, FDP, LOC	LOC, NOLV

Table 5. Other code smell order

Code smell order	Improved metrics			
	Data class	God class	Feature envy	Long method
God class Vs other Code smells	No impact	WMC/WMCNAMM, LOC/LOCNAMM, Negative improvement on (ATFD, CBO, RFC)	No impact	LOC, MAXNESTING

```
1 //Customer.java
2 public class Customer{
3    private String firstName;
4    private String lastName;
5    private String title;
6    private Address address;
7
8 public String getFirstName(){
9    return firstName;
10 }
11 public String getLastName(){
12    return lastName;
13 }
14 public String getTitle(){
15    return title;
16 }
17 public Address getAddress(){
18    return address;
19 }
20 public void setFirstName(String firstName){
21    this.firstName=firstName;
22 }
23 public void setLastName(String lastName){
24    this.lastName=lastName;
25 }
26 public void setTitle(String title){
27    this.title=title;
28 }
29 public void setAddress(Address address){
30    this.address=address;
31 }
32 }
```

```
1 //Address.java
2 public class Address{
3    private String city;
4    private String postcode;
5    private String country;
6    public String getCity(){
7    return city;
8    }
9    public String getPostcode(){
10    return postcode;
11    }
12    public String getCountry(){
13    return country;
14    }
15    public void setCity(String city){
16    this.city=city;
17    }
18    public void setPostcode(String postcode){
19    this.postcode=postcode;
20    }
21    public void setCountry(String country){
22    this.country=country;
23    }
24 }
```

```
1 //God Class
2 public class CustomerSummaryView{
3         private Customer customer;
4         public CustomerSummaryView
(Customer customer){
5                 this.customer=customer;
6         }
7         //Long Method
8         public String getCustomerSummary
(){
9         Address
address=customer.getAddress();
10         //Featur Envy
11         return customer.getTitle() + " "
+ customer.getFirstName() + " " +
customer.getLastName()+
12                 "," + address.getCity()+
"," + address.getPostcode() + "," +
address.getCountry();
13         }
14 }
```

Fig. 2. Example of simple java source code

Table 6. Results of applied refactoring on suggested order

Suggested Refactoring order	Affected Class	Size					Complexity					Coupling				
		Metric	Before Refactor		After Refactor		Metric	Before Refactor		After Refactor		Metric	Before Refactor		After Refactor	
			Class Level	Method Level	Class Level	Method Level		Class Level	Method Level	Class Level	Method Level		Class Level	Method Level	Class Level	Method Level
Dataclass-> Featureenvy-> Longmethod-> Godclass	Customer Summary View	LOC	10	4	9*	3*	NOAV	-	9		1*	ATFD	7	7	0*	0*
		LOC NAMM	10	-	9*	-	NOLV	-	1		0*	FANOUT	2	2	1*	1*
												CBO	2	-	1*	-
												FDP	-	2	-	0*
												RFC	9	-	3*	-
												CINT	-	7	-	0*

Table 7. Continuation of Table 6

Suggested Refactoring order	Affected Class	Size					Data Abstraction				
		Metric	Before Refactor		After Refacotr		Metric	Before Refactor		After Refactor	
			Class Level	Method Level	Class Level	Method Level		Class Level	Method Level	Class Level	Method Level
Dataclass-> Featureenvy-> Longmethod-> Godclass	Customer Summary View, Address, Customer	TCC	0.14 (Customer) 0.20 (Address)	-	0.16* 0.24*	-	LAA	-	0.125	-	1*

4.5 Code Smell Order

Below, the suggested order is evaluated on simple java project.

Data Class → Feature Envy → Long Method → God Class: Figure 2 shows that, the customer summary view class, lines 11–12 has data class code smell. So, if we refactor line number 11 first using move method, then that line is moved to the customer class. After that line 12 is moved to address class using same refactoring method. From these refactorings, the method getSummary and customersummaryview class dispels the other existing code smell. With this suggested order, number of refactoring needed to remove all the smells is 2.

Let us, consider the other order, **God Class → Long Method → Feature Envy → Data Class:** Figure 2 shows that, Because of line numbers 7–13, the customer summary view class has affected with god class. So, if we refactor line numbers 7–13 are extracted has a new class using extract class refactoring technique. After that extracted new class is solved with extract method to remove long method. Then after, we solve feature envy method with 2 refactoring techniques. With this order, number of refactoring needed to remove all the smells is 4.

5 Experimental Results

In this section, we have evaluated proposed approach with the help of simple java source code shown in Fig. 2. The suggested refactoring order is applied on the source code and shown results in the Table 6. In this paper, we theoretically proved that other orders may not be recommended. In this section, we are practically showing the results of suggested order in Table 6. In this table, * represents metrics improvement, - for not applicable. From the table, we can observe that with our approach, God class metrics (LOC/LOCNAMM, ATFD, CBO, RFC), long method metric (NOLV, LOC) and feature envy metrics (FDP, ATFD) are improved. The suggested order is more impacted on design quality dimensions than other orders i.e., The improved metrics (highlighted) in the table which cover all aspects of quality dimensions are listed in appendix Fig. 3. These results suggest that, the developer can reduce their efforts by using the proposed approach (Tables 6 and 7).

6 Conclusion and Future Work

From the analysis, through this paper, we tried to order the code smells for refactoring based on two factors: (1) finding relevant metrics for each code smell dataset with the help of feature selection technique (2) analyzing the internal relation among the code smells with those relevant metrics. The results suggest the developers to save their effort by using our approach.

We have chosen 4 code smells and a simple java project to evaluate our approach. The proposed approach is not only limited to 4 code smells in future, we want to investigate other code smells which are there in the literature.

Appendix

Quality dimension	Metric Name	Metric Label	Level or Granularity
Coupling	FANOUT	-	Class, Method
	FANIN	-	Class
	ATFD	Access to foreign data	Method
	FDP	Foreign Data Providers	Method
	RFC	Response For Class	Class
	CBO	Coupling Between Objects Class	Class
	CFNAMM	Called Foreign Not Accessor or Mutator Methods	Class, Method
	CINT	Coupling Intensity	Method
	MaMCL	Maximum Message Chain Length	Method
	MeMCL	Mean Message Chain Length	Method
	NMCS	Number of Message Chain Statements	Method
	CC	Changing Classes	Method
	CM	Changing Methods	Method
Size	LOC	Lines of Code	Project, Package, Class, Method
	LOCNAMM	Lines of Code Without Accessor or Mutator Methods	Class
	NOPK	Number of Packages	Project
	NOCS	Number of Classes	Project, Package
	NOM	Number of Methods	Project, Package, Class
	NOMNAMM	Number of Not Accessor or Mutator Methods	Project, Package, Class
	NOA	Number of Attributes	Class
Inheritance	DIT	Depth of Inheritance Tree	Class
	NOI	Number of Interfaces	Project, Package
	NOC	Number of Children	Class
	NMO	Number of Methods Overridden	Class
	NIM	Number of Inherited Methods	Class
	NOII	Number of Implemented Interfaces	Class
Encapsulation	NOAM	Number of Accessor Methods	Class
	NOPA	Number of Public Attribute	Class
	LAA	Locality of Attribute Accesses	Method
Complexity	CYCLO	Cyclomatic Complexity	Method
	WMC	Weighted Methods Count	Class
	WMCNAMM	Weighted Methods Count of Not Accessor or Mutator Methods	Class
	AMW	Average Methods Weight	Class
	AMWNAMM	Average Methods Weight of Not Accessor or Mutator Methods	Class
	MAXNESTING	Maximum Nesting Level	Method
	CLNAMM	Called Local Not Accessor or Mutator Methods	Method
	NOP	Number of Parameters	Method
	NOAV	Number of Accessed Variables	Method
	ATLD	Access to Local Data	Method
	NOLV	Number of Local Variable	Method

Fig. 3. Code metric abbrevations

References

1. Martin, F., Beck, K., Brant, J., Opdyke, W., Roberts, D.: Refactoring: improving the design of existing programs (1999)
2. Opdyke, W.F.: Refactoring: a program restructuring aid in designing object-oriented application frameworks. Ph.D. dissertation, Ph.D. thesis, University of Illinois at Urbana-Champaign (1992)
3. Rasool, G., Arshad, Z.: A review of code smell mining techniques. J. Softw.: Evol. Process **27**(11), 867–895 (2015)

4. Liu, H., Ma, Z., Shao, W., Niu, Z.: Schedule of bad smell detection and resolution: a new way to save effort. IEEE Trans. Softw. Eng. **38**(1), 220–235 (2012)
5. Ouni, A., Kessentini, M., Bechikh, S., Sahraoui, H.: Prioritizing code-smells correction tasks using chemical reaction optimization. Softw. Qual. J. **23**(2), 323–361 (2015)
6. Vidal, S.A., Marcos, C., Díaz-Pace, J.A.: An approach to prioritize code smells for refactoring. Autom. Softw. Eng. **23**(3), 501–532 (2016)
7. Fontana, F.A., Mäntylä, M.V., Zanoni, M., Marino, A.: Comparing and experimenting machine learning techniques for code smell detection. Empirical Softw. Eng. **21**(3), 1143–1191 (2016)
8. Booch, G.: Object-oriented Analysis and Design. Addison-Wesley, Boston (1980)
9. Demeyer, S., Ducasse, S., Nierstrasz, O.: Object-Oriented Reengineering Patterns. Elsevier (2002)
10. Pietrzak, B., Walter, B.: Leveraging code smell detection with inter-smell relations. In: Abrahamsson, P., Marchesi, M., Succi, G. (eds.) XP 2006. LNCS, vol. 4044, pp. 75–84. Springer, Heidelberg (2006). https://doi.org/10.1007/11774129_8
11. Quinlan, J.: Induction of decision trees. Mach. Learn. **1**(1), 81–106 (1986)
12. Fontana, F.A., Ferme, V., Zanoni, M.: Towards assessing software architecture quality by exploiting code smell relations. In: Proceedings of the Second International Workshop on Software Architecture and Metrics, pp. 1–7. IEEE Press (2015)

Optimal Feature Selection of Technical Indicator and Stock Prediction Using Machine Learning Technique

Nagaraj Naik[✉] and Biju R. Mohan

Department of Information Technology, National Institute of Technology, Karnataka, Surathkal, India
{it16fv04.nagaraj,biju}@nitk.ac.in
http://www.nitk.c.in

Abstract. Short-term trading is a difficult task due to fluctuating demand and supply in the stock market. These demands and supply are reflected in stock prices. The stock prices may be predicted using technical indicators. Most of the existing literature considered the limited technical indicators to measure short-term prices. We have considered 33 different combinations of technical indicators to predict the stock prices. The paper has two objectives, first is the technical indicator feature selection and identification of the relevant technical indicators by using Boruta feature selection technique. The second objective is an accurate prediction model for stocks. To predict stock prices we have proposed ANN (Artificial Neural Network) Regression prediction model and model performance is evaluated using metrics is Mean absolute error (MAE) and Root mean square error (RMSE). The experimental results are better than the existing method by decreasing the error rate in the prediction to 12%. We have used the National Stock Exchange, India (NSE) data for the experiment.

Keywords: Boruta feature selection · ANN · Stock prediction

1 Introduction

Predicting stock price for short-term trade is a challenging task due to the volatile nature of the stock market. This leads an attraction for researchers and academicians to enhance the prediction model. Interpreting stock critical information earlier gives the profitable trading [1]. Efficient Market Hypothesis stated that all stock related information impact on stock prices rather than technical trading [7].

There are two ways of stock analysis can be performed to gain huge profits [10]. First, the fundamental analysis and next is technical analysis. Fundamental analysis is about of evaluating an individual company share price by looking at company past performances. A fundamental analyst studies anything that affects the company share prices, including company earnings, the condition of

© Springer Nature Singapore Pte Ltd. 2019
A. K. Somani et al. (Eds.): ICETCE 2019, CCIS 985, pp. 261–268, 2019.
https://doi.org/10.1007/978-981-13-8300-7_22

Table 1. Technical indicators and its formulas [5]

Technical indicator name	Calculation	Number of days(n)
Simple Moving Average (SMA)	$(C_t+C_{t-1}+....C_{t-n+1})/n$	5,10,14,30,50,100,200
Exponential Moving Average (EMA)	$(C_t\text{-}SMA(n)_{t-1})$ * (2/n+1) $+SMA(n)_{t-1}$	5,10,14,30,50,100,200
Momentum Indicator (MOM)	$C_t\text{-}C_n$-9	5,10,14
Stochastic oscillator (STCK)	100*$((C_t$ - $L_t(n))/(H_t(n)$ - $L_t(n))$	14
Stochastic oscillator (STCD)	$(100*((C_t$ - $L_t(n))/(H_t(n)$ - $L_t(n)))/3$	14
Moving Average Convergence/Divergence (MACD)	$SMA(n) - SMA(n)$	26,13,19,45,25,15
Relative Strength Index (RSI)	100 - (100/(1 + Avg(Gain)/Avg(Loss)))	14
Williams R (R)	$((H_t\text{-}C_t)/(H_n\text{-}L_n))$x 100	14,28,50,100
Accumulation/distribution index(A/D)	$H_t\text{-}C_{t-1}/H_t\text{-}L_t$ $((C_t/L)/(H/C))/(H/L)$	14
Commodity Channel Index (CCI)	$(H+L+C/3)$ - SMA)/ (0.015 * Mean deviation)	14,50,100

company macroeconomic factors such as the overall economy, industry conditions and company management. If a particular stock is undervalued and company fundamentals are strong, then it indicates buy signal to investors. If the stock is overvalued and stock fundamentals are weak, then it indicates sell signals to investors [4].

Technical analysis is a method of measuring stock. It involves many statistical analysis of market data, such as stock open price, close price, day high price, day low price and volume of the stock. Technical analysis is different from fundamental analysis. Technical analysis does not attempt to measure a stock by looking at company financial data. Technical analysis is an attempt of finding traders or investors sentiments. By looking at the technical chart, the technical analyst may predict the trend of market sentiments [5, 10]. To realize trading signals detection, several methods have been developed, among which artificial intelligence methods have drawn more and more attention by both investors and researchers [3].

The benefits involved in predicting trading signals to gain more returns have motivated researchers to design newer and more advanced tools and prediction techniques. The paper has two objectives, first is the technical indicator feature selection and identification of the relevant technical indicators by using Brouta feature selection techniques. The second objective is an accurate prediction model for stocks. The proposed ANN Regression prediction model outperform compared to existing work.

2 Related Work

Existing trading rules were not gainful for future periods when market condition changes dynamically. Chourmouziadis and Chatzoglou [4] proposed short-term technical trading strategy by considering the daily price of the stock using fuzzy systems. An automatic way of buying and selling financial securities without the help of portfolio managers has been discussed. The combination of technical trading indicators like moving average, alpha, beta and volatility of the stock over a period of time has been proposed [2]. Nakano et al. [9] proposed a method in which non-linear financial time-series data are considered and machine learning techniques were used for predicting stock prices. Mousavi et al. [8] proposed generalized Exponential Moving Average technical indicator model to predict the stock prices. The future performance of stock indices has been studied using fuzzy time series modeling [12]. Return and risk are important objectives for managing a portfolio. Macedo et al. [6] proposed a model to enhance technical trading rule indicator based on Moving average convergence/divergence, Relative Strength Index, Bollinger Bands and Contrarian Bollinger Bands. Artificial Neural Network (ANN) has been widely used in predicting stock for financial markets. Zhang and Wu [17] proposed optimization technique with back propagation ANN to predict stock prices and indices.

Technical indicators like moving average, moving average convergence and divergence, relative strength index and commodity channel index have been used to predict the stock price [15]. Performing feature extraction could help to reduce the redundant features, which can reduce the measurements, storage requirements and the running time of classifiers. It also avoids the curse of dimensionality and improves prediction performance as well as facilitate data visualization and understanding [14]. Ticknor [13] proposed artificial neural network approach to improve the prediction performance. Preis et al. [11] hypothesized that investors may use a Google hits ratio of pages are used to take the decision to predict stock price. Macroeconomic factors are believed to influence stock market movements. Machine learning methods, which are data-driven and assumption free have become more popular in stock market prediction [16].

3   Research Data

In this paper, stock data are collected from http://www.nseindia.com. The data contains information about stock such as stock day open price, day low price, day high price and day close price. We have considered ICICI Bank and State bank of India stocks for the experiment. The dataset range is obtained from the year 2008 to 2018.

In Table 1 the abbreviations used are as follows:
n $\rightarrow$ Total number of days.
H_t $\rightarrow$ day high stock price at period t.
L_t $\rightarrow$ day low stock price at period t.
C_t $\rightarrow$ day close stock price at period t.

4   Proposed Work

Overall Framework of the proposed model is described in Fig. 1. The data are retrieved from NSE. We have considered 33 different combinations of technical indicators and evaluated based on formulas which are described in Table 1. We have used Boruta feature selection method to select the optimal feature of technical indicators. The step by step proposed Boruta feature selection algorithm is stated below.

Step 1: Create duplicate copies of technical indicators feature.

Step 2: Do the random Shuffle original technical indicators and duplicate copies of technical indicator to remove their correlations with the outcome variable.

Step 3: Apply random forest algorithm to find important technical indicator feature based on higher mean values.

Step 4: Calculate Z score by using Mean/Std deviation.

Step 5: Find the maximum Z score on duplicates technical indicator feature.

Step 6: Remove technical indicator feature if Z is less than Technical indicator feature.

Step 7: Repeat the above steps till iteration completes.

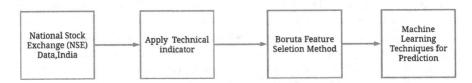

Fig. 1. Overall framework of proposed prediction model

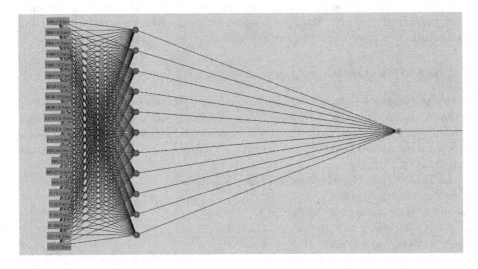

Fig. 2. The proposed ANN regression prediction model

Table 2. Stock ICICI Bank after feature selection.

Technical indicator feature	meanImp	medianImp	minImp	maxImp	normHits	Decision
SMA 5 days	5.905820	5.805600	3.5773017	8.330396	1.00000000	Confirmed
SMA 10 days	4.412296	4.450179	2.5664467	6.381897	0.94949495	Confirmed
SMA 14 days	3.904888	3.865653	1.7642646	5.652071	0.86868687	Confirmed
WMA 5 days	5.108720	5.116738	1.9129550	7.403865	0.94949495	Confirmed
WMA 10 days	4.715484	4.789195	2.7125061	6.435136	0.95959596	Confirmed
WMA 14 days	4.357254	4.259337	1.9938488	6.654934	0.91919192	Confirmed
MOM 5 days	18.692034	18.659231	16.9670413	20.344669	1.00000000	Confirmed
MOM 10 days	10.406941	10.374657	9.1582195	11.926020	1.00000000	Confirmed
MOM 14 days	9.074784	9.058466	7.0379555	10.879017	1.00000000	Confirmed
STCK1 9 days	21.269620	21.339074	18.4123667	23.683564	1.00000000	Confirmed
STCK2 9 days	25.729514	25.897507	22.7613795	28.723707	1.00000000	Confirmed
RSI 14 days	15.487287	15.438833	13.8918381	16.732855	1.00000000	Confirmed
RSI 28 days	8.608997	8.602433	7.0550301	10.168651	1.00000000	Confirmed
MACD1 (26,13,9) days	16.152372	16.184757	14.2102377	18.254305	1.00000000	Confirmed
MACD SIGNAL	9.164158	9.182269	7.2546778	10.961601	1.00000000	Confirmed
MACD (15,25,15)	6.046035	5.977311	3.7503675	8.219024	1.00000000	Confirmed
R 14 days	41.157543	41.103670	38.5168811	44.780878	1.00000000	Confirmed
R 28 days	24.057435	24.017315	22.3425494	26.501515	1.00000000	Confirmed
R 50 days	19.071730	19.049960	17.2172337	21.637499	1.00000000	Confirmed
R 100 days	12.560288	12.563706	10.8720538	14.235407	1.00000000	Confirmed
CCI 14 days	17.287549	17.246896	15.4191912	19.279900	1.00000000	Confirmed
CCI 50 days	9.203503	9.272982	7.1159672	11.020916	1.00000000	Confirmed
CCI 100 days	6.259018	6.366987	3.7899522	8.369810	0.98989899	Confirmed

Prediction Model

ANN Regression Model. The ANN can provide a more accurate prediction model for the larger amount of data, nonlinear data, nonstationary and has been a popular approach for stock market prediction. Feature selection performed on technical indicator using Boruta algorithm and selected technical indicator feature is given as input to the ANN regression model and it is described in Fig. 2. In this work, the ANN is used to predict the stock prices. ANN has three layers, each layer is connected to the other. The neurons represent the technical indicators. The sigmoid function activation function is used in the ANN regression model. The threshold value 0.5 has been set. A gradient descent momentum parameters are considered to determine the weights and to reduce the global minimum.

5   Experimental Results and Discussion

There are two phases, the first phase of experiment considered feature selection technique. The second phase of the experiment is the prediction model. In the

Table 3. Stock State Bank of India after feature selection.

Technical indicator feature	meanImp	medianImp	minImp	maxImp	normHits	Decision
SMA 5 days	5.4606723	5.5251714	3.1273954	7.653046	1.00000000	Confirmed
SMA 10 days	3.6382503	3.6526033	0.8677842	6.344588	0.81818182	Confirmed
WMA 5 days	4.8639987	4.9798041	2.1810807	7.662506	0.96969697	Confirmed
WMA 10 days	4.2772306	4.2978051	2.0651554	8.159167	0.91919192	Confirmed
WMA 14 days	3.7222162	3.9236859	0.8468174	6.263128	0.87878788	Confirmed
MOM 5 days	19.2684238	19.1960889	17.7018349	21.289981	1.00000000	Confirmed
MOM 10 days	8.5963907	8.6058484	6.7912856	9.896962	1.00000000	Confirmed
MOM 14 days	9.1535623	9.1827829	7.7671871	10.944682	1.00000000	Confirmed
STCK1 9 days	22.8154629	22.8720799	20.0837955	25.518795	1.00000000	Confirmed
STCK2 9 days	21.6583364	21.6522237	19.0834711	23.759105	1.00000000	Confirmed
RSI 14 days	15.1007526	15.0648255	13.6517412	17.075092	1.00000000	Confirmed
RSI 28 days	9.0990375	9.0610509	6.6856226	10.547252	1.00000000	Confirmed
MACD1 (26,13,9) days	14.6987380	14.6169198	12.7518900	16.634019	1.00000000	Confirmed
MACD SIGNAL	8.3677429	8.4367255	6.3049964	10.389434	1.00000000	Confirmed
MACD (15,25,15)	6.1898880	6.2728086	4.1248029	7.845896	1.00000000	Confirmed
R 14 days	37.9879805	37.9176749	34.9232519	40.963526	1.00000000	Confirmed
R 28 days	25.7820969	25.7832651	23.5718441	27.630123	1.00000000	Confirmed
R 50 days	18.4300230	18.4008101	16.4257141	20.670375	1.00000000	Confirmed
R 100 days	14.0062758	13.9214321	11.6389235	16.356005	1.00000000	Confirmed
A/D	17.2844799	17.2619585	15.0965778	19.177774	1.00000000	Confirmed
CCI 14 days	7.1505888	7.1446862	5.2038171	9.449850	1.00000000	Confirmed
CCI 50 days	8.7992939	8.8263404	6.7923020	10.430373	1.00000000	Confirmed

first phase of the experiment, we have used the Boruta method to select the best feature based on the Z Score. ICICI Bank and State bank of India stocks after feature selection are described in Tables 2 and 3. These selected features of the stock technical indicator are given as input to the ANN regression model. The MAE and RMSE are used to evaluate the performance of the prediction model and it is described in Eqs. 1 and 2. The proposed ANN Regression prediction model performance is compared with existing work and it is shown in Table 4. All tests are trained with a 10 fold cross-validation based model. From this study, we can conclude that technical indicator feature is important to predict the stock prices.

$$MAE = \frac{1}{n} \sum_{t=1}^{n} |e_t| \tag{1}$$

$$RMSE = \sqrt{\frac{1}{n} \sum_{t=1}^{n} e_t^2} \tag{2}$$

Table 4. Result comparison with existing work.

Stock Name	Prediction Model	MAE	RMSE
ICICI Bank	ANN [5]	27.0583	36.0444
ICICI Bank	**Proposed ANN Model**	**15.1221**	**19.9444**
State Bank of India	ANN [5]	27.7392	36.4834
State Bank of India	**Proposed ANN Model**	**17.4341**	**23.1585**

6 Conclusion

The paper focused on stock prediction for short-term trading. The stock data is collected from the National Stock Exchange (NSE). The paper has two objectives, first is the technical indicator feature selection and identification of the relevant technical indicators by using Boruta feature selection techniques. The second objective is an accurate prediction model for stocks. The performance for stock ICICI Bank MAE is 15.12 and stock State Bank of India MAE is 14.4. The experimental prediction model ANN outperforms compares to existing work by decreasing the error rate in the prediction. The future work can be identified the microeconomics, macroeconomics factor and fundamental analysis find the quality of stocks.

Acknowledgment. This work has been supported by the Visvesvaraya Ph.D Scheme for Electronics and IT (Media Lab Asia), the departments of MeitY, Government of India. The Task carried out at the Department of Information Technology, NITK Surathkal, Mangalore, India.

References

1. Alexander, S.S.: Price movements in speculative markets: trends or random walks. Ind. Manag. Rev. **2**(2), 7 (1961). (pre-1986)
2. Berutich, J.M., López, F., Luna, F., Quintana, D.: Robust technical trading strategies using GP for algorithmic portfolio selection. Expert Syst. Appl. **46**, 307–315 (2016)
3. Chen, Y., Hao, Y.: Integrating principle component analysis and weighted support vector machine for stock trading signals prediction. Neurocomputing **321**, 381–402 (2018)
4. Chourmouziadis, K., Chatzoglou, P.D.: An intelligent short term stock trading fuzzy system for assisting investors in portfolio management. Expert Syst. Appl. 43, 298–311 (2016)
5. Kara, Y., Boyacioglu, M.A., Baykan, O.K.: Predicting direction of stock price index movement using artificial neural networks and support vector machines: the sample of the Istanbul stock exchange. Expert syst. Appl. **38**(5), 5311–5319 (2011)
6. Macedo, L.L., Godinho, P., Alves, M.J.: Mean-semivariance portfolio optimization with multiobjective evolutionary algorithms and technical analysis rules. Expert Syst. Appl. **79**, 33–43 (2017)
7. Malkiel, B.G., Fama, E.F.: Efficient capital markets: a review of theory and empirical work. J. Financ. **25**(2), 383–417 (1970)
8. Mousavi, S., Esfahanipour, A., Zarandi, M.H.F.: A novel approach to dynamic portfolio trading system using multitree genetic programming. Knowl.-Based Syst. **66**, 68–81 (2014)
9. Nakano, M., Takahashi, A., Takahashi, S.: Generalized exponential moving average (EMA) model with particle filtering and anomaly detection. Expert Syst. Appl. **73**, 187–200 (2017)
10. Patel, J., Shah, S., Thakkar, P., Kotecha, K.: Predicting stock market index using fusion of machine learning techniques. Expert Syst. Appl. **42**(4), 2162–2172 (2015)
11. Preis, T., Moat, H.S., Stanley, H.E.: Quantifying trading behavior in financial markets using Google trends. Sci. Rep. **3**, 01684 (2013)
12. Rubio, A., Bermúdez, J.D., Vercher, E.: Improving stock index forecasts by using a new weighted fuzzy-trend time series method. Expert Syst. Appl. **76**, 12–20 (2017)
13. Ticknor, J.L.: A Bayesian regularized artificial neural network for stock market forecasting. Expert Syst. Appl. **40**(14), 5501–5506 (2013)
14. Tsai, C.-F., Hsiao, Y.-C.: Combining multiple feature selection methods for stock prediction: union, intersection, and multi-intersection approaches. Decis. Support Syst. **50**(1), 258–269 (2010)
15. Tsai, C.-F., Lin, Y.-C., Yen, D.C., Chen, Y.-M.: Predicting stock returns by classifier ensembles. Appl. Soft Comput. **11**(2), 2452–2459 (2011)
16. Vaisla, K.S., Bhatt, A.K.: An analysis of the performance of artificial neural network technique for stock market forecasting. Int. J. Comput. Sci. Eng. **2**(6), 2104–2109 (2010)
17. Zhang, Y., Lenan, W.: Stock market prediction of S&P 500 via combination of improved BCO approach and BP neural network. Expert Syst. Appl. **36**(5), 8849–8854 (2009)

Using Social Networking Sites: A Qualitative Cross-Cultural Comparison

Prateek Kalia[1]([⊠]), Norchene Ben Dahmane Mouelhi[2],
Sana Tebessi Hachana[3], Faten Malek[4], and Mehdi Dahmen[5]

[1] Sant Baba Bhag Singh University, Jalandhar, Punjab, India
kalia.prateek@gmail.com
[2] IHEC Carthage, Unicar, Université de Carthage, Tunis, Tunisia
[3] Université de Manouba, ISCAE, Manouba, Tunisia
[4] ESSCA, Angers, France
[5] Université Laval, Quebec City, Canada

Abstract. The purpose of this study was to determine the motives, the brakes and the ways of using Social Networking Sites (SNS) based on dimensional models that describe national cultural values explaining and predicting behavior. The study used a qualitative method, we conducted face-to-face in-depth interviews with 32 Tunisian, 29 Indian, 20 Pakistani, 20 French and 20 Canadian participants. The data was collected and transcribed in each of the five countries by independent researchers for each country. However, the analysis of the data was carried out by a single researcher to eliminate bias. We concluded that cultural environment influences the way people communicate and use social networking sites. The sources of motivation, the brakes and the reasons for the non-use of SNS also depend on cultural aspects of each society. We conducted an intercontinental study where the countries studied represent the four continents of the World, namely North America, Africa, Asia and Europe.

Keywords: Social networking sites (SNS) · Motivation · Brakes · Constraints · Culture

1 Introduction

Social networking sites (SNS) are of considerable importance for both businesses and individuals. Wide audiences can communicate, cooperate and connect through second-generation web applications like Facebook, Instagram, Linkedin, etc. Morever, companies can take decisions on the basis of customer online profiles, build relationship, interact and connect with the customer, as well as carry out online marketing, observe and analyze user behavior through SNS [1–4]. However, the diversity of audiences due to different demographics, nationalities, languages, interests and cultures can lead to variations in preference and use of SNS [1]. Previous studies in the past tried to decipher social media in a cross cultural/country context [5, 6]. For example, Goodrich and de Mooij [7] tried to examine the influence of culture on the role of social media in the consumer decision-making process and found that Hofstede's cultural dimensions explain cross-cultural differences in both online and offline purchasing decision

© Springer Nature Singapore Pte Ltd. 2019
A. K. Somani et al. (Eds.): ICETCE 2019, CCIS 985, pp. 269–285, 2019.
https://doi.org/10.1007/978-981-13-8300-7_23

influence. However, researchers have acknowledged shortage of knowledge about social media and its international or cross cultural applicability [7]. Therefore, the purpose of this study is to determine the motives, the brakes and the ways SNS is used based on cultural environment and to understand how culture influences the way people communicate and use media, based on dimensional models that describe national cultural values which explain and predict behavior [8]. The study contributes to the marketing literature by examining the influence of culture on the use of online information sources across five countries: Tunisia, France, India, Canada and Pakistan. Additionally, it identifies the brakes, sources of avoidance and the reasons for the non-use of its networks. Our major research questions are:

RQ 1. If members of different cultures are attracted to the same social networks and have the same attitudes to SNS.

RQ 2. What are the differences in the reasons for using these networks and the sources of motivation?

RQ 3. To know the reasons for avoidance, the brakes and the reasons for the non-use of these networks.

RQ 4. Which cultural variables best explain these differences?

2   Theoretical Background and Research Questions

2.1   Social Media Perception

Past researchers have highlighted how users' attitudes can affect their intention to use SNS. Alarcón-del-Amo et al. [9] found that user-friendly websites and trust in SNS can reduce perceived risk and creates positive attitudes. Positive attitude and perception can be developed by influencing individuals (i.e. enhancing their moods, memories, motivations or abilities), by improving contextual clues (e.g. classical conditioning) or through persuasive messages, like two-way communication, message credibility and memory messages etc. [10]. As compared to perceived usefulness and ease of use, attitude can change quickly, but its temporary and unstable nature can be managed through continuous effort to maintain it [11]. While divulging online behavior, previous studies have confirmed that demographic variables like city residence, age, occupation, family income, gender and marital status can significantly affect information search behavior [12, 13] and influence growth/conservation of perception and attitudes towards use of technology [10]. Using the Social Exchange Theory as a basis, Keating et al. [14] propose that perceptions about social media vary according to psychosocial (anger, self-esteem, job satisfaction, family support, friend support) and demographic (gender, age, children, work hours, health problems) variables and found that younger individuals reported positive perceptions, whereas younger individuals with health problems reported negative perceptions of social media. They also found that angry individuals with strong friend support reported positive perceptions, but angry individuals with low self-esteem reported negative perceptions of social media. They admitted that social media has been used by individuals as a kind of coping mechanism i.e. release of anger, gathering social support; however, it also results in social criticism, reduced self-esteem and a feeling of being left out. Lin and Lu [15] observed that gender can contribute to a difference in the effect of perceived benefit and network externalities can affect the continued intention to use SNS. They also highlight that, out of three sources of network externalities, the number of peers using the SNS that has the greatest impact on enjoyment and usefulness of SNS for women. However, for men the number of peers and perceived complementarity are most important and men do not enjoy SNS which have greater numbers of members. Joinson [16] found that females visit Facebook more frequently than males. With respect to occupational categories, Kalia et al. [17] reported a significant difference within business-service and business-student groups in terms of their perception of the service quality of web-based services. Smith [18] analysed undergraduate perceptions of social media technologies (SMTs) in the learning process and discovered that students consider social media as a 'double-edged sword' which can distract or assist learning. They also found that student-student and student-content interactions were predominant, rather than faculty-student interactions through social media for learning purposes.

2.2 Social Media Motivation

Lin and Lu [15] apply arranged externalities and inspiration hypotheses to investigate factors influencing people joining SNS and discovered that enjoyment is the most crucial factor, followed by peer number and usefulness. While investigating the link between webographics and use of web services by internet users, Kalia [19] found a significant link with webographic characteristics of respondents, such as internet experience, computer experience, usage level (time spent), access device, access place, web skill and type of data plan. During analysis to check the distinction in Chinese users' satisfactions with various social media and their implications for planning brand strategies via web-based networking media, Gao and Feng [20] observed that different motives underpin the use of the different social media. They observed, for instance, that SNS are used for social interaction whereas microblogs are mostly used for self-expression, entertainment and information. Similarly, Smock et al. [21] observed that users seeking relaxing entertainment, expressive information sharing and social interaction reliably indicate the general use of Facebook. Kaplan and Haenlein [22] stated that people use social media for self-presentation with the desire to control the impressions that other people form of them. This objective can be achieved with the goal of gaining rewards or of creating an image that is consistent with one's personal identity. Such a presentation is done by revealing personal information either consciously or unconsciously. Zhu and Chen [23] argued that people use SNS to seek information, advice, solutions and answers to resolve their personal problems (e.g., legal, tax, finance, day-care). They also mentioned that people use SNS to build relationships (e.g., Facebook, LinkedIn), to connect to 'self-media' kinds of SNS (e.g. Twitter, Weibo), to share their interests, creativity, and hobbies (YouTube, Instagram, Flickr, Pinterest), to cooperate (e.g., Quora, Reddit) or to satisfy their human needs (autonomy, competence, relatedness, physical thriving, security, self-esteem, self-actualization, pleasure-stimulation, money-luxury, and popularity-influence) through social media.

2.3   Social Media Constraints

One of the very strong barriers which prevents active participation and use of SNS by users can be culture. For example, values such as modesty and the desire to save face in Asian cultures. A user may decide not to post an online query or suggest because they are worried by the threat of ridicule; or they may avoid being too active on SNS for fear of appearing indecent or pretentious. Similarly, information hoarding may also vary depending on the country in question. For instance, Michailova and Husted [24] note significant information hoarding and lack of information sharing among employees in Russian organizations and attributed this behavior to the need to cope with uncertainty (i.e. in countries with unstable economic conditions, high power distance and strong hierarchies, knowledge sharing may be inhibited). Another reason why people avoid using SNS could be a cultural preference for face-to-face communication. Even organizational culture can act as significant organizational barriers to knowledge sharing; for example, in 'vertical' culture top managers can block the flow of vital information. Furthermore, higher-level managers may avoid participating in online communities. In the case of collectivist cultures, due to sharper distinction made between in-group and out-group members, there may be natural inhibition of inter-group knowledge sharing. Another significant barrier to SNS usage could be the generation gap. For example, in situation where one is at risk of "losing face", younger people are more tolerant and bolder than older people [25]. Apart from demographic and cultural factors, Web proficiency can also affect use of Web related services. For example, Kalia [26] found significant differences within webskill and data plan categories, while conducting research with respect to perceived service quality (PSQ) in online context. Another important factor restricting SNS use is the issue of trust, which can further influence perceived usefulness, perceived ease of use, perceived risk and attitude [9]. Trust has two facets i.e. trust in the other party and trust in control mechanisms [27]. SNS can boost trust considerably by properly maintaining infrastructure, facilitating transactions, installing firewalls, using authentication mechanisms and ensuring the protection of privacy and information [28, 29].

2.4   Cultural Dimensions

Hofstede [8] defined culture as, *"collective programming of the mind that distinguishes the members of one group or category of people from others"*. This author identified five dimensions of the culture that contribute to the understanding of national cultural values: Individualism/collectivism, uncertainty avoidance, power distance, masculinity-femininity and long/short-term orientation.In their literature review, Soares et al. [30] described culture as *"the complex whole which includes knowledge, belief, art, morals, custom and any other capabilities and habit acquired by man as a member of society"*.

Both definitions highlight the importance of behavioral changes due to culture. Against the backdrop of exponential growth and use of SNS, researchers have observed that the reasons for and way of using SNS differ due to social and cultural differences. Kim et al. [5] examined motives and patterns of SNS use among US and Korean students and reported that major motives like seeking friends, social support, entertainment, information, and convenience are similar between the two countries, but the emphasis placed on each of these motives differs. For example, Korean students focus on obtaining social support through existing social relationships, while US students seek entertainment. They also found difference regarding development and management of relationships through SNS in the two countries. Alexandre et al. [25] carried out qualitative research on cultural factors (degree of collectivism, competitiveness, the importance of saving face, in-group orientation, attention paid to power and hierarchy, and culture-specific preferences for communication modes) and observed differences in knowledge-sharing strategies in virtual communities among participants in three different countries. For example, the study revealed that modesty and competitiveness between employees were major obstacle to information sharing in China as compared to Russia and Brazil. Goodrich and de Mooij [7] used cultural dimensions to compare the use of social media and other sources of information for consumer decision-making in 50 countries and observed considerable variations in the use of information sources that influence online purchasing decisions across cultures. They also concluded that there are major differences in online complaint behavior by country.

3   Research Questions

- If members of different cultures are attracted to the same SNS and have the same attitudes, then what are the differences in their behavior when using these SNS?
- What are their sources of motivation for using SNS in different cultures?
- What are the reasons of non-usage of SNS in different cultures?
- Which cultural variables best explain these differences?

4   Research Method

4.1   Participants

The study used a qualitative methodology and was based on 121 in-depth interviews based on the concept of data saturation [31]. We conducted face-to-face qualitative research with 32 Tunisians, 29 Indian, 20 Pakistanis, 20 French participants and 20 Canadians. We considered three key features of Branthwaite and Patterson [32] "*that make qualitative research a unique and valuable marketing tool to consumers*", namely conversation, active listening to the underlying dialogue and "fusion of minds". Table 1 shows the detailed sample demographics.

Table 1. Sample demographics

Measure	Item	Tunisia		India		Pakistan		France		Canada	
		n	%	n	%	n	%	n	%	n	%
Gender	Male	13	40.6	14	48.3	13	65	9	45	11	55
	Female	19	59.4	15	51.7	7	35	11	55	9	45
Age (years)	18–24	16	50	17	58.6	7	35	6	30	3	15
	25–29	3	9.4	2	6.9	8	40	4	20	5	25
	30–34	4	12.5	9	31	2	10	6	30	5	25
	35–39	4	12.5	1	3.4	1	5	2	10	6	30
	40–44	5	15.6	0	0	2	10	2	10	1	5
Academic level	Secondary	1	3.1	0	0	1	5	1	5	3	15
	Bachelor	13	40.6	16	55.1	10	50	14	70	7	35
	Masters	17	53.1	9	31	9	45	3	15	5	25
	Ph. D.	1	3.1	4	13.8	0	0	2	10	5	25
Occupation	Student	16	50	15	51.7	7	35	8	40	6	30
	Employee	8	25	8	27.6	9	45	5	25	9	45
	Senior executive	4	12.5	0	0	3	15	2	10	2	10
	Liberal profession	3	9.3	5	17.2	0	0	5	25	2	10
	Unemployed	1	3.2	1	3.4	1	5	0	0	1	5

4.2   Materials and Procedure

In order to explore the motives, the brakes and the reason for non-use of SNS it was obvious to undertake qualitative exploratory research [32]. Five countries were covered in this survey: Tunisia, France, India, Canada and Pakistan. The research was based on a semi-structured interview guide containing four topics.

We started by asking the participants whether they had a social media account. If the answer was yes, they proceeded to topic one and if the answer was no, they proceeded to the second topic. Under topic one participants were asked for their general information regarding social networking i.e. uses and gratification they seek from networking sites (i.e. social media used, opinion of each of them, the reason for using social media, motives to use social media). Subsequently, we examined the extent to which respondents use the SNS of their choice (i.e., how many times per day they connect to social media, for how many hours, at which moment of the day, and how many contacts they have).

Under second section we enquired respondents for the reasons they avoid SNS, possible threats and the factors holding them back from network use (i.e., is it possible that you avoid connecting to social media? If yes, why? Do you think that using social media can be threatening? What are the constraints for using social media? Participants were also asked to predict why other people avoid creating an account on SNS).

To frame our study and examine the effect of national cultural differences on use of social network in Tunisia, France, Canada, Pakistan and India, we applied an universal criteria for making international comparisons of cultures assessed through the five cultural dimensions namely: Individualism/collectivism, uncertainty avoidance, power distance, masculinity-femininity and long/short-term orientation developed by Hofstede [8].

The last part of the interview guide contained demographic questions about age, gender, race, school classification and socio-professional category. The data was collected and transcribed in each of the five countries by independent researchers for each country. During the final stage, data from five countries was collected, coded and analyzed entirely by single researchers to detect commonalities and differences between the countries. The content analysis was conducted using the data analysis software Sphinx IQ2.

5 Results and Discussion

5.1 Preliminary Results

We obtained descriptive information on SNS usage among survey respondents from Tunisia, France, India, Pakistan and Canada. Results and discussions are as below:

Do You Have SNS Account? Among the participants surveyed, 27 Tunisians (84.4%), 14 Pakistanis (70.0%), 29 (100%) Indian, 20 French (100%) and 16 Canadians (80%), indicated that they have a SNS account; the rest did not. Among the participants who had accounts, for Tunisia, 55.6% were women, whereas for Pakistanis 44% were women and 56.3% were men.

Preferred SNS. Of the social networks most used in the countries studied, Facebook came out as the most popular SNS used by Tunisian participants most frequently (92.6%), followed by Instagram (44.4%), Linkedin (29.6%), Twitter (22.2%), and others such as Snapchat (22,2%) and WhatsApp (18,5%). In Pakistan, Facebook ranked as the most popular SNS(100.0%). It was used by all interviewees, followed by WhatsApp (68.4%), Instagram (21.1%), Linkedin (10.5%) and Twitter (10.5%). In India, we observed similar preference for SNS as in Pakistan i.e. Facebook (100.0%), WhatsApp (48.3%), Linkedin (34.5%) and Instagram (31.0%). In France, we found the following figures: Facebook (100.0%), Snapchat (50.0%), Instagram (45.0%), Twitter (40.0%), Linkedin (35.0%). Among the Canadian respondents, Facebook was cited first by 93.8% of the respondents, followed by Linkedin (37.5%), Instagram (37.5%) and Twitter (18.8%). To synthesize these results, we ranked the order of preference of the different social networks according to the countries and proposed a general classification (Table 2). Facebook was at first place, followed by WhatsApp, Instagram, and Twitter, with Snapchat as least preferred SNS.

Table 2. Preference of social network

Country	Rank 1	Rank 2	Rank 3	Rank 4	Rank 5
Tunisia	Facebook	Instagram	Linkedin	Twitter	Snapchat
Pakistan	Facebook	WhatsApp	Instagram	Linkedin	Twitter
India	Facebook	WhatsApp	Linkedin	Instagram	Snapchat
France	Facebook	Snapchat	Instagram	Twitter	Linkedin
Canada	Facebook	Linkedin	Instagram	Twitter	Snapchat

Time Spent on SNS. With respect to average time spent on SNS, Tunisian participants spent between 3–4 h, Indian used social networks for 1 h, Pakistanis spent 3 h. On the other hand, the French and the Canadian remain connected to their accounts for 2 h (These data were recoded according to the most frequent values).

Login Frequency on SNS. Respondets were asked about number of times they login to their SNS account per day and for how long. We have received following results: the Pakistanis connect to their SNS accounts for an average of three times a day, during night, morning and afternoon. The French connect more frequently i.e. four times a day and they do so at night (70.0%), in the morning (30.0%), in the afternoon (25.0%), and when they have free time (20.0%). Canadians connect 10 times a day: in the evening (43,8%), at night (37.5%), in morning (37,5%) at no time (12.5%), all day (6.3%), and when they have free time (6.3%). Meanwhile, the Tunisian respondents connect nine times or more, in the evening, morning, during break, before sleep, during the day, when the alarm clock goes off, in the early morning and afternoon; however, this high frequency can be explained by the fact that in these two countries people connect to social networks for a short time. Additionally, Indian do not have specific times when they login thier social networks. They check them "all the time" – at night, in the morning, throughout the day, when they have free time, at the weekend and before sleeping.

SNS Contact List. Respondets were asked about the number of contacts they have. It was observed that, respondents in Tunisia had highest number of contacts; in fact, most Tunisian participants had more than 1000 contacts (40.7%), followed by India with approximately 500 contacts (55.2%), Pakistan with more than 200 contacts, French respondents had contact list varying between 200 and 250 (40%). In case of Canadian respondents contact list averaged between more than 800 contacts for 25% of respondents and 250 and 300 contacts for 25%.

5.2 Motives for Using SNS

One of the investigative aims of this research was to ascertain the perception users have of social networks. These are discussed below:

Use and Gratification. In Tunisia, Internet users perceived that social networks are "very useful" (26.7%), "very interesting" (26.7%), "provide benefits" (26.7%), "essential to our lives" (23.3%), "practical" (16.7%), "provide distraction" (13.3%), "economic benefits"(10.0%). Few respondents stated negative comments such as "unnecessary content" (13.3%), "no interest" (10.0%), "intrusive" (10.0%), "voyeuristic" (6.7%), "wastage of time" (3.3%), "I do not like" (3.3%).

Indian respondents stated that, SNS are "very helpful" (32.1%), "useless" (21.4%), "good" (7.1%), "risk spreading false information" (7.1%), "are a integral part of our daily schedule" (7,1%), "connect people in the world" (7.1%), "a way of passing time" (7.1%), "getting calmed by sharing of feelings" (7.1%), "circumstances demands" (3.6%), "help connectivity" (3.6%), "gives inspiration"(3.6%), "provides services" (3,6%), "makes a person more social" (3.6%), "risk of addiction" (3.6%) and "fun to use" (3.6%).

In Pakistan, users responded that social networks "helps in connectivity" (30.0%), "fun to use" (15,0%), "are integral part of our daily schedule" (15,0%), "risk of addiction" (15,0%), "is wasteful" (5,0%), "provides services" (5,0%), "all is fake and artificial World" (5,0%), "getting calmed by sharing of feelings" (5,0%), "time saving" 5,0%, "gives inspiration" (5,0%), "for time pass" (5,0%).

The French perceived social networks as "very useful" (55.0%), "create some connectivity with the world" (50.0%), "very interesting" (15.0%), "convenient" (10.0%) but they also think that SNS "disclose our privacy" (25.0%), "there is a bad side" (20.0%), and SNS can be "very dangerous" (15.0%).

The Canadians considered social networks as "all easy to access" (15.8%), "essential" (10.5%) and "easy to use" 5.3%. On the darker side they responded that SNS are "unnecessary networks" (5.3%), you "lose a lot of time" (5.3%) and "one becomes addicted quickly" (5.3%).

Motivation. In terms of the motivations for using social networks, Tunisians preferred connecting to SNS to "follow the news" (85.2%), "contact friends" (66.7%), "be up to date" (37.%), "search for information" (22.2%), "meet again" (22.2%); "distraction" (22.2%), "to discover new offers" (18.5%), "for photo and video sharing" (18.5%), as a "means of communication" (18.5%), "to follow the news of friends" (14.8%), out of "Curiosity" (11.1%), "to share information" (11.1%), "to maintain family ties" (7.4%) and to "keep my pictures" (3.7%).

For Canadians, motivations to use SNS were similar to the Tunisians i.e "to follow the news" (50.0%), and "to connect with friends" (37,5%). The other motives were "to follow the news of my friends" (25.0%),"to seek professional opportunities" (6.3%), and "to share videos" (6.3%). These results are consistent with Castro and Marquez [6] on motivation for the use of social networks in Colombia.

In India, SNS users were motivated by the desire "to connect with people" (63.0%), "to get information" (29.6%), "to share information" (25.9%), "to pass the time" (22.2%), "to know what my friends are doing" (14.8%), "to express views" (11.1%), "to find friends" (3,7%), "to download videos" (3.7%), "to contact new people" (3.7%), "to show support for brand" (3,7%), and "to obtain information" (3.7%).

Pakistanis respondents replied that they use SNS "to be in touch with friends and family" (70.0%), " to get information from around the globe" (20.0%), "for entertainment" (20.0%), "to get information about new trends" (20,0%), "to share my life with others" (10.0%), "to get information about my friends" (10.0%), "for commercial use" (10.0%), "sharing pictures" (5,0%), "to pass time" (5,0%), "to adopt every new technology" (5.0%), "to meet new people" (5.0%).

Socialisation and maintaining relationships was found to be important for French respondents. Analysis revealed that they use SNS "to share photos" (55.0%), "establish relationship with friends" (50.0%) and "professional goals" (40.0%). We also observed the desire "to share information" (40.0%), "to talk about one's privacy" (40.0%), to "keep in touch with people who are far away" (20.0%), "to share moments" (20.0%), "for personal development" (15.0%), and to "find friends" (10.0%).

In view of the above results, the first motive for having a social account in India was "to be in touch with friends and family". In Pakistan and France, it was "to connect with people" and "to establish a relationship with friends". Nevertheless, for Tunisian and Canadian participants, it was more of "collecting information".

Realtive analysis of findings revealed that users of social networks with motivation to stay in touch with friends have limited number of contacts. Whereas, those having need for information have greater number of friends in their contact list. Indeed, Pakistanis, French and Indian have fewer friends and their needs are more oriented towards social support and seeking friends whereas the Tunisians and Canadians have higher number of friends and they use SNS to follow the news.

5.3   Avoidance, Threats and Obstacles in SNS Use

The third objective of this research was to analyze the reasons for the avoidance, threats and obstacles perceived by respondents towards SNS use.

Avoidance. Eighty percent of Tunisian interviewees avoided using social networks for the following reasons: "when I am busy" (30.8%), "when I feel I'm wasting my time" (26.9%), "when I want to avoid people" (19.2%), "when I want to isolate myself" (19.2%), "when I feel I am becoming addicted" (19.2%), "source of conflict" (11.5%), "instructive" (11.5%), "when I am depressed" (7.7%), "when I abuse" (7.7%), "obligation to respond to messages" (3.8%), "vacancy" (3.8%).

In Pakistan, 72.2% avoid SNS and explained that they do this "to have a control" (33.3%), "not really" (27.8%), "to avoid some contacts" (22.2%), "to get a practical life" (16.7%), "very unproductive" (5.6%), and "to save my time" (5.6%).

Seventy percent of the French sometimes avoided SNS use as compared to 30.0% respondents who never avoided. The reasons for these avoidances were: "during the holidays" (25.0%), "when I'm very busy" (10.0%), "to focus on what I do" (10.0%), "a few evenings" (5.0%), "we are cut off from real life" (5.0%), "to sleep and relax" (5.0%), "when publications annoy me" (5.0%), and "to avoid some contacts" (5.0%).

Sixty five percent Canadians avoided SNS to "concentrate on work" (26.3%), "avoid somebody" (15.8%), "when I'm busy" (15.8%), "it's a waste of time" (10.5%), and "avoid threats related to the disclosure of personal data" (5.3%).

Around sixty percent of Indian respondents avoided SNS as compared to remaining 39.3%. Reasons for avoidance were "excessive time consumption" (28.6%), "circumstances demand" (28.6%), "when I am busy in personal affairs" (14.3%), "my studies and create a lot of distraction" (14.3%), "and some abusive people" (14,3%).

Threat. Different types of threats perceived by respondents from five countries have been summarized below.

All Tunisian respondents reported threats such as, "account piracy" (41.4%), "bad influence" (20.7%), "exposure of his or her privacy" (20.7%), "associability in relation to off-line life" (17.2%), "espionage" (17.2%), "no security" (13.8%), "lying" (13.8%), "torque problem" (3.4%).

Almonst 48.3% Indian respondents perceived threat, and 51.7% replied that social networks pose no threat. For those who thought it does, the reasons were: "sometimes privacy is in danger" (19.2%), "Internet access" (19.2%), "addiction" (15.4%), "cyber bullying" (7.7%), "pushes us away from real life" (7,7%), "we know the real personality of the other" (3.8%), "rumors" (3,8%), "security" (3,8%), "making a perception about a person" (3,8%), "time constraints" (3,8%), "work pressure" (3,8%), "waste of time" (3.8%).

Around 68.4% Pakistanis agreed that there is a threat translating into "violation of right to privacy" (15.8%), "if used in a wrong way" (10.5%), "information is not secure" (5.3%).

The Canadians perceived that SNS can harm (44.4%), because you can "become dependent" (22.2%), there is a "virus transmission threat" (5.6%) and the threat of "piracy of a personal account" (5.6%).

The threats perceived by the French were related to "posting of inappropriate and shocking content" (31.6%), "the breach of privacy" (31.6%), "disclosure of rumors" (21.1%), "harassment" (15.8%), "piracy of accounts" (10.5%) and the risk that "we forget true communication" (10%).

Obstacle. We observed that some respondents (except Indian and French) were not having SNS accounts. The Tunisian non-users explained this in the following ways: "it is my private life" (50.0%), "there is no security" (21.9%), "I am totally uninterested" (21.9%), "low technological knowledge" (21.9%), "resistance to change because I like real life" (21.9%), and finally "lack of time" (15.6%).

For Pakistanis, the reasons were "privacy theft" (47.4%), "security" (26.3%), "it wastes time" (10.5%), "it is very distracting" (5.3%), "it is too social" (5.3%), and "it is very addictive" (5,3%).

Some Canadians didn't use SNS because they "fear for their private lives" (25.0%), "they are afraid of addiction" (15.0%), "does not adhere to technology" (15.0%), "not interested" (15.0%), "do not prefer to communicate by these means". Table 3 gives examples of verbatims that illustrates perception, motivation, avoidance, threats perceived by respondents during use and non-use of SNS.

5.4   Interpretation of Results Across Cultural Dimensions

Individualism/Collectivism. The number of social friends through SNS was higher among the Tunisian respondents. They reported having an average of more than 1000 contacts; the figure stood at 800 or more for Canadian, whereas Indians had an average of 500 contacts; for the French the number was between 200 and 250 contacts and Pakistanis had only 200. This suggests that Tunisian and Canadian place more importance on creating social networking. This is due to the fact that individualistic and low-context cultures highlight more casual, practical relationships [33], whereas Pakistani may be more exclusive in their network building, connecting with people they are involved in, through social bonds and commitment [34].

Tunisian, Indian and Canadian participants demonstrated a different profile as compared to Pakistanis and French in terms of their SNS uses. Tunisian, Indian and Canadian respondents reported that they use SNS 9 or 10 times per day, at different times of the day, where the Indian respondents were connected all day. Among Pakistanis there was a different pattern: they connected to SNS only three times, while the French respondents were connected for four times a day. This is because the former only connect quickly, for very short moments, and used social networks only for a personal purpose such as to "search for information" while the French and Pakistani connected during "specific moments for a social purpose" such as to "stay in touch with friends".

Concerning the gratification, in most countries, when it comes to work the participants were motivated by information (first or second most common answer). However, according to previous research, seeking information suggests belonging to an individualistic culture, whereas sharing information means that they fall under the collectivist culture.

We examined the interest in obtaining information country wise in our study and observed that Tunisians seek information for the following reasons: "ease" (52.2%), "speed" (39.1%), "accessibility" (17.4%), and "practicality" (13.0%), and "utility" (4.3%), "availability of information" (4.3%). This underlines that the behavior of the Tunisians is guided by an individualistic culture. However, 70.4% were willing to share information, while 29.6% were not.

Pakistanis were also interested in obtaining information for their own knowledge i.e. for the desire "to get new information" (5.0%), the fact that "it provides first-hand experience of people" (20.0%),"it is authentic" (5.0%), it "is convenient" (5.0%), "it is fast" (5,0%), and "widely available" (30.0%). Up to 90% of Pakistanis shared information with their contacts, which is a higher rate than Tunisians.

For French people, the search for information via social networks was explained by the fact that "information is available rapidly" (26.3%), "information is accessible" (21.1%) and "information is authentic, without filtering" (15.8%).

The Indian respondents mention that they do not use social networks for information purposes. 20.5% say, "I don't look for information on social media and for those who are interested in obtaining information", "do so to get to know the views of others" (18.5%), "to obtain different kinds of information" (14.8%), "as there is more personal information available" (11.1%), there are posts by the people (3.7%). Similar results were observed in Canada. In fact, 43.8% of Canadians "do not look for information on social media".

We noticed that the arguments of Indian and Canadian respondents were more oriented towards a social approach and concern for others. Indeed, the social relationship of individuals provides social support online [35]. They fall under the collectivist culture. People who belong to the collectivist culture attach importance to the objectives of the group [8]. In addition, in our case, the interest of looking at the information lies rather in the interest we have for the group and concerns about what it thinks. In collective and high-power distance cultures, social media tend to strengthen the sharing of feelings and ideas [36].

Uncertainty Avoidance. When it comes to trusting people, which is an aspect of the uncertainty avoidance [7], we noticed that Tunisians followed by Canadian and Indian, trust people more than institutions and experts – the exact figures for people who feel this way was 61.5% for Tunisians, 55.2% for Indian and 56.3% for the Canadians, while the Pakistanis were divided i.e. 50% respondents trusted people and remaining 50% had faith in institutions and experts. In contrast, percentage of French respondents having trust in institutions was 75%. This underscores the notion that SNS users in

Tunisia, Canada and India have low uncertainty avoidance, while French respondents belong to high uncertainty avoidance group.

According to the results of the survey, Canada, France, Tunisia and Pakistan indicate that their friends are part of their identity whereas Indians think otherwise. Indeed, the respective percentages of Canadian, French, Tunisian and Pakistani respondents who think like this stand at 81.3%, 60%, 59.3% and 60,0% respectively. For 58.6% of Indian this was not the case.

Members of individualistic cultures see themselves as independent from others (Indian), whereas collectivists see themselves as interdependent on other members (Canadians, French, Pakistanis and Tunisians).

Concerning contacts on social networks, Tunisian easily accept invitations and encounters via SNS (66.7% of participants). The same result was found in case of Pakistanis (65.0% of participants). In contrast, Indian, Canadian and French do not accept invitations easily and avoid meeting new people (51.7%, 62.5% and 80% respectively). Internet users who use social networks in individualistic, low-context cultures (i.e., Tunisia and Pakistan) have a tendency to use this means of communication to meet new people and thus accept a new invitation to get to know each other, SNS users in collectivistic, high-context cultures (i.e., India, Canada and France) prefer to keep this means of communication to communicate with their closest contacts and friends and do not use this means to make new encounters.

Power Distance. The study indicated that information is essential, and the Tunisian and Canadian respondents wanted to obtain it by all possible means. In individualistic and low power distance cultures, information is an important tool for this purpose and people search for information more actively, including from social networks [7, 37].

Masculinity/Femininity. Our analysis revealed that respondents are more concerned about their personal success than the quality of social life. Indeed, this is true in most countries under this study (Tunisia 74.1%, India 65.5%, Pakistan 65.5%, Canada 56.3%). In other words, these findings indicate that these countries are closer to the masculinity dimension [8]. On the other hand, France attach considerable importance to the quality of social life (60%) which reveals the femininity dimension among the French.

Long/Short Term Orientation. In the same way, people who employ a discreet and reserved style of communication are related to long-term-oriented collectivistic cultures and they sometimes prefer not to disclose their identities. This is the case for Tunisians (74.1% discreet, 25.9% expressive) Pakistanis (60.0% discreet; 40.0% expressive) and the French (65% discreet, 35% expressive). While in short-term-oriented collectivistic cultures, Internet users communicate in an expressive way and presents themselves explicitly, which holds true for the Indian participants (48.3% discreet; 51.7% expressive). Canadians fall somewhere between the two cultures, namely, long-term-oriented collectivistic cultures and short-term-oriented collectivistic cultures because 50% communicate on social media through an expressive manner and 50% communicate on social media using a discreet communication style.

Table 3. Synthesis of the most cited verbatims of Perception, Motivation, Avoidance, Threat for using SNS and the reason of the non-use.

	Perception	Motivation	Avoidance	Threat	Reason of non use
Tunisia	*Very useful; Very Interesting; Provide benefits*	*Follow the news; Contacting friends; Be up to date*	*When I am busy; When I feel I'm wasting my time; When I want to avoid people*	*Account piracy; Bad influence; Expose his or her privacy*	*It is my life's privacy; There is no security; I am totally disinterested*
France	*Very useful; Create some connectivity with the World; Very interesting*	*To share photos; Establishing relationship with friends; Professional goal*	*During the holidays; When I'm very busy; To focus on what I do*	*Post inappropriate and shocking content; Breach of privacy; Disclosing rumors*	
Pakistan	*Helps in connectivity; Fun to use; integral part of our daily schedule*	*To be in touch with friends and family; To get information around the globe; For entertainment*	*To have a control; Not really; To avoid some contacts*	*Violation of right to privacy; If used in a wrong way; Information is not secure*	*Privacy theft; Security; It wastes time; Very distracting*
Canada	*All easy to access; Are essential; Easy to use*	*To follow the news; To connect with friends; Follow the news of my friends*	*To concentrate in the work; To avoid somebody; When I'm busy*	*Becomes dependent; Virus transmission threat; Piracy of a personal account*	*Fear for their private lives; They are afraid of addiction; Does not adhere to technology*
India	*Very helpful; Useless; Good; Risk of spreading false information*	*To connect with poeple; To get information; To share information*	*Excessive time consumption; Circumstances demands; When I am busy in personal affairs*	*Sometimes privacy at danger; Internet access; Addiction*	

6 Conclusion, Managerial Implications and Future Research Directions

As outlined by this paper, the principal objectives of determining the motives, the brakes and the ways of using SNS (social networking sites) have been achieved. The results indicate that members of different cultures do not have same attitudes to social networking sites (SNS).

The sources of motivation, the brakes and the reasons for the non-use of SNS depend on the cultural aspects of the community. Several cultural variables like individualistic or collectivist orientation contribute towards creating these differences. For example, in individualistic, low-context cultures (i.e., Tunisia and Pakistan), users tend to use SNS to meet new people, while users in collectivistic, high-context cultures (i.e., India, Canada and France) use SNS to communicate with their closest contacts and friends and do not use this means of communication to have new encounters. As Maschio [38] explains, individualism is *"the notion that richness of experience, of sensation and of desire should be major life goals"*. Knowledge of the cultural variables that influence users' attitudes to SNS helps in understanding users' reaction and their online behavior. Thus, the results of this study highlight the role of culture as one of the factors affecting use or non-use of social networks. This study widens the prospects of using and satisfying social networks of new media in an intercultural context and to a certain extent in an intercontinental context too. The results confirm the influence of Tunisian, French, Indian, Pakistani and Canadian culture on the various motivations of use of social networks and for some of its brakes. In practical terms, the results of this research suggest that social networking specialists, defined as "community managers", should apply culturally specific strategies to attract more users. More specifically, community managers need to develop more features [6] to attract Internet users from individualist cultures, while in collectivist cultures they may need to incorporate features that facilitate the exchange of social support through relationships and social benefits. In this study, we just considered one country per continent. However, it will be interesting to compare more than two countries in differents continents to highlight differences by cultures related to the country or the continents.

References

1. Boyd, D.M., Ellison, N.B.: Social network sites: definition, history, and scholarship. J. Comput. Commun. **13**, 210–230 (2008). https://doi.org/10.1111/j.1083-6101.2007.00393.x
2. Constantinides, E., Fountain, S.J.: Special issue papers web 2.0: conceptual foundations and marketing issues. J. Direct Data Digit. Mark. Pract. **9**, 231–244 (2008). https://doi.org/10.1057/palgrave.dddmp.4350098
3. Donath, J.: Identity and deception in the virtual community. In: Kollock, P., Smith, M. (eds.) Communities in Cyberspace. Routledge, London (1998)
4. Tikkanen, H., Hietanen, J., Henttonen, T., Rokka, J.: Exploring virtual worlds: success factors in virtual world marketing. Manag. Decis. **47**, 1357–1381 (2009). https://doi.org/10.1108/00251740910984596

5. Kim, Y., Sohn, D., Choi, S.M.: Cultural difference in motivations for using social network sites: a comparative study of American and Korean college students. Comput. Hum. Behav. **27**, 365–372 (2011). https://doi.org/10.1016/j.chb.2010.08.015

6. Castro, L., Marquez, J.: The use of Facebook to explore self-concept: analysing Colombian consumers. Qual. Mark. Res. **20**, 43–59 (2017). https://doi.org/10.1108/QMR-12-2015-0086

7. Goodrich, K., de Mooij, M.: How 'social' are social media? A cross-cultural comparison of online and offline purchase decision influences. J. Mark. Commun. **20**, 1–14 (2014). https://doi.org/10.1080/13527266.2013.797773

8. Hofstede, G.: Dimensionalizing cultures: the Hofstede model in context. Online Read. Psychol. Cult. **2**, 1–26 (2011). https://doi.org/10.9707/2307-0919.1014

9. Alarcón-Del-Amo, M.D.C., Lorenzo-Romero, C., del Chiappa, G.: Adoption of social networking sites by Italian. Inf. Syst. E-bus Manag. **12**, 165–187 (2014). https://doi.org/10.1007/s10257-013-0215-2

10. Yang, H.D., Yoo, Y.: It's all about attitude: revisiting the technology acceptance model. Decis. Support Syst. **38**, 19–31 (2004). https://doi.org/10.1016/S0167-9236(03)00062-9

11. Thompson, R.C., Hunt, J.G.: Inside the black box of alpha, beta, and gamma change: using a cognitive- processing model to assess attitude structure. Acad. Manag. Rev. **21**, 655–690 (1996)

12. Kalia, P., Singh, T., Kaur, N.: An empirical study of online shoppers' search behaviour with respect to sources of information in Northern India. Prod. A. Q. J. Natl. Prod. Counc. **56**, 353–361 (2016)

13. Odell, P.M., Korgen, K.O., Schumacher, P., Delucchi, M.: Internet use among female and male college students. CyberPsychol. Behav. **3**, 855–862 (2000). https://doi.org/10.1089/10949310050191836

14. Keating, R.T., Hendy, H.M., Can, S.H.: Demographic and psychosocial variables associated with good and bad perceptions of social media use. Comput. Hum. Behav. **57**, 93–98 (2016). https://doi.org/10.1016/j.chb.2015.12.002

15. Lin, K.Y., Lu, H.P.: Why people use social networking sites: an empirical study integrating network externalities and motivation theory. Comput. Hum. Behav. **27**, 1152–1161 (2011). https://doi.org/10.1016/j.chb.2010.12.009

16. Joinson, A.N.: 'Looking at', 'looking up' or 'keeping up with' people? Motives and uses of Facebook. In: CHI 2008 Proceedings of Online Social Networks 1027–1036 (2008). https://doi.org/10.1145/1357054.1357213

17. Kalia, P., Law, P., Arora, R.: Determining impact of demographics on perceived service quality in online retail. In: Khosrow-Pour, M. (ed.) Encyclopedia of Information Science and Technology, 4th edn, pp. 2882–2896. IGI Global, Hershey (2017). https://doi.org/10.4018/978-1-5225-2255-3.ch252

18. Smith, E.E.: A real double-edged sword: undergraduate perceptions of social media in their learning. Comput. Educ. (2016). https://doi.org/10.1016/j.compedu.2016.09.009

19. Kalia, P.: Determining effect of webographics on customer's purchase frequency in e-retail. J. Internet Bank. Commer. **21**, 1–24 (2016)

20. Gao, Q., Feng, C.: Branding with social media: user gratifications, usage patterns, and brand message content strategies. Comput. Hum. Behav. **63**, 868–890 (2016). https://doi.org/10.1016/j.chb.2016.06.022

21. Smock, A.D., Ellison, N.B., Lampe, C., Wohn, D.Y.: Facebook as a toolkit: a uses and gratification approach to unbundling feature use. Comput. Hum. Behav. **27**, 2322–2329 (2011). https://doi.org/10.1016/j.chb.2011.07.011

22. Kaplan, A.M., Haenlein, M.: Users of the world, unite! The challenges and opportunities of social media. Bus. Horiz. **53**, 59–68 (2010). https://doi.org/10.1016/j.bushor.2009.09.003

23. Zhu, Y.-Q., Chen, H.-G.: Social media and human need satisfaction: implications for social media marketing. Bus. Horizons **58**, 335–345 (2015). https://doi.org/10.1016/j.bushor.2015.01.006

24. Michailova, S., Husted, K.: Knowledge sharing in Russian companies with foreign participation. Manag. Int. **6**, 19–28 (2003)

25. Alexandre, A., Martin, M., Wei, L., et al.: Cultural influences on knowledge sharing through online communities of practice. J. Knowl. Manag. **10**, 94–107 (2006). https://doi.org/10.1108/13673270610650139

26. Kalia, P.: Webographics and perceived service quality: an Indian e-retail context. Int. J. Serv. Econ. Manag. **8**, 152–168 (2017). https://doi.org/10.1504/IJSEM.2017.10012733

27. Pavlou, P.A.: Consumer acceptance of electronic commerce: integrating trust and risk with the technology acceptance model. Int. J. Electron. Commer. **7**, 101–134 (2003)

28. Cassell, J., Bickmore, T.: External manifestations of trustworthiness in the interface. Commun. ACM **43**, 50–56 (2000). https://doi.org/10.1145/355112.355123

29. Shin, D.H.: The effects of trust, security and privacy in social networking: a security-based approach to understand the pattern of adoption. Interact. Comput. **22**, 428–438 (2010). https://doi.org/10.1016/j.intcom.2010.05.001

30. Soares, A.M., Farhangmehr, M., Shoham, A.: Hofstede's dimensions of culture in international marketing studies. J. Bus. Res. **60**, 277–284 (2007). https://doi.org/10.1016/j.jbusres.2006.10.018

31. Boddy, C.R.: Sample size for qualitative research. Qual. Mark. Res. Int. J. **19**, 426–432 (2016). https://doi.org/10.1108/QMR-06-2016-0053

32. Branthwaite, A., Patterson, S.: The power of qualitative research in the era of social media. Qual. Mark. Res. Int. J. **14**, 430–440 (2011). https://doi.org/10.1108/13522751111163245

33. Hall, E.T.: Beyond Culture. Anchor Books (1976)

34. Parks, M.R., Floyd, K.: Making friends in cyberspace. J. Commun. **46**, 80–97 (1996). https://doi.org/10.1111/j.1460-2466.1996.tb01462.x

35. Bugshan, H.: Co-innovation: the role of online communities. J. Strateg. Mark. **23**, 175–186 (2015). https://doi.org/10.1080/0965254X.2014.920905

36. Pornpitakpan, C.: The persuasiveness of source credibility: a critical review of five decades' evidence. J. Appl. Soc. Psychol. **34**, 243–281 (2004)

37. Boase, J., Horrigan, J.B., Wellman, B., Rainie, L.: The strength of internet ties: the internet and email aid users in maintaining their social networks and provide pathways to help when people face big decisions. In: Pew Internet American Life Project (2006). http://www.pewinternet.org/files/old-media/Files/Reports/2006/PIP_Internet_ties.pdf.pdf. Accessed 29 June 2018

38. Maschio, T.J.: Culture, desire and consumer culture in America in the new age of social media. Qual. Mark. Res. Int. J. **19**, 416–425 (2016)

POS Tagging and Structural Annotation of Handwritten Text Image Corpus of Devnagari Script

Maninder Singh Nehra[1]([⊠]), Neeta Nain[1], Mushtaq Ahmed[1], and Deepa Modi[2]

[1] Malaviya National Institute of Technology, Jaipur, India
maninder4unehra@yahoo.com
[2] Swami Keshvanand Institute of Technology, Management and Gramothan, Jaipur, India

Abstract. Natural Language Processing (NLP) germaneness required a large benchmark annotated dataset. Handwritten and impressed text corpus plays a momentous role in pattern recognition algorithm for benchmarking. Part-of-speech tagging is very recurrent and subjugated types of annotation. Because POS tagging is significant to many linguistic annotations like lemmatization, syntactic parsing, semantic annotation, etc. Part-of-Speech tagging together with the structural annotations of handwritten text image corpus of Devnagari script of 1300 handwritten form collected from different geographical location and demographics are narrating in this paper.

Keywords: Corpus · Hindi · Handwritten · POS and annotation

1 Introduction

Linguistic research along with pattern recognition requires a standard benchmark dataset or corpus [21–23]. As reported by Swaran [1] There are one hundred twenty two (122) major languages in India and two thousand three hundred seventy one (2371) dialects, in which constitutionally 22 are standard languages. Adequate research has been done in the area of NLP for many languages, including English, but by virtue of typical character style and modifiers ample research is not done for Devnagari script.

The collection of texts is a corpus, texts are written (handwritten or printed) or speech, that are recorded in a digitized form [2]. A corpus may be quite small or very large, containing many millions of words. The corpus text is analyzed with the help of frequency list. The corpus linguistic plays an important functionality in the research of a language. According to Tim et al. [3] the corpus should be representative, widely available, should only contain public domain material, should not be larger than necessary and should be perceived to be valid and useful. Based on the nature of text and application, the corpus can be classified as Generalized corpus, Specialized corpus, Monitor corpus etc. According to the category of text corpus is Monolingual corpus, Bilingual corpus or Multilingual corpus. These corpora may be Un-annotated corpus or Annotated corpus. The un-annotated corpus is just a group of text with no portrayal or markup of text. On the hand annotated ones is a collection of text in a organized and technical manner in a Encoding Standard of corpus [4]. The different levels of corpus annotation are: Morphology, Lexicon, Phrase, Syntax, Semantic and Discourse Analysis. These annotations can be applied in sequence or separately [5]. POS Tagging (POST) is the procedure of mark-up a token in a sentence with a particular POS tag like Noun, Pronoun, Adverb, Adjectives etc. [5]. That are the input for several other annotation method like syntactic parsing, lemmatization, semantic annotation etc. Examples of POS tagging for Devnagari script text are:

Input Hindi Text

पेरिस आतंकी हमलों के मास्टरमाइंड के खिलाफ कार्रवाई के दौरान गई महिला फिदायीन के बारे में नए खुलासे हुए हैं और कुछ सामने आए हैं |

Part-of-Speech Tagging of Text

पेरिस_NNC आतंकी_NN हमलों_NN के_PREP मास्टरमाइंड_NN के_PREP खिलाफ_PREP कार्रवाई_NN के_PREP दौरान_PREP गई_VAUX महिला_NNC फिदायीन_NN के_PREP बारे_NN में_PREP नए_JJ खुलासे_NN हुए_VAUX हैं_VAUX और_CC कुछ_QF सामने_NLOC आए_VFM हैं_VAUX |_PUNC.

2 Literature Review

The first corpus was created for English language in 1961, known as Brown corpus of size 1 million words [6]. In the last decade of 20th-century various corpus of handwritten and printed text, of different languages and categories were developed. Marti and Bunke [7] developed a handwritten database for English language known as IAM database. The data set contains of approximately 556 written forms written by 250 different writers, total 4881 lines of text and 43751 handwritten words. Raza and Abidi [8] design handwritten Urdu benchmark data set. In this handwritten corpus dataset, there is 400 filled form written with the help of 200 writers. In this corpus sports, entertainment, blogs, religion, editorials and science fiction categories are selected. The database consists of 23833 printed words, 23812 handwritten words, 2068 Urdu sentences with 2051 Urdu lines and 783 numerals. Matthias and Bunke [9] developed handwritten off-line data set for English (IAM). The database consists 417 forms, 3703 text lines, and 25000 handwritten words. Johansson and Leech [10] developed corpus named as LOB. It is the foremost texts corpus of modern English language in computer readable form. A different version of this corpus released. This corpus consists 11827 sentences and 134740 words. Methasate et al. [11] developed a handwritten character corpus for Thai language. The corpus contains of online and offline handwritten characters. The corpus is separated into two parts, offline and online handwritten character corpus of approximately 30,000 images for the online corpus, 14,000 images for offline corpus by 63 writers and 143 writers for the offline corpus. Francisco, Jesus etc. [12] developed a handwritten signature corpus known as GPDS-960 corpus. It is a corpus of signature twenty four authentic signature and 30 forgery of 960 individual. IRESTE [13] is a handwritten dual data set of French and English. Prakash et al. [14] designed an Urdu corpus with annotation at four tiers. These corpora are detailed in Table 1.

Table 1. Widely used standard handwritten database

Name of database	Language/Script	Size of database	No. of writers
IAM [7]	English	556-Forms 4881-Lines 43751-Words	250
URDU CORPUS [8]	Urdu	400-Forms 2068-Sentences 783-Numerals	200
Matthias and Bunke [9]	English	417-Forms 3703-Lines 25000-Words	-
LOB [10]	English	11827-Sentences 134740-Words	-
Thai [11]	Thai	30000-Online images 14000-Offline images	63
GPDS [12]	English	24-Genuine signature 30-Forgeries	960
IRESTE [13]	French English	28657-French words 7689-English words	-
CALAM [14]	Urdu	1000-Forms	500

There exists many of implementation for syntactic and morphological annotation for Devnagari script. Deepa et al. [5] developed POS tagger for Devnagari Hindi using a Rule-based technique with a precision of 91.84%. Garg et al. [15] proposed a Hindi POS tagger based on defined rule with 87.55% precision. Savant et al. [16] developed Devnagari Hindi POS tagging and chunking system, based on maximum entropy and Markov model with a precision result of 88.4%. Shrivastav et al. [17] designed Hindi tagger by Naive Stemming: Harnessing morphological information with no extensive linguistic knowledge based on Hidden Markov model approach with 93.12% precision result. Kuhoo et al. [18] developed a POS tagger for Hindi by CN2 algorithm with a precision of 93.45%. A Dalal et al. [19] designed POS tagger of Devnagari Hindi based on maximum entropy and Markov model with the 94.38% precision. These tagging methods are detailed in Table 2.

3 Motivation

The mother of approximately all Indian scripts is Devnagari. The official and national language of India is Hindi. In Indian constitution, there are twenty two are scheduled languages and Hindi is one of them. The 39.29% of total population (422 million) speak Hindi language and Hindi is the mother tongue of 25.071% population [1]. Throughout world fourth largest speaking community is Hindi-Urdu after English, Mandarin Chinese and Spanish [14]. So, it is useful to develop the Hindi corpus with part-of-speech tagging and structural annotation.

Table 2. Different tagging methods

Method used	Model used	No. of Tags	Training set (no. of words)	Testing set (no. of words)	Perfo. Result
Part-of-Speech Tagging of Hindi Corpus [5]	Rule-based model	29	9,000	3,189	91.84%
Rule-Based Hindi Part of Speech Tagger [15]	Rule-based model	30	18249	26149	87.55%
Hindi Part-of-Speech Tagging and Chunking [16]	Maximum entropy Markov model	29	35000	8750	88.4%
Hindi POS Tagger Using Naive Stemming [17]	Hidden Markov model	NA	53400	13500	93.12%
Morphological - Constructing a POS Tagger for Hindi [18]	CN2 Algorithm	27	15562	3890	93.45%
Building Feature Rich POS Tagger [19]	Maximum entropy Markov model	27	15,562	3890	94.38%

4 Proposed Work Hindi Handwritten Text Image Corpus

To develop a proficient dataset, it should have different genres. The proposed corpus development is divided into 7 genres and 19 sub-genres. The layout of data collection form with constraints are designed in such a way that it is straightforward, evident and comfortable for writers. A significant information on a single form makes the availability of corpus in the multi-disciplinary research areas of NLP. The form is divided into four parts. The top block is the header, for writers to write their own information. Then printed text block of 3 to 5 lines or 50 to 70 words approximately. Space is given for handwritten text and footer block for signature, address of form generators and UID of the form is given. The proposed handwritten text images corpus is designed by collecting approximately 1300 handwritten text images that include all domains out of which 140 forms are discarded. Total no. of words are 63800 out of which 7.83% are discarded. Proposed Devnagari script handwritten dataset with POS tagging and structural annotation possess 58800 words. A sample of structural annotation and uploaded form are depicted in Fig. 1. Which also includes GT information of the uploaded data and some the other as:

- Category and subcategory of the form.
- Unicode corresponding to the handwritten text.
- Handwritten lines in the form.

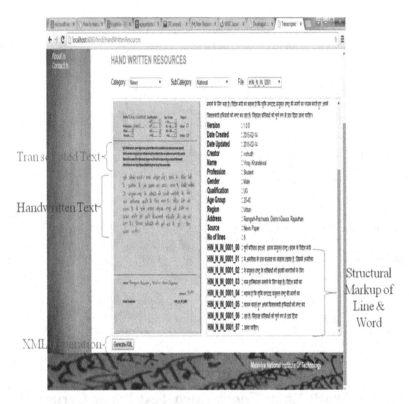

Fig. 1. Sample structural annotation of the handwritten text image

5 Proposed Part-of-Speech Tagging

The three main functions of POS dividing the input text in sentences, words and assign tags (POS) for input text.

5.1 Experimental Setup and POS Tagging Method

In the proposed part-of-tagging method the unigram and rules-based approaches are used with 29 tag set. A manually developed dataset of Hindi is used in the system. The system is trained over 11,600 Hindi words with their individual POS tags. The proposed system work in two phases. In first phase it tag known words according to trained dataset and in second phase it tag unknown words. In the first phase, according to unigram model, POS tags for a word don't depend upon its previous tag [19]. The proposed system follows this model and assigns a most probable tag to all known words based on the already trained dataset. The unigram probability of a word w_i with tag t_i are calculated as:

$$P(t_i/w_i) = \text{frequency}(w_i/t_i)\text{frequency}(w_i). \ldots \qquad (1)$$

In the second phase the unknown words are tagged using the rules base on lexical and contextual features.

5.1.1 Rules Based upon Lexical Features

The lexical feature based rule includes rules based prefixes, suffixes, and rules based on regular expression. These rules find particular patterns in the input text. Lexical features are do not effected from the context of a current word.

1. Rules-Based on Regular Expressions: This rules search patterns like punctuation mark (. : ; I), special symbols (# $ %), numerical data (930 735 211) etc. in the input text and assign tag accordingly.
2. Rules-Based on Prefixes: Several Hindi words that starts with prefixes like अव, अप etc. These words can be tagged with high probability tag like Noun, Adjective etc.
3. Rules-Based on Suffixes: In this rules, the words are also tagged with high probability as in prefixes based rules.

5.1.2 Rules Based upon Contextual Features

These rules are chosen from Hindi grammar and based on various combination previous tag, current tag and next tag depends upon the words context The words are tagged according.

The Rules are as Follow

1. If current tag is postposition then the previous tag will be probably the noun.
   Example: सोहन ने पानी मे फुल देखा।
   In this example, "मे" is postposition and "पानी" is noun.
2. If current tag is adjective then the next tag will be probably the noun.
   Example: सोहन खुबसुरत लङ्क़ा है।
   In this example, "खुबसुरत" is adjective and "लङ्क़ा" is noun.
3. If current tag is the verb like Finite Main, Nonfinite Adjective, Nonfinite Adverbial or Nonfinite Nominal, then the previous tag will be probably the noun.
   Example: वह शहर जा रहा है।
   In this example, "जा" is verb and "शहर" is noun.
4. If current tag is pronoun then next tag will be probably the noun.
   Example: यह हमारी कार है ।
   In this example, "हमारी" is pronoun and "कार" is noun.
5. If current tag is noun and the next tag is proper noun then the current tag will be probably compound proper noun.
   Example: संजवि अग़वाल जा रहा है ।
   In this example, "संजवि" is compound proper noun and "अग़वाल" is proper noun.
6. If current tag is auxiliary verb then the previous tag will be probably the finite main verb.
   In the previous example "रहा" is auxiliary verb and "जा" is main verb.

7. If two consecutive tag are noun then the first tag will be probably compound common noun.

Example: केन्द्र सरकार ने बेरोजगारो के लिए काम नही कयिा।

In this example, "केन्द्र" is compound common noun and "सरकार" is noun.

8. If current tag is the verb and previous is noun, adjectives or adverb then previous tag is changed to a noun in kriya mula, adjective in kriya mulla or adverb in kriya mulla respectively.

Example: हमने फल लाल होते ही तोड़ लयिा।

In this example, "होते" is verb and "लाल" is adjective in kriya mulla.

5.1.3 Ambiguity Removing

A word can have more than one meaning and according to its meaning in the given context it can have more than one POS tags in different sentences. So the ambiguity removing rules are implemented in the POS tagger. The process is depicted in Fig. 2. The ambiguity removing process checks the context of the ambiguous word and assigns or re-assigns tag to the word accordingly. Example: आम आदमी आम बहुत पसंद करते हैं | In this sentence the word "आम" appears twice but have different POS tag at both places. The ambiguity removing process checks the context of first appearance of "आम", as next tag is noun so "आम" is tagged as adjective. While second appearance of "आम", next tag is not noun so "आम" is tagged as noun.

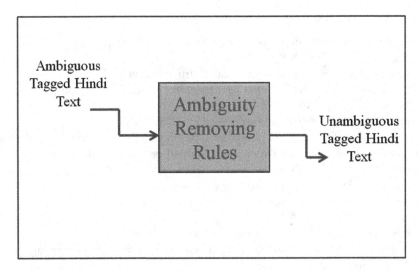

Fig. 2. Ambiguity removing process in proposed POS tagging approach

The complete experimental setups of POS tagging are depicted in Fig. 3. Sample results of splitting and tagging are shown in Figs. 4 and 5 respectively. The proposed system split input text into sentences with 100% accuracy.

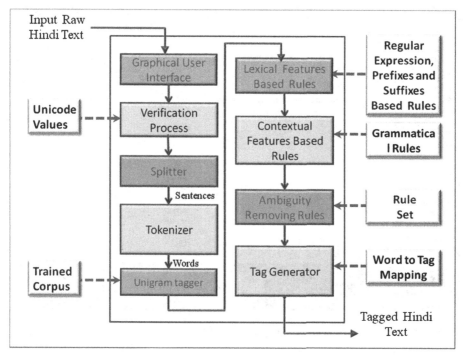

Fig. 3. Experimental framework of proposed POS tagging approach

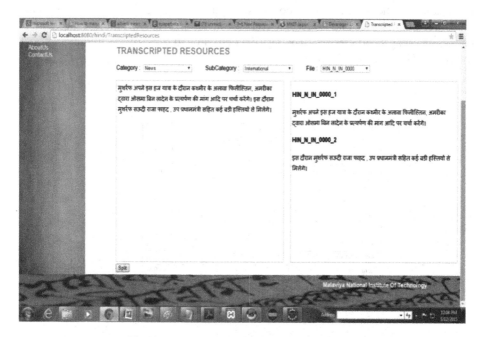

Fig. 4. Sample result of splitting the input text

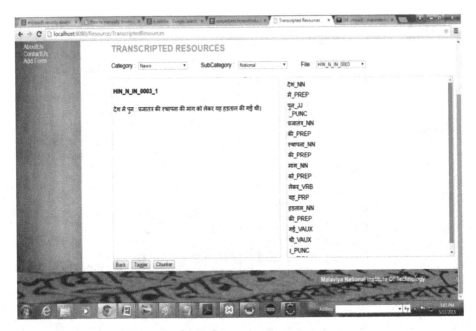

Fig. 5. Sample result of POS tagging

5.2   Performance Result of POS Tagging Method

Hindi text is collected of various domains like news, science & technology, history, literature, politics etc. and collected from recourses like news, online data, stories, books, newspaper, articles etc. The test data are around 24% of the training set and follow the Gaussian distribution.

Table 3. Performance result of POS tagging

Domain	No. of Words	No. of Tagged Words	Correctly Tagged Words	P	A	S
Science & Technology	340	327	311	95.10	91.47	96.17
Literature	500	463	435	94.38	87.4	92.60
News	662	558	538	96.42	81.26	84.29
Politics	516	487	471	96.71	91.27	94.37
History	382	360	341	94.72	89.26	94.24
Architecture	226	168	157	94.64	70.35	74.33
Economy	236	201	187	95.02	80.93	85.16

The system has validated on data sets as shown in Fig. 6, through holdout method of cross-validation. In this method the complete dataset is divided into two sets, the bigger set is training set, and smaller one is test data set. Rules are derived from the training set, and the upcoming data are tagged according to these rules. The system performance are evaluate as precision and accuracy and sensitivity detailed as Fayyad et al. [20].

$$\text{Precision} = \text{Correctly tagged words} / \text{Total tagged words} \ldots \ldots \tag{2}$$

$$\text{Accuracy} = \text{Correctly tagged words} / \text{Total words} \ldots \ldots \ldots \ldots \ldots \tag{3}$$

$$\text{Sensitivity} = \text{Total tagged words} / \text{Total words} \ldots \ldots \ldots \ldots \ldots \tag{4}$$

The performance of POS tagging is shown in Table 3, by P- Precision, A- Accuracy, and S-Sensitivity. The proposed system achieved 95.28% average precision, 84.56% average accuracy and 88.73% average sensitivity. Comparative results of precision, accuracy, and sensitivity are graphically depicted in Fig. 7. Yield precisions by system are uppermost with superior accuracies while having data set in this range (Up to best knowledge).

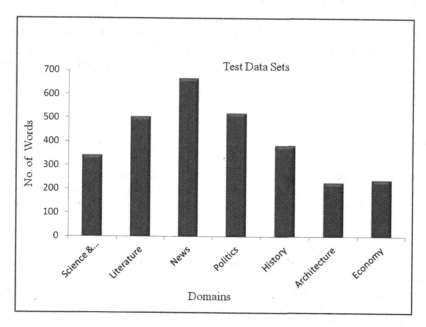

Fig. 6. Test data set

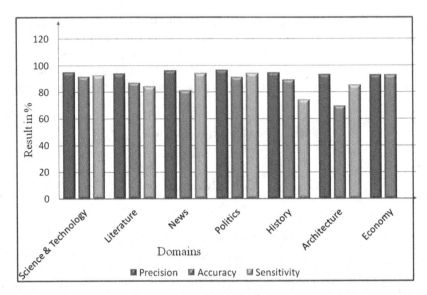

Fig. 7. Comparative analysis of precision, accuracy and sensitivity for various genre

6   Conclusion

A corpus has been designed for Devnagari script which having a large scaled data set with a different number of lines and tokens. The corpus belongs to seven diverse categories and 19 sub categories. The handwritten text images corpus is designed by collecting approximately 1300 handwritten text images and 58,800 words that include all domains. Structural annotation and POS tagging, Using unigram and rule-based method with a precision of 95.28%. The yield precisions by the model are utmost with high-quality accuracies. For linguistic related research data set would be useful on the Devnagari Hindi.

References

1. Swaran: Challenges of Multilingual Web in India: Technology Development & Standardization Perspective. Department Information technology Government of India
2. McCarthy, M.: From Corpus to Course Book. Cambridge University Press (2004). http://www.cambridge.org
3. Tim, B., et al.: A corpus for the evaluation of lossless compression algorithms. In: IEEE Conference, pp. 201–210 (1997)
4. Prakash, C., et al.: An annotated urdu corpus of handwritten text image and benchmarking of corpus. In: MIPRO 2014, Croatia (2014)
5. Deepa, M., et al.: A survey of techniques for two-level corpus annotation for Hindi. Int. Bull. Math. Res. **2**, 194–206 (2015)
6. Francis, W.N.: Brown Corpus Manual. Brown University, July 1979

7. Marti, U.-V., Bunke, H.: A full English sentence database for off-line handwriting recognition. In: Proceedings of the 5th ICDAR, pp. 705–708 (1999)
8. Raza, A., Abidi, A.: An unconstrained benchmark urdu handwritten sentence database with automatic line segmentation. In: ICFHR, pp. 489–494 (2012)
9. Marti, U.-V., Bunke, H.: The IAM-database: an English sentence database for offline handwriting recognition. In: ICDAR, pp. 35–39 (2002)
10. Johansson, S., Leech, G.: The Tagged Lob Corpus. Norwegian Computing Centre for the Humanities, Bergen (1986)
11. Sutat, S., Methasate, L.: Thai handwritten character corpus. In: ISCIT, pp. 486–491 (2004)
12. Francisco, V., Jesús, A.: Off-line handwritten signature GPDS-960 corpus. In: Proceedings of Ninth International Conference on Document Analysis and Recognition, pp. 764–768 (2007)
13. Christian, V.-C., et al.: The IRESTE On/Off (IRONOFF) dual handwriting database. In: Proceedings of 5th ICDAR (1999)
14. Choudhary, P., Nain, N.: A four-tier annotated urdu handwritten text image dataset for multidisciplinary research on urdu script. ACM Trans. Asian Low-Resour. Lang. Inf. Process. **15**(4), 26 (2016)
15. Garg, N., et al.: Rule- based Hindi part-of-speech tagger. In: COLING 2012, pp. 163–174 (2012)
16. Sawant, U., et al.: Hindi part-of-speech tagging and chunking: a maximum entropy approach. In: Proceeding of the NLPAI Machine Learning, Mumbai, India (2006)
17. Shrivastava, M., et al.: Hindi pos tagger using naïve stemming: harnessing morphological information without extensive linguistic knowledge (2008)
18. Kuhoo, G., et al.: Morphological richness offsets resource poverty-an experience in building a pos tagger for Hindi. In: COLING, pp. 779–786 (2006)
19. Dalal, A., et al.: Building feature rich pos tagger for morphologically rich languages: experiences in Hindi. In: ICON 2007
20. Fayyad, U.M., et al.: Advances in Knowledge Discovery and Data Mining. American Association for Artificial Intelligence, Menlo Park (1996)
21. Mittal, N., Agarwal, B., Chouhan, G., Bania, N., Pareek, P.: Sentiment analysis of hindi reviews based on negation and discourse relation. In: The Proceedings of 11th Workshop on Asian Language Resources, (in conjunction with IJCNLP-2013), pp. 45–50 (2013)
22. Yadav, M., Purwar, R.K., Mittal, M.: Handwritten hindi character recognition-a review. IET Image Proc. **12**(11), 1919–1933 (2018). https://doi.org/10.1049/iet-ipr.2017.0184
23. Mittal, N., Agarwal, B., Chouhan, G., Pareek, P., Bania, N.: Discourse based sentiment analysis for hindi reviews. In: Maji, P., Ghosh, A., Murty, M.N., Ghosh, K., Pal, S.K. (eds.) PReMI 2013. LNCS, vol. 8251, pp. 720–725. Springer, Heidelberg (2013). https://doi.org/10.1007/978-3-642-45062-4_102

Automatic Parallelization of C Code Using OpenMP

Gaurav Singal[1]($\boxtimes$), Dinesh Gopalani[2], Riti Kushwaha[2], and Tapas Badal[1]

[1] CSE Department, Bennett University, Greater Noida 201310, India
gauravsingal789@gmail.com, tapas.badal@bennett.edu.in
[2] Department of CSE, MNIT, Jaipur 302017, India
dgopalani.cse@mnit.ac.in, riti.kushwaha07@gmail.com

Abstract. Automatic parallelization is necessary for all system. Every person wants the program to execute as soon as possible. Now Days, programmer want to get run faster the sequential program. Automatic parallelization is the greatest challenge in now days. Parallelization implies converting the sequential code to parallel code to getting better utilization of multi-core processor. In parallelization, multi-core use the memory in sharing mode or massage passing. Now day's programmers don't want to take extra overheads of parallelization because they want it from the compiler that's called automatic parallelization. Its main reason is to free the programmers from manual parallelization process. The conversion of a program into parallelize form is very complex work due to program analysis and an unknown value of the variable during compile time. The main reason of conversion is execution time of program due to loops, so the most challenging task is to parallelize the loops and run it on multi-core by breaking the loop iterations. In parallelization process, the compiler must have to check the dependent between loop statements that they are independent of each other. If they are dependent or effect the other statement by running the statement in parallel, so it does not convert it. After checking the dependency, test converts it into parallelization by using OpenMP API. We add some line of OpenMP for enabling parallelization in the loop.

Keywords: OpenMP · Automatic parallelization · Data dependency · C programming · Lex and Yacc · Language

1 Introduction

Today's multi-core processor dominate the market of computer processors. Traditional sequential program not utilize the processing power of multi-core architecture, not able to achieve speed up through multi-core processor. Availability of multi-core architecture enable execution of program in parallel and result in great speed up, which can't be achieved in single core architecture. Parallelization of program can be done in two ways, either programmer can develop program with adding parallel constraints in a program by him self or parallelization is

© Springer Nature Singapore Pte Ltd. 2019
A. K. Somani et al. (Eds.): ICETCE 2019, CCIS 985, pp. 298–309, 2019.
https://doi.org/10.1007/978-981-13-8300-7_25

done with automatically parallelize compiler. Second approach ease task of programming, thus programmer doesn't need to pay a lot attention on parallelizing construct while development of program. In sequential program loops consume a large amount of complete execution time. So our main aim is to break the loop to different processes and parallelization [1] of loop can reduce execution time significantly, thus great performance is gained and maximum utilization of core in processors [2].

Parallel processing is used to increase performance of systems. We can apply parallelization on single processor as well as multiprocessor. Applying it on multiprocessor is achieving more speedup. Automatic parallelization of loop will give us high performance computers.

Now days, the speed of processor is increased and now multiprocessor are shipped in the every personal computer. In the last decade most of compilers are working on sequential programming model. This is very time consuming process. Now, User doesn't want to wait for long time to run a program. Parallelization is more necessary in the systems. Programmers want to write a code in sequential manner and don't want to take overhead of parallelization [3] so it is want from compilers. So due to parallelization we get speedup in our program execution time. So implementation based the main purpose is to break the loop in to different thread so we achieve the maximum utilization of multi processors. In parallelization shared memory is also important concern.

Objective of this work is to generate a automatic compiler that automatically convert C code to parallelize one. Automatic parallelization is used to achieve proper utilization of resources or processors. In C programming, parallelization achieving is more tough so we add some thread to easy the process. Automatic compiler takes c program and apply OpenMP construct to given loop to parallelize it. Our main concern is to break the loop and send it onto different threads so, each thread perform their execution separately, but we have to take care of variable that are used in loop and shared to each thread a copy. We give the automatic conversion of serial program to parallel without taking the knowledge of parallelization to the user. So user easily write a program in sequential language but it run on parallel compiler and give output with better speed-up.

We are trying to develop a compiler that automatically generate a parallel code for sequential program. First, in this we generate the equation from loop statements, after that we reduce the problem to simpler way and apply dependence test to find the dependence between loop statements to convert it on parallelize. For dependency we apply gcd test for check the coefficients and after that we apply banergee test to check their lower bound or upper bound of loop to find it in range. So we add OpenMP construct to the code for parallelization to achieve maximum speedup. In this we take a sequential C program as a input and parallelize code is generated that can run on multi-core processors. So we achieve some level of processor utilization. In this work we use lex and yacc [4] tool to generate token and parsed the token to meaning full statements. Using this tool we easily get the position where we have to put OpenMP construct and check the dependence. So our compiler is getting input of C program then convert it on parallelize form where is requirement and give maximum speedup. We implement some as to shared variable and some as private for as per use in run time Whereas, array variable is always declared in shared. We use some mathematical forms to test the dependence in loop statements like echelon and uni-modular matrix. Using mathematical forms it easy to understand the loop statements and easily perform action on statements. So at last we make a compiler that covert automatically sequential program to parallelize without knowledge of this to user.

In this section, we describe the problem of sequential program and requirement of parallelization. In the second section, we discuss about OpenMP language. OpenMP constructs, how to used it and what about threads? In third section we classifies the dependency and different test of dependency. In fourth section, we discuss about this purpose work, algorithm, flow graph of process and give a conversion of sequential program to parallelize program. In last section, we write conclusion and future work of project.

2 OpenMP

OpenMP (Open Multi Processor) OpenMP is shared memory API (Application Program Interfaces) that facilitate shared memory parallelism in programming [5]. It is used to create shared memory parallel programs. It's not a new programming language. It's notation can be bind with different languages like FORTRAN, C, C++, etc. The used of OpenMP code at desired position it is very beneficial for parallel with minimal modification in the code. The advantage of OpenMP is that the responsibility of working out in parallel is up to the compiler. OpenMP have to declare both shared and private variable. In this we don't required to change the most of the program but more important work is to identifying the parallel code.

OpenMP execute a program by groups of threads. But in multiple threads is working then we had to share the resources. Each individual thread also needs their own resources. Multiple threads run on a single processor to execute a code in parallel with context switching. We have to shared a copy of variable to each threads. The program run with single thread as master thread then the multiple thread is fork when parallel code is begin and at the end it convert it again initial thread. Figure 1 shows, the team of threads activated when the code is able to run in parallel and after completion of threads it combined it on master thread. In OpenMP program, we write a parallel code without changing the sequential code that used if we not getting better performance then we can use sequential program. The omp_get_thread_num value 0 shows that its master thread. And if our compiler is not supported for OpenMP Api, so we used this statement that not give error [5].

3 Introduction to Data Dependence's

In Sequential program their are group of statement. Their is an action that perform on certain sequence. After converting it onto parallelization their is no

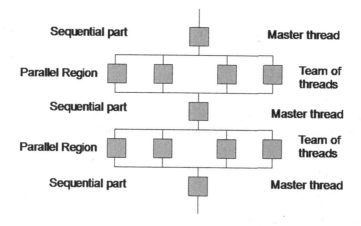

Fig. 1. Fork join programming model

sequence because any sentence can execute before any other statement. So we have to check that their is no reflect in the output after this [6] So we check the variable that modifying or accessing the another variable because any variable in first statement is getting read and in second statement is writing to the same memory location [7]. If we interchange the order of these statements then reading variable statement gives incorrect output. So, this is called Dependence. So after changing the order of statements does not reflect the output of main program means there is no dependence [8].

In this we classified different type of dependency like flow, anti, output, etc. We also discuss about dependency test that used to check the dependency for loop statements.

3.1 Data Dependence

There are two statements S1 and S2. Then S2 is depends on S1 if they came in one of three condition.

- S1 is reading and S2 is writing
- S1 is writing and S2 is reading
- S1 and S2 both are in writing

and there is a path from S1 to S2 in the block. It classified in three categories

- Flow Dependence(RAW)
  Statement S1 is writing into a memory location and later on is read by S2 statement.

```
S1 :    arr[] = ....
S2 :    .... = arr[]
```

– Anti Dependence(WAR)
Statement S1 is reading into a memory location and later on is writes by S2 statement.

```
S1 :      .... = arr[]
S2 :      arr[] = ....
```

– Output Dependence(WAW)
Statement S1 is writing into a memory location and later on is also write by S2 statement.

```
S1 :      arr[] = ....
S2 :      arr[] = ....
```

When executing a parallel program than the order of execution is change according to sequential execution. So the result of execution may also change, so we have to check all the case of statements. This is known by dependence analysis [9].

3.2 GCD Test

GCD (Greatest Common Divisor) test is used for finding the dependence between loop statements. In this test we take the gcd of all coefficients in the equation and then it should be divide to constant part of equation. If it's not divisible then we say their is no dependency. Below we show gcd test in two part for one dimensional and two dimensional equations problem [10, 11].

3.3 Banergee's Test

Banergee's test find the dependency using intermediate value. In this we compute the lower bound and the upper bound of the equations [10].

3.4 Omega Test

The omega test is used for find the exact dependency between the statement. This test determine whether their is integer solution or not. In this our input is linear equations in the form of linear equalities and inequalities [12]. This process is goes on different steps.

– Normalizing constraints
– Equalities Constraints
– Dealing With Inequalities Constraints

4   Proposed Work

In the starting chapters we introduce some data dependency tests, information about OpenMP. In this we show the full technique of my project that we have applied in my project from starting to the end. In this chapter we are explaining flow graph, algorithm, example of Dependence test and conversion of sequential c code to parallelize C Code. In this work, we have used some tool like Lex and Yacc for parsing of input c program after then we used mathematical form to solve out the big equations and then apply dependence test for parallelization. In the last we generate output file of parallelize code.

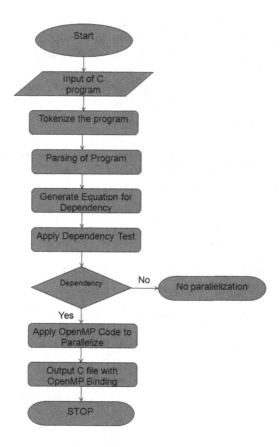

Fig. 2. Flow chart of my project

4.1   Flow Diagram

Flow graph is used for showing the flow of program. In Fig. 2, we show the flow of whole process of the project. In the starting we generate the token of input file using Lexical analyzer. Then we parser for parsing the token stream into meaning full sentence. So, now we can find out the loop statements where we have separate the code for parallelization and generate equations for applying dependence test. In the above chapter we have explained a dependence test and find the dependency of loop after that we check that is true or false. If dependence is exist then we add the code of OpenMP into input file and also defined the shared and private variable during run time of code. At last we display the output file with OpenMP binding.

4.2   Algorithm

We implement an algorithm for whole process that we done in my project. It takes an input file of C code that we have to parallelize. In the output we generate the parallelize code of input file to achieve maximum utilization of processor. Inside the algorithm we have used some more function like echelon reduction method for simplified the equations. We use dependence algorithm word in algorithm that is basically telling that its applying dependence test that is describe on last chapter.

Algorithm 1. Process_ of_ project()

1: **Input:** file : Input C program
2:             Dep: Dependency variable that store the result of dependency test
3: **Output:** file 1 : Output C program with OpenMP
4: Generate token form Lexical Analyzer
5: Parsed the token to grammar using Yacc
6: Recognize Loop from Input File
7: Generate mathematical Equations from loop statements
8: Dep=Dependence_algorithm()
9: **if** $dep == 1$ **then**
10:     Add OpenMP threads for parallelize.
11: **else**
12:     No Output is generated
13:     Exit to program.
14: **end if**
15: file1=Output_file

4.3   Example of Dependency Test

Now we are taking an loop with one nested level and check the dependency using data dependence algorithm by gcd test and banergee test. In this first we generate input matrix, continue with generation of Echelon and Uni-modular matrix and then apple test to find the dependency of statements.

4.4   Example of Conversion of C Code Using OpenMP

Here we explaining the conversion of sequential C code to parallelize version using OpenMP api. In this we taking a two matrix value multiplication in input and after parallelize we used thread to do sum in shared mode. In this program first we check where we have to apply parallelization. So using lex and yacc tool we find the loop in the program and apply dependence test to loop statements. After checking dependence we add OpenMP constraints in sequential program before each loop. At last we check that the variable are in which mode like sharing or personal copy of each variable to each thread.

Sequential Dot-Product

```c
#include<stdio.h>
void main()
{
    for (i = 0; i < n; i++)
    {
        a[i] = i * 1;
        b [i] = i * 2;
    }
}
```

Parallelize Dot-Product with OpenMP

```c
#include<stdio.h>
#ifdef_OPENMP
    #include<omp.h>
#else
    #define omp_get_thread_num() 0
#endif
void main()
{
 int tid, nthreads;
 //Adding OpenMP parallel Direcitves
 #pragma omp parallel for shared
 (a_arr, b_arr) private(i)
 for (i = 0; i < n; i++)
 {
  tid = omp_get_thread_num()
  if (tid == 0)
  {
   nthreads = omp_get_num_threads();
   printf("Threads =%d \n", nthreads);
  }
 a_arr [i] = i * 1;
 b_arr [i] = i * 2;
 }
}
```

In above example we add OpenMP directive in both loop because in upper loop we just initialize the array variable with integer value so there is no dependence and both array are in sharing mode during multiple thread run. Example: Consider the problem of finding the dependence between S1 and S2 in the loop nest

```
L1: for i = 1 to 100 {
L2:   for j = 1 to i {
S1:     arr[2i-1,i+j] = .........
S2:     ........... = arr[3j,i+2]
      }
    }
```

caused by the two variables shown.

The output variable of $S1$ and the input variable of $S2$ are elements of a two-dimensional array arr. The variable $arr(2i - 1, i + j)$ can be written as $arr(IA + a_0)$ where $I = i + j$,

$$A = \begin{pmatrix} 2 & 1 \\ 0 & 1 \end{pmatrix} \text{ and } a_0 = (-1, 0)$$

since

$$arr(2I_1 - 1, I_1 + I_2) = (I_1 + I_2) \begin{pmatrix} 2 & 1 \\ 0 & 1 \end{pmatrix} + (-1, 0)$$

Similarly, the variable $arr(3I_2, I_1 + 2)$ can be written as $arr(IB + b_0)$, where

$$B = \begin{pmatrix} 0 & 1 \\ 3 & 0 \end{pmatrix} \text{ and } b_0 = (0, 2).$$

The instance of $arr(IA + a_0)$ for an index value i is $arr(iA + a_0)$, and the instance of $arr(IB + b_0)$ for an index value j is $arr(jB + b_0)$. They will represent the same memory location iff

$$iA + a_0 = jB + b_0$$

$$iA - jB = b_0 - a_0$$

This is the dependence equation for the above loop. It also be write as

$$(i; j) \begin{pmatrix} A \\ -B \end{pmatrix} = (1, 2)$$

where (i;j) is the result of adding the components of j to i, and $\begin{pmatrix} A \\ -B \end{pmatrix}$ is the result of adding the rows of $-B$ to lower part in the rows of A. after generating it can use for input in echelon reduction algorithm, we get the system:

$$(i_1, i_2, j_1, j_2) \begin{pmatrix} 2 & 1 \\ 0 & 1 \\ 0 & -1 \\ -3 & 0 \end{pmatrix} = (1, 2)$$

There are n equations in $2m$ variables, where n is the number of dimensions of the array X, and m is the number of loops in the loop nest; here $n = 2$ and $m = 2$.

To solve this system, we apply Algorithm 1, find two matrices

$$U = \begin{pmatrix} 1 & 0 & 0 & 1 \\ 0 & 0 & 1 & 0 \\ 3 & 0 & 3 & 2 \\ 0 & 1 & 1 & 0 \end{pmatrix} \text{ and } S = \begin{pmatrix} -1 & 1 \\ 0 & -1 \\ 0 & 0 \\ 0 & 0 \end{pmatrix}$$

such that U in unimodular, S is echelon, and $U\begin{pmatrix} A \\ -B \end{pmatrix} = S$. The system has an (integer) solution in the variable i_1, i_2, j_1, j_2 iff the system shown below has an (integer) solution in v_1, v_2, v_3, v_4 to multiply with echelon matrix and get the vector (1,2)

$$(v_1, v_2, v_3, v_4).S = (1, 2)$$

If the equation has no solution, then two instances of the variables of S1 and S2 can never represent the same memory location, and hence they cannot cause a dependence between S1 and S2. However, we find that equation does have a solution. The general solution is given by to solving the matrix.

$$- v_1 = 1 \tag{1}$$

$$v_1 - v_2 = 2 \tag{2}$$

$$\Rightarrow v_2 = -3$$

Note that the number of integer parameters (v_1, v_2, v_3, v_4) whose values are determined by equation is ρ where

$$\rho = \text{rank}\begin{pmatrix} A \\ -B \end{pmatrix} = \text{rank}(S) = 2.$$

The general solution to system can be written as

$$(i_1, i_2, j_1, j_2) = (v_1, v_2, v_3, v_4).U$$
$$= (-1, -3, t_3, t_4).U$$
$$= (-1 + 3v_3, v_4, -3 + 3v_4, -1 + 2v_3).$$

Let U_1 denote the matrix formed by the first two columns of U and U_2 the matrix formed by the last two columns. Then we have

$$(i_1, i_2) = (-1, -3, v_3, v_4).U_1$$
$$(j_1, j_2) = (-1, -3, v_3, v_4).U_2$$

Next, consider the constraints of the dependence problem. The index variables I_1 and I_2 must satisfy the inequalities:

$$1 \leq 100$$
$$1 \leq I_1$$

or

$$1 \leq I_1$$
$$1 \leq I_2$$
$$I_1 \leq 100$$
$$-I_1 + I_2 \leq 0.$$

In matrix notation, we can write

$$p_0 \leq IP$$
$$IQ \leq q_0$$

where $p_0 = (1,1), P = I_2, q_0 = (100,0)$, and $Q = \begin{pmatrix} 1 & -1 \\ 0 & 1 \end{pmatrix}$ Since (i_1, i_2) and (j_1, j_2) are values of $I = (I_1, I_2)$, they must satisfy

$$p_0 \leq (i_1, i_2)P$$
$$(i_1, i_2)Q \leq q_0$$
$$p_0 \leq (j_1, j_2)P$$
$$(j_1, j_2)Q \leq q_0$$

we generate the equation for (v_3, v_4) follow as.

$$-3v_3 \leq -2$$
$$-v_4 \leq -1$$
$$-3v_3 - v_4 \leq -4$$
$$-2v_3 \leq -2$$
$$-3v_3 \leq 101$$
$$-3v_3 + v_4 \leq -1$$
$$3v_3 + v_4 \leq 103$$
$$-v_3 - v_4 \leq -2$$

of using graph problem we can find out the value of v_3 and v_4.

For each solution (i_1, i_2, j_1, j_2) to system that satisfies above, the corresponding integer vector (v_3, v_4) must satisfy just above, Conversely, for each (integer) solution (v_3, v_4) to (5.1), we get from (5.3) and (integer) solution (i_1, i_2, j_1, j_2) to (5.1) satisfying the constraints (5.5). If a solution to (5.6) exists, then that would indicate the existence of two iterations (i_1, i_2) and (j_1, j_2) of the loop nest, such that the instance $S1(i_1, i_2)$ of $S1$ and the instance $S2(j_1, j_2)$ of $S2$ reference the same memory location.

5   Conclusion and Future Work

This work is to achieve automatic parallelization, maximum utilization of resource, time saving from running program in sequential mode. In this we use

OpenMP API for C language to achieve automatic parallelization. In our work, we generate a output file that consist of sequential code and OpenMP Statement that add threads in a program. In this our main aim to relive the programmer to overhead of parallelization because we are generating automatic compiler that convert automatically sequential program to parallelize one. Using OpenMP api we get maximum utilization of multi-core architecture systems, because sequential code never get proper utilization of multi-core process. We convert sequential C program to parallelize one which containing input C program with OpenMP directive to add parallelism. To achieve the parallelization we first find out where we apply parallelization or achieve maximum utilization of resource using lex and yacc tool to parse the input program. Then we apply OpenMP directive before for loop starting after check the dependency between loop statements. Our future work is to exceed the parallelization by adding advanced form of OpenMP to reduce the running time. In this we can increase the level of parallelization by applying it to on functions, pointer etc.

References

1. Li, Z., Yew, P.C., Zhu, C.Q.: An efficient data dependence analysis for parallelizing compilers. IEEE Trans. Parallel Distrib. Syst. **1**(1), 26–34 (1990)
2. Eigenmann, R., Hoeflinger, J., Padua, D.: On the automatic parallelization of the perfect benchmarks. IEEE Trans. Parallel Distrib. Syst. **9**(1), 5–23 (1998)
3. Banerjee, U.K.: Loop Parallelization. Kluwer Academic Publishers, Norwell (1994)
4. Levine, J.R., Mason, T., Brown, D.: lex & yacc O'Reilly & Associates. Sebastopol (1990)
5. Chapman, B., Jost, G., Van Der Pas, R.: Using OpenMP: portable shared memory parallel programming, vol. 10. MIT press, Cambridge (2008)
6. Pugh, W., Wonnacott, D.: Eliminating false data dependences using the omega test. In: ACM SIGPLAN Notices, vol. 27, pp. 140–151. ACM (1992)
7. Kyriakopoulos, K., Psarris, K.: Efficient techniques for advanced data dependence analysis. In: Proceedings of the 14th International Conference on Parallel Architectures and Compilation Techniques, PACT 2005, pp. 143–156, Washington, DC, USA. IEEE Computer Society (2005)
8. Burke, M., Cytron, R.: Interprocedural dependence analysis and parallelization. SIGPLAN Not. **21**(7), 162–175 (1986)
9. Psarris, K., Kyriakopoulos, K.: An experimental evaluation of data dependence analysis techniques. IEEE Trans. Parallel Distrib. Syst. **15**(3), 196–213 (2004)
10. Pugh, W.: A practical algorithm for exact array dependence analysis. Commun. ACM **35**(8), 102–114 (1992)
11. Psarris, K.: The Banerjee-Wolfe and GCD tests on exact data dependence information. J. Parallel Distrib. Comput. **32**(2), 119–138 (1996)
12. Pugh, W.: The omega test: a fast and practical integer programming algorithm for dependence analysis. In: Proceedings of the 1991 ACM/IEEE Conference on Supercomputing, Supercomputing 1991, pp. 4–13, New York, NY, USA. ACM (1991)

Expert System in Determining the Quality of Superior Gourami Seed Using Forward Chaining-Based Websites

M. Sivaram[1], B. Bazeer Ahamed[2], D. Yuvaraj[3], V. Manikandan[1],
Nabila Ghassani Karlus[4], Andri Sahata Sitanggang[5],
Aliza Abdul Latif[6], and Andino Maseleno[7(✉)]

[1] Department of Computer Networking, Lebanese French University, Erbil, Iraq
[2] Department of Computer Science and Engineering,
Balaji Institute of Technology and Science, Laknepalli, India
[3] Department of Computer Science, Cihan University, Duhok,
Kurdistan Region, Iraq
[4] Department of Information Systems, STMIK Pringsewu, Pringsewu,
Lampung, Indonesia
[5] Department of Information Systems, Faculty of Engineering,
Universitas Komputer Indonesia, Bandung, Indonesia
[6] College of Computer Science, Universiti Tenaga Nasional, Kajang, Malaysia
[7] Institute of Informatics and Computing Energy, Universiti Tenaga Nasional,
Kajang, Malaysia
andimaseleno@gmail.com

Abstract. Gourami (Osphronemusgouramy) is the original fish of Indonesian waters that has spread to Southeast Asia and China. Taxonomically, based on The fish labirinth, it is including the family Osphronemidae. Gourami fish is one of many commodities cultivated by the farmers, this is because the market demand is high enough. This fish is one of well-known consumption fish species and has many demands in Indonesia. This is because Gourami has superiorities, there are good taste of meat, easy maintenance and relatively stable prices. This fish has long been known and has been widely cultivated. To determine the type of gourami seed we used forward chaining method, so we easily know the quality of the seed of the Gourami by using this application.

Keywords: Expert system · Gourami · Forward chaining

1 Introduction

1.1 Background

Gourami is one of the types of fish that has a high economic value, this is due to the taste of the meat chewy [1]. Cultivation of Gourami becomes one of the commodities selected by fish farmers. To support the activities of this cultivation needs to be improved in a sustainable seed procurement [2]. In morphology, this fish has a lateral line of a single, complete and uninterrupted, flaky stenoid and has teeth on its lower

A. K. Somani et al. (Eds.): ICETCE 2019, CCIS 985, pp. 310–321, 2019.
https://doi.org/10.1007/978-981-13-8300-7_26

jaw, rounded tail fin. Weak fingers first belly fin is long threads that serve as a means of sensing with height of 2.0–2.1 times that of the standard length [3]. There are 8 to 10 pieces young fish and on the base of the tail, there is a round black dots. Gourami also has distinctive physical form their flat, somewhat long and wide. The body is covered with scales rather rough edges, small mouth, sloping proper not under the tip of the snout. Bottom mouth look stand out a little than the upper mouth. The tip of the mouth can toss that it looks like a muzzle [4].

The appearance of different adult Gourami is minor. The differences can be observed based on body size, color, shape of the head and forehead. The colour and behaviour of young Gourami are far more interesting than the adult Gourami [5]. While in young fish there are eight straight arms. The dark spot with the fringe is yellow or silvery found on parts of the body above the anal fin and pectoral fins, there are black spots at the base [6]. Gourami belongs to the order of the Labirynthici which has an extra breathing tool called labyrinth, it folds the respiratory epithelium which is derived from the first sheet of the gills, so the fish can take oxygen directly from the air. The existence of this extra breathing tool allows gourami can live in the low oxygen water [7].

1.2 Formulation of the Problem

By looking at the background of the above, author takes a summary of problem as follow: How can we determine the best seed of Gourami fish?

1.3 Limitation of the Problem

Based on the above problems then the authors can conclude the limitation of the problem in this research, that this is only a limited scope of Gourami atone village and this application was developed using Forward Chaining-based Website.

1.4 Research Benefits

a. The creation of expert system makes it easy to determine the quality of Gourami seed.
b. As a means of exercise and the development of insights for the author in applying the theory as the basis for research.
c. Acquiring more knowledge and insight about Gourami (Osphronemusgouramy).

2 Literature Review

2.1 Expert System

An expert system is a system that strives to adopt human knowledge to the computer so that the computer can solve problems such as commonly done by members, and a good planned expert system in order to solve a particular problem with copying work from members [8, 9]. An expert (expert system) in General is a system that strives to adopt human knowledge to the computer planned and implemented using a particular computer language, so as to be able to program problem solving such as those made by members [10, 11]. Currently the middleexpert systems developed in various kinds of fields, one of them in the field of agriculture or plantation.

An expert system is a computer program to replicate the ability in solving problems from an expert. An expert is a person who has the ability or understand in facing problem [12, 13]. Based on the experience, an expert develops who made it unable to solve problems with good results and efficient. An expert system is a system of knowledge, namely system-based that imitates the reasoning from an expert in a particular field to solve a problem or to give advice. This system uses human knowledge to solve a problem that requires the expertise of an expert [14, 15].

2.1.1 Components of an Expert System

Components on expert system as follows:

(a) User Interface
User interface is a mechanism used by users and expert system for communicate. Interface is receiving help from the users and turn it into a form that can be accepted by the system. In addition to the interface is receiving from the system and presents a form that could be understood by users.

(b) Knowledge Base
Knowledge base is knowledge for understanding, formulation, and solution of problem.

(c) Knowledge Acquisition
It is always Knowledge Test, accumulation, transfer, and membership in the transformation of knowledge sources in solving the problem into a computer program. In this Knowledge tries to embed more knowledge transfer into to the knowledge base. Knowledge acquires from experts, equipped with books, database, reports, observations and user experiences.

(d) Machine/motor inferential (inference engine)
This component contains the mechanism of thought patterns and reasoning used by experts in solving a problem. Inference is a computer program that provides a methodology for reasoning about information in the knowledge base and in the workplace, and to formulate conclusion

(e) Workplace/blackboard
Workplace is the area of a set of working memory, it is used to record ongoing events including the decision while.

(f) Facility description
Component description is an additional facility that will improve the ability of the expert system, it is used to track the response and give an explanation about the conduct of expert system interactively through the question.

(g) Knowledge improvement
Experts have the ability to analyze and improve the performance and the ability to learn from its performance. This ability is important in the computerized study, so that the program will be able to analyze the cause of success and failure of it and also evaluate whether knowledge is suitable for the future or not.

2.2 Website

Website is a collection of web pages associated with other files related. In a website there are a page known as Home Page. The Home Page is the first page a times when someone visits a website. From Home Page, visitors can click the hyperlink to view the first move other contained in the website [16, 17].

2.3 MYSQL

MySQL is an opensource SQL database management system and most popular at this time. MySQL is used to access database. MySQL Database System features like support some multithreaded, multi-user, database and SQL management system (DBMS) [18, 19]. A database is created for the purposes of database systems, it must be reliable and easy to use. MySQL is actually a derivative of one of the main concepts in the database since a long time, namely the SQL (Structured Query Language). SQL is a database operation, the concept is especially for the election or selection of data entry, and that ease data operations.

2.4 PHP

PHP is a server-side scripting language with HTML in creating dynamic web pages. The intent of a server-side scripting is syntax and commands that will be executed fully served but included on the HTML document. The creation of this web is a combination of PHP as the programming language and HTML as a web page Builder [20, 21].

2.5 Definition of Gourami

Gourami with the latin name "Osphronemusgouramy" is one of the cultivated freshwater fish [22]. Fish is included in the family labyrinthici, due to the use of additional respiratory fish such as labyrinth. Gourami fish has elongated body and body shape reaches 65 cm, height and flattened laterally. It has a relatively small mouth size, tilted and has muzzle. Gourami fish also have a lateral line of a single, uninterrupted and complete [23]. As well as having a very slimy scales and coarse and stenoid-shaped (rounded). Gourami fish also has teeth on the bottom, its mouth has a tail with characteristics like the Moon that is black or dark. In addition, the tail of the fish also has a rounded tail fin as well as in the complete pair of fins that also looks good. In General, Gourami has brownish to blackish with Mark spots – black and white also at the pectoral fins. However, the thick meat on fish gourami reaches 1–2 cm and also has a very fine scales. Therefore, many people like this fish because it has a very thick meat. Beside having thick meat, Gourami can be also used for delicious food [24].

1. The qualified seed. The seed reliable in terms of trusted because The seed is emphasized on quality. The reference about a trusted seed can be obtained from fish farmers information. Qualified seed is produced by the hatchery (a porch seeds) that uses high standard in seeding because people in the area of seeding as well as reliable because of experience and academic background as well as the apply a good seed production system.

2. Healthy and not Handicapped visually conditions. Seed that will be cultivated can observe. Seed to be cultivated must be healthy and not defective, because the derivative or disability due to be injured by the disease. Criteria of healthy seed will be actively move and will respond when given a stimulus. The stimulation we can do with feeding. When the feed directly in the sea then the seed can be sure either.

3. Same size seed. Same size seed is the seed that has the same size, it can be utilized the food more efficient because there is no fish competing in obtaining food. If the seed is not the same then became alarmed at the large portion it will get to eat a lot while the smaller one will be left growth.

4. Response to Feeding. Response to Feeding and identifying that the seed quickly get the stimulus. Qualified seed will respond to feeding by striking when feeding.

5. Free of disease organisms. Qualified seed is free from the organism diseases such as parasites, bacteria, fungi, or viruses.

6. In accordance with the standards of qualified seed it can be identified by their nature qualitative or quantitative criteria. Qualitative criteria are conditions indicated by the seed based on the origins and the observations are visible. Good seed is a result of spawning that not a single offspring with normal body shape with active movement, the tub against the current of water as well as the stimulus from the outside. Quantitative criteria known from the data age, length uniformity size, minimum weights, as well as the uniformity of the agility of his movement towards stimuli from the outside and against the flow of the water.

3   Research Methods

3.1   Methods of Data Collection

(a) Observations
   Observation is collecting data or information which must be executed by conducting observation directly to the place to be investigated [25]. Author conducted observation, by surveying locations in Tanggamus directly on the Gourami's farm in the region in order to collect authentic and specific data.

(b) Literature Research Study
   Literature research study is a study from various references like books as well as the results of previous studies of its kind that is useful to get the base theories regarding the issue that will be examine [26]. In this research, it was conducted the browsing as well as learning from a variety of references in existing libraries, namely the journal about expert system of Gourami farm using a web application, a web-based expert system for selecting Gourami seed and help expert system in determining the quality of Gourami seed use website application.

(c) Interview
   In this case researchers conducted face-to-face interview and conducted question and answer directly to the farmer. The question that the Gourami farmer still did not know about Gourami seed quality. Therefore required an expert system-based websites to determine Gourami seed in order to make it easier for farmer.

3.2   Systems Development Methods

3.2.1   SDLC Method

The system development method of SDLC (System Development Life Cycle) or often referred to as the approach of waterfall (waterfall) method. The waterfall method was first introduced by W. Royce in 1970. Waterfall is the classic model of a simple linear systems with output flow of each stage is the input to the next stage.

3.2.1.1 Planning Stages

At this point the author made planning by collecting materials as well as a feasibility study requirement. What goals wish to be accomplished, the implementation period as well as considering the availability of fund.

3.2.1.2 Analysis System

At this stage the author analyzed the entire system requirements for the proposed information systems. By specifying the necessary skills for the needs of the end user information, which is useful for completing the stage design of the system [27].

3.2.1.3 System Design

The author tried to conduct the design of systems that can meet the needs of the users. The design consists of the design of the logic and physical that can produce specific system that meets the system requirements developed at this stage of the analysis system.

3.2.1.4 Code

Translating the results of the design process into a form of a computer program that is understandable by the engine computer.

3.2.1.5 Test Program

Free trial software is a critical element of the SQA (Software Quality Assurance) and presenting a thorough review towards the specification, design and coding. UEFA presented the lack of abnormalities that occur in software development. During the initial definition and development phase, development seeks to build software from abstract concept to implementation allowed.

3.2.1.6 System Implementation

Implementation stage is a stage where all of the elements and the activities of the system was merged with the following measures: setting up Physical Facilities.

Prepared, among others, computers and its peripheral, including security physical to keep the equipment in a long time.

User is prepared by first i.e. procedurally by providing training or tutorial about functions appropriate information systems. The goal is to let users understand and colonised system operations and the workings of the system as well as what is retrieved from the system.

3.2.1.7 System Maintenance

There are several stages performed by author:

- Fix design errors in the information system program
- Then modify the system to adapt to environmental change system to resolve the new issues
- Keep the system from possible problems in the days to come.

3.3   Research Framework

Research frame work is a form of research from the overall research process in the form of diagrams that describes the outline of the logic flow passage of a study based on website-based information systems journal. The following description by using the flowchart below:

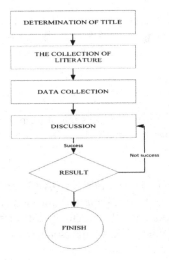

Fig. 1. Flowchart

Based on Fig. 1. It started with the determination of the title, if the title had been specified then the next step was the collection of literature such as the theories needed in this study after that the author collected data by means of interviews, observations and literature study.

After all the next step is to assemble the discussion of this study by designing and implementing successfully then the eligible to use. But if the research did not get results that match then it is not worth the research to be used. But if the research did not get results that match then it was not feasible to research to be designed and implemented.

4 Discussion

After analyzing existing data and information, the next step was designing a system that illustrated the true system concept.

4.1 Design Rule

From some of the criteria that wouldbe taken the result in an outcome that illustrated whether Gourami was preferably in use or not.

Rule 1: IF

- trusted the seed (already has a name) = Yes
- healthy and not defects = Yes
- same seed Size = Yes
- response to feeding = Yes
- free from disease organisms = Yes
- complies with Standard = Yes

THEN Your Gourami was Qualified Seed; Else Your Gourami was not Qualified Seed.

4.2 DFD and the Design of the Interface

Data Flow Diagram (DFD) was a data flow diagram that functioned to map a environmental model, presented with a single circle represented the whole system. This DFD expert system was illustrated on the Fig. 2.

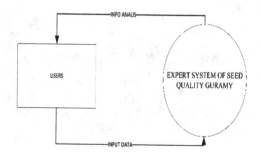

Fig. 2. Data Flow Diagram

Interface design expert system to determine seed quality using nutmeg application website is shown in Fig. 3.

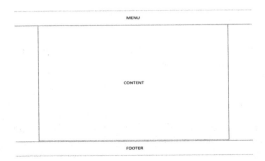

Fig. 3. The design of the main page of the web-based interface

4.3 System Implementation

Implementation of the system is the transformation of the design of the interface to the form results so that the form of a web page that connects users with the website.

4.4 Main Page Display

Main page display was the initial display when application was opened showing information expert system as shown in Fig. 4.

Fig. 4. Main page display

4.5 Display Input Page

Display input page is the page for the user to input the criteria with the criteria that user has prepared in accordance with standards of quality seeds as shown in Fig. 5.

Fig. 5. Display input page

5 Conclusion and Future Direction

5.1 Conclusion

Based on the above explanation can be concluded as follows: gourami used seed that meet the criteria as trusted seed (already had a name), healthy and not disabled, the same seed size, the response to the grant packed, free from disease organisms, in accordance with the standards.

5.2 Future Direction

Suggestions for further development on this app are:

1. Application of decision support system for the selection of gourami seed must be used.
2. Application should be able to provide information on the overall gouramiseed.

References

1. Akash, M., Hossain, M.A.R.: A Southeast Asian species in the Ganges Delta: on spreading extent of non-native croaking gourami Trichopsis vittata (Cuvier, 1831) in Bangladesh. Bioinvasions Rec. **7**(4), 447–450 (2018)
2. Masoudi Asil, S., Abedian Kenari, A.M., Van Der Kraak, G.: Effect of different levels of dietary arachidonic acid on calcium, thyroid hormone, and cortisol levels in vitellogenesis and maturation stages of female blue gourami (Trichopodus trichopterus, Pallas, 1770). Fish. Sci. Technol. **7**(2), 109–116 (2018)
3. Nuryanto, A., Amalia, G., Khairani, D., Pramono, H., Bhagawati, D.: Molecular characterization of four giant gourami strains from Java and Sumatra. Biodiversitas J. Biol. Divers. **19**(2), 528–534 (2018)
4. Abedian Kenari, A., Van Der Kraak, G.: The effects of dietary arachidonic acid on the calcium, cortisol and thyroid hormone levels in vitellogenesis and maturation stages of female Blue gourami (Trichopodus trichopterus, Pallas, 1770). Fish. Sci. Technol. **7**(2), 17–18 (2018)

5. Slembrouck, J., et al.: Gender identification in farmed giant gourami (Osphronemus goramy): a methodology for better broodstock management. Aquaculture **498**, 388–395 (2019)

6. Prayogo, N.A., Siregar, A.S., Sukardi, P., Nugrayani, D., Ekasanti, A., Bessho, R.Y.: Identification and Expression of vitellogenin gene in the Gouramy (Osphronemous Gourammy) under photoperiods manipulation. In: E3S Web of Conferences, vol. 47, p. 02001. EDP Sciences (2018)

7. Ramadhani, A.W., Samik, A., Madyawati, S.P.: Vitellogenesis of Giant gourami (Osphronemus goramy) examined by the measurement of Estradiol-17β, vitellogenin concentration and the size of ovary. Aquac. Stud. **18**(2), 39–51 (2018)

8. Abadi, S., et al.: Hazard level of vehicle smoke by fuzzy multiple attribute decision making with simple additive weighting method. Int. J. Pharm. Res. **10**(4) (2018)

9. Fauzi, M., et al.: The design of fuzzy expert system implementation for analyzing transmissible disease of human. Int. J. Pharm. Res. **10**(4) (2018)

10. Susilowati, T., et al.: Sucipto, determination of scholarship recipients using simple additive weighting method. Int. J. Pure Appl. Math. **119**(15), 2231–2238 (2018)

11. Sugiyarti, E., Jasmi, K.A., Basiron, B., Huda, M., Shankar, K., Maseleno, A.: Decision support system of scholarship grantee selection using data mining. Int. J. Pure Appl. Math. **119**(15), 2239–2249 (2018)

12. Putra, D.A.D., Jasmi, K.A., Basiron, B., Huda, M., Maseleno, A., Shankar, K., Aminudin, N.: Tactical steps for e-government development. Int. J. Pure Appl. Math. **119**(15), 2251–2258 (2018)

13. Kurniasih, D., Jasmi, K.A., Basiron, B., Huda, M., Maseleno, A.: The uses of fuzzy logic method for finding agriculture and livestock value of potential village. Int. J. Eng. Technol. (UAE) **7**(3), 1091–1095 (2018)

14. Adela, H., Jasmi, K.A., Basiron, B., Huda, M., Maseleno, A.: Selection of dancer member using simple additive weighting. Int. J. Eng. Technol. (UAE) **7**(3), 1091–1107 (2018)

15. Muslihudin, M., et al.: Prediction of layer chicken disease using fuzzy analytical hierarchy process. Int. J. Eng. Technol. (UAE) **7**(2.26), 90–94 (2018)

16. Muslihudin, M., Fauzi, Susanti, T.S., Sucipto, Maseleno, A.: The priority of rural road development using fuzzy logic based simple additive weighting. Int. J. Pure Appl. Math. **118** (8), 9–16 (2018)

17. Irviani, R., Dinulhaq, I., Irawan, D., Renaldo, R., Kasmi, K., Maseleno, A.: Areas prone of the bad nutrition based multi attribute decision making with fuzzy simple additive weighting for optimal analysis. Int. J. Pure Appl. Math. **118**(7), 589–596 (2018)

18. Fauzi, N., Noviarti, T., Muslihudin, M., Irviani, R., Maseleno, A.: Optimal dengue endemic region prediction using fuzzy simple additive weighting based algorithm. Int. J. Pure Appl. Math. **118**(7), 473–478 (2018)

19. Susilowati, T., Anggraeni, E.Y., Fauzi, Andewi, W., Handayani, Y., Maseleno, A.: Using profile matching method to employee position movement. Int. J. Pure Appl. Math. **118**(7), 415–423 (2018)

20. Muslihudin, M., Trisnawati, Latif, A., Ipnuwati, S., Wati, R., Maseleno, A.: A solution to competency test expertise of engineering motorcycles using simple additive weighting approach. Int. J. Pure Appl. Math. **118**(7), 261–267 (2018)

21. Oktafianto, M.R.A.A., Fitrian, Y., Zulkifli, S., Wulandari, A.M.: Dismissal working relationship using analytic hierarchy process method. Int. J. Pure Appl. Math. **118**(7), 177–184 (2018)

22. Mendez-Sanchez, J.F., Burggren, W.W.: Hypoxia-induced developmental plasticity of larval growth, gill and labyrinth organ morphometrics in two anabantoid fish: The facultative air-breather Siamese fighting fish (Betta splendens) and the obligate air-breather the blue gourami (Trichopodus trichopterus). J. Morphol. (2018)

23. Nurhuda, A.M., Samsundari, S., Zubaidah, A.: Pengaruh perbedaan interval waktu pemuasaan terhadap pertumbuhan dan rasio efisiensi protein ikan gurame (Osphronemus gouramy). Acta Aquat.: Aquat. Sci. J. 5(2), 59–63 (2018)

24. Pratiwi, H.C., Manan, A.: Teknik Dasar Histologi pada Ikan Gurami (Osphronemus gouramy)[The Basic Histology Technique of Gouramy Fish (Osphronemus gourami)]. Jurnal Ilmiah Perikanan dan Kelautan 7(2), 153–158 (2019)

25. Waziana, W., Irviani, R., Oktaviani, I., Satria, F., Kurniawan, D., Maseleno, A.: Fuzzy simple additive weighting for determination of recipients breeding farm program. Int. J. Pure Appl. Math. 118(7), 93–100 (2018)

26. Mukodimah, S., Muslihudin, M., Fauzi, Andoyo, A., Hartati, S., Maseleno, A.: Fuzzy simple additive weighting and its application to toddler healthy food. Int. J. Pure Appl. Math. 118(7), 1–7 (2018)

27. Manikandan, V., Porkodi, V., Mohammed, A.S., Sivaram, M.: Privacy preserving data mining using threshold based fuzzy cmeans clustering. ICTACT J. Soft Comput. 9(1) (2018)

Proactive Cache Placement and Optimal Partitioning in Named Data Networking

Ganesh Pakle[(⊠)] and Ramchandra Manthalkar

Shri Guru Gobind Singhji Institute of Engineering and Technology,
Nanded, MH, India
gkpakle@sggs.ac.in

Abstract. In-network caching plays an important role in improving network performance in Named Data Networking. The effective cache management in NDN involves cache placement, replacement and eviction policies. This paper presents low complexity, efficient cache placement strategy by using cache-span concept. The popular objects are stored within cache-span radius of Autonomous system to increase the cache hit ratio. The proactive cache placement identifies router locations to store the popular contents to serve large proportion of requests within cache-span. An optimal partitioning based on cache-span level is also presented. The partitioning is required to store content objects and information of cached content of nearby routers defined in cache-span. Our simulations show that proactive caching increases cache hit ratio and reduces round trip latency as compared to existing approaches. The methods are evaluated using analysis model and also by simulation using ndnSIM. The simulation study shows that our method outperforms default NDN caching strategies and also more sophisticated approaches in terms of cache hit ratio and round-trip latency.

Keywords: Named Data Networking · Caching · Optimal cache partitioning

1 Introduction

The current internet model has become more content centric i.e. users are interested in content and not in the location where it is stored. Users want reliable and fast content retrieval. Due to sophisticated developments in internet technology and increase in user base throughout the world, the demand for content access is also increased manifold. The internet traffic today is experiencing explosive growth. This growth also poses challenges in modernizing internet technology and increasing capacity for growing demand. To tackle with ever growing demand, P2P networking and CDN were introduced as an overlay to current internet architecture. But these solutions are unscalable. For scaled content dissemination, Future internet architectures have proposed various models based on name-based content delivery.

Most promising architecture based on named-content is Named Data Networking (NDN) [1]. NDN supports name-based forwarding, routing and in-network caching. Each router in NDN is provisioned with content store which acts as cache to store the content object on forwarding path, which then serves subsequent requests for same

© Springer Nature Singapore Pte Ltd. 2019
A. K. Somani et al. (Eds.): ICETCE 2019, CCIS 985, pp. 322–335, 2019.
https://doi.org/10.1007/978-981-13-8300-7_27

objects. NDN routers, by caching most popular items in their content store, can reduce access latency and network load. The subsequent requests are served by data object from their content store and need not be forwarded to source.

If requested content object is not found in local CS, router checks entry in local PIT, if the entry exists, it discards the interest packet. PIT records only non-served requests. If PIT does not contain entry for interest packet, it records new entry and forwards the interest packet on one of the forwarding interfaces according to forwarding strategy. as soon as the router receives data packet for corresponding interest packet, it deletes the entry from PIT, stores the copy of data packet in content stores and forwards data packet to requesting consumer [2].

1.1   Caching in NDN

Each router in NDN can select what to cache according to its cache insertion and eviction policy. NDN also supports distributed caching in the network. For every received content object, router makes a copy and stores it in its content store and forwards a copy to requester. As the required content is available at nearest site, Publishers are not obliged to send separate copies of same content for each request thereby saving bandwidth in upstream routers.

NDN supports Leave Copy Everywhere as default caching strategy. The NDN routers caches all the content en-route to delivery. This makes user content available from any intermediate routers. This results in lowering load on content servers, reduced redundant transmission, no or little congestion on network path and decreased response time for fetching content for user requests.

Caching all content is unpractical. En-route caching in all routers wastes the cache space. Early cache replacement may result in cache miss for same content requested by other users and router may need to run cache replacement operations often. Some content may never be requested by any users. Thus, an efficient and better caching strategy is needed for NDN.

The performance of good caching policy is decided by two main factors. Firstly, cache policy should improve quality of experience of end users through minimized content access delay and high hit ratio and Secondly, improved network performance measured through reduced network traffic load and lessening burden on end servers.

Cache policy is used to determine placement of named content objects at cache nodes or routers called cache insertion policies. The parameters for good cache policy required are content popularity, policy complexity, cooperation of caching nodes with other nodes in network, caching granularity and cache redundancy.

In this paper we propose effective cache placement strategy to minimize the access latency. The placement of every cached object is a periodical decision in any one of routers along the forwarding path based on cache-span selection of router. This means, the required data object is always cached in any one of router within the AS with maximum hops equal to cache-span. This limits the latency to be equal to average one-way path length from requesting consumer to router with cached object in its content store. We also propose an optimal cache partitioning for storing location information of neighbor router within cache-span. The content store is partitioned into object cache and location cache. Content store partition stores the location information of all routers under cache-span level.

The main contributions of this paper are outlined as follows:

- Design of an effective cache placement strategy to minimize the latency in NDN.
- Optimal partitioning of content store for placement of content objects and location information of neighbor cache.
- Performance evaluation of proposed strategy to determine effectiveness in terms of cache hit ratio, latency, cache content diversity etc.
- Comparison of proposed methods with default methods such as LCE, LCProb and more sophisticated methods like EMC, etc.

2 Related Work

Extensive research has been done all over the world to explore and investigate effective forwarding and caching in NDN.

NDN architecture provides Leave Copy Everywhere (LCE) as default cache placement policy [1, 2]. In LCE, every data packet is cached in all intermediate routers between publisher to consumer. This results in excessive cache redundancy and lacks diversity in cached objects in system. For reducing redundancy, variations of LCE are proposed as Leaving Copies with Probability (LCProb) which caches content objects at each router with uniform probability equal to 1/hop count [3, 12].

Coordinated caching and routing scheme is proposed in [4]. Each node in NDN is assigned with specific chunk of available name space. A publisher ID based on hash value, instead of publisher name for each object is used to cache objects in designed RRs (Responsible Routers). RRs then advertise PID prefixes in its AS. Consumers forward interest packet to corresponding RR based on PID of publisher. The scheme has advantage that requested content is not duplicated within AS and saves on additional signaling overhead. But the RR results in indirect routing which can cause longer delivery paths for interest packets. Every RR performs tunneling for defined set of publishers which incurs considerable computational costs at each designated responsible router.

Effective multipath caching [5] is a popularity based coordinated caching strategy proposed for reducing Inter-ISP traffic. In initialization stage, EMC aggregates all name-based routing paths to extract caching topology. All NDN routers then periodically measures incoming data packet request rate for every object. List of most popular items is then shared with subordinate routers. EMC avoids object duplication by removing cached objects from downstream routers. Although EMC is effective in reducing Inter-ISP traffic, it stores popular contents in upstream routers instead of downstream routers. The downstream routers cannot store popular objects because objects are removed from its list. This may result in being duplicating unpopular objects in downstream routers.

For an effective in-network cache management, cache placement, replacement policy and caching location are most important. The caching scheme based on this combination is proposed in [6] The scheme is named as RPL. A data packet is cached only when it satisfies constraints such as hop reduction gain, local content popularity, replacement penalty and cache space contention. A cached content achieves maximum gain when it is popular and cached for longer duration at a location far from publisher. For every interest packet, all intermediate routers on forwarding path calculates its contribution gain based on caching contribution. Every router also creates trails to forward future requests toward cached copy. Hop count and max contribution fields are added in RPL to avoid need of additional signaling. The scheme has additional computational overhead due to computation of maximal contribution for every incoming interest packet at every intermediate router on forwarding path. Routers also need to maintain additional forwarding table to record trails for up-stream routers, rendering complex forwarding process. As cached content in forwarding paths are frequently changed, intermediate routers need to update trails for every replacement cycle.

When router cache is full, it needs to identify which cached objects need to be replaced first. Cache replacement strategies such as Least Frequently used (LFU), Least Recently used (LRU), First In First Out (FIFO) and randomized replacement are commonly used in web caches [7]. LRU replaces content objects which are not used recently. Oldest content in cache is replaced first by FIFO. LFU replaces cached object based on latest popularity. The cached objects having least number of requests are replaced first in LFU. LFU is best strategy to store most popular items but it converges slowly to popularity changes. Randomized replacement strategy arbitrarily removes a random cached object. NDN architecture supports both LRU and LFU caching strategies. A more sophisticated approaches for cache placement strategy based on SDN is proposed in [13].

3   Methodology

3.1   Effective Cache Placement Policy

We consider an NDN network of graph topology represented by $\mathcal{G} = (\mathcal{V}, \mathcal{L})$ where $\mathcal{V}$ denotes set of caches in every NDN node on interest forwarding path and $\mathcal{L}$ denotes transmission links interconnecting the NDN nodes. For generality, we assume that every data packet received has same size and every slot in content store can accommodate at most one data packet. Request for data arrive in Poisson process. Interest packet is first routed towards publisher via shortest path and matching data packet always traverses reverse path and intermediate routers may cache data packet.

NDN architecture supports that only the first interest is forwarded on upstream routers and subsequent request for same data are served from content store. This results in request aggregation. i.e. upstream routers see only one interest request.

We divide the NDN topology as collection of domains $\mathcal{D} = (d_1....d_n)$ and $D \subseteq \mathcal{G}$. Each domain represents group of nodes connected by links. The domain has two types, Interdomain traffic is carried by *Intra-AS* domain and *AS* domains carry intra-domain traffic. The NDN routers are represented as nodes in domains arranged in directed graph. Intra-domain path connects nodes only in AS domain. The shortest path connecting a node in AS to another node in another AS follows the path with one or more

Intra-AS domains. The connecting link from one AS to another AS does not pass through any other AS domains. For closely connected ASs, the connecting path between two ASs may avoid Intra-AS domains. The general objective for any good caching strategy in NDN is to minimize the communication cost from node u to node v for requested data item which is cached at node v measured in delay, number of hops, etc. Figure 1 shows the general model of NDN topology showing the path of user request and its routing through the AS and Intra-AS domains. To achieve lower access latencies and saving bandwidth usage, network nodes should have sufficient caching space. In NDN, each router needs to make a decision of caching for every data packet which is received as response to user's interest packet. If objects are cached in all intermediate routers, it will result in very low latencies but this strategy will result in replication of same objects in multiple routers in network. If caching of objects is too sparse, the access latency will not tend to decrease and caching performance will be sub-par.

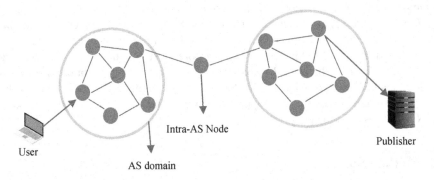

Fig. 1. Arrangement of AS and Intra-AS nodes in NDN arbitrary topology

In this paper we are proposing proactive caching scheme to address the NDN caching problem outlined above by proactively organizing the caching decision for every data packet arriving at router. The proactive caching scheme counteract the need of deciding where to place cached object in NDN router. We assume that each requested data item must be cached in one of content stores along the path to publisher. And Each node has finite cache which cannot be exceeded.

To avoid excessive caching replication, we introduce the concept of *cache-span*. An object is cached only once in given *cache-span* (measured in terms of hops in an AS). Thus, cache span mechanism distributes caching object throughout the network evenly thereby avoiding making ad-hoc decisions about placement of caching objects near to consumer or publisher. This also eliminates designating edge or core routers for caching objects. If objects are cached closer to publishers, cache hit ratio is always high and overall latency is not reduced. If we cache objects near to consumers which in-turn leads to very low cache hits.

Table 1. Notations used for proactive cache modeling

Symbol	Notations
C	No of cache objects
R	No of caches
S	Cache size
X	Fraction of requests served by caches
Q	Variable with value 1 if object is cached, 0 otherwise
p_i	Probability that object i is requested
α	Zipf exponent of object request frequency distribution

For modeling, we assume that item requests follow a uniform probability distribution over each item being requested and independent of previous requests. Table 1. summarizes notations and symbols used for proactive cache modeling. Request for cache objects are issued for items $1....C$ of fixed catalogue size of C and probability distribution in $\{p_1, p_2, ..., p_N\}$. The topology consists of R number of caches with $R < C$. Each cache has same amount of caching space of $S < \lceil C/R \rceil$ number of objects. Caching objects are mapped uniformly to routers in AS according to hash function $f_m : [1....C] \rightarrow [1...R]$. Therefore, we model assignment of cache object i to a router in cache-span using a Bernoulli random variable X_i such that

$$f_{X_i}(x) = \begin{cases} \frac{1}{R}, & x = 1 \\ 1 - \frac{1}{R}, & x = 0 \end{cases} \tag{1}$$

For given cache-span, we apply modulo function for evenly distributing cached objects along the path to publisher. Every object in NDN is identified with unique name identifier. We apply modulo function to unique identifier to obtain specific location as per Eq. (1), along consumer publisher path of length equal to cache-span. Thus, any NDN router, contains a specific set of non-overlapping objects resulting in storing more diverse cache objects in content store.

We can now express the total number of objects each router can receive over unit time for caching without considering object popularity given as

$$X = \sum_{i=1}^{C} Q_i p_i \tag{2}$$

Where p_i is probability that an object i is requested. Since Q_i is an i.i.d. random variable, using Eq. 2 the variance of X is calculated as

$$Var(X) = Var\left(X = \sum_{i=1}^{C} Q_i p_i\right) = \sum_{i=1}^{C} Var(Q_i) p_i^2 = \frac{R-1}{R^2} \sum_{i=1}^{C} p_i^2 \tag{3}$$

Using Eq. 3 we derive expression for total load across cache in cache-span radius as

$$X_c = \sqrt{R-1} \sqrt{\sum_{i=1}^{C} p_i^2} \tag{4}$$

From Eq. 4 we conclude that load distribution is directly proportional to square root of number of caches in cache-span and also on skewness of object popularity defined as $\sum_{i=1}^{C} p_i^2$. When one object has probability equal to 1 and other objects in cache has probability 0 then value of X_c is maximum.

We now derive expression for load across caches with popular objects distributed according to zipf distribution [9]. The zipf distribution defines the probability of an object being requested as

$$p_k = \frac{k^{-\alpha}}{\sum_{m=1}^{C} k^{-\alpha}} \tag{5}$$

Where α is skewness parameter with value $\alpha > 0$. For value of $\alpha = \frac{1}{2}$, Load across caches is derived from Eq. 5 as

$$X_c = \sum_{i=1}^{C} p_i^2 = \sum_{i=1}^{C} \left(\frac{k^{-1}}{\sum_{m=1}^{C} m^{-1}} \right) \approx \frac{\log(C+1)}{4(\sqrt{C+1}-1)^2} \tag{6}$$

For $\alpha = 1$,

$$X_c = \sum_{i=1}^{C} p_i^2 = \sum_{i=1}^{C} \left(\frac{k^{-1}}{\sum_{m=1}^{C} m^{-1}} \right) \approx \frac{C}{(C+1)\log^2(C+1)} \tag{7}$$

From Eqs. 6 and 7 we conclude that for low values of α, load imbalance in caches is close to 0, and increases with increase in value of α. In general, we note that skewness in object popularity effects load distribution across caches.

When a NDN router stores an object in cache, the object has multiple home locations as router cache and at the publisher. Proactive caching allocates part of available cache space to store information about location of neighboring items. When NDN router receives interest packet, the router search for its corresponding data packet in its cache and also in the list of neighboring cached items. If hit is detected for neighboring cache, router sends request to nearby cache. Adaptive caching scheme limits neighboring routers to single domain as either Intra-AS or AS domain. i.e. neighboring caches are checked only for members of same Intra-AS or AS domains.

3.2   Analysis of Proactive Caching

We now present analysis model for determining latency in retrieving data packet corresponding to interest packet which is cached in one of the routers in AS domain. We partition the cached objects in two categories as popular and unpopular objects. Every router content store cache only popular content object. Popularity of an object is determined simply by examining PIT entries for historical access request for the object. We assume that every object cached in content store is unique and cached objects encountered from one consumer access to next are independent.

Let's consider a case where a consumer sends interest packet for required data. If the item is unpopular, it may not be cached in any AS routers and request will be forwarded to publisher. In this case, the latency will be average consumer-publisher path. Let T denote average one-way latency in number of hops for every cache hit. Let p denote probability that the cached item is popular. So, the probability of cached item being unpopular is $(1 - p)$. Let T_u be average path length from requesting consumer to publisher and T_q be average length from consumer to popular item which may be stored in content store of router along forwarding path. Then

$$T = pT_q + (1 - p)T_u \tag{8}$$

Let's derive the expression for T_u. The average path length from a AS node to Intra-AS node inside a AS domain is represented as D_s. The average path length across Intra-AS domains is represented as D_t. Then

$$T_u = D_t + 2D_s \tag{9}$$

since path to nearest availability of object traverses two ASs and single Intra-AS domain.

Now we need to derive expression for T_p. Let H_i denote hit for an object in an intermediate cache store occurred at nearest router at distance i from consumer. Let M_i denote miss occurred for a requested object from a consumer at a distance up to i from request node. Let the total distance from requesting node to publisher be d in number of hops. Then

$$T_p = \sum_{i=1}^{d} iP_r\{H_i \cap M_{i-1}\} \tag{10}$$

$$= \sum_{i=1}^{d} iP_r\{H_i|M_{i-1}\}P_r\{M_{i-1}\} \tag{11}$$

Probabilities in Eq. 11 are affected by cache policy, location of caches and client access patterns of popular objects.

We assume that objects stored in content store of every router are independent. This assumption requires that client access patterns for requested data objects are uniform. Probability of caching a popular item is equally likely for all NDN routers. As caches are independent, we are interested in calculating latency for popular items. The probability that a hit occurs for a popular object in cache of size S is S/P. Let S_i denote size of content store at distance i from requesting client then

$$L_p = \sum_{i=1}^{d} iP_r \{H_i \cap M_{i-1}\} \tag{12}$$

3.3   Optimal Cache Partitioning Framework

For implementation of cache-span caches we need to partition router content store in to two parts as *object cache*, to store data packets received from publisher and *location cache* to store the shared information from neighboring router content store. When a router receives an interest packet on one of its interfaces, it first checks its content store for corresponding data packet. If there is miss for requested object, it looks into its location cache. If it results in hit at location cache pointing to nearby router, it redirects interest packet to nearby router. In this scenario, two cases may occur: either requested object is at nearby router or it is not available as router has replaced this object with another object. In later case, the request will then be forwarded to publisher for latest copy of requested content object. This results in use of minimum content store space.

For optimal cache partitioning we need to associate cache-span with level. The level l means router stores content objects information of router at distance l. This helps router to search required data object within a distance of modulo l from itself. Routers with cache-span level one corresponds to storing of data object information of direct neighbors and routers with content stores with only cache has cache-span level zero.

Table 2. Parameters definitions and notations for cache partitioning

Router Content Store size	S
Number of objects cached at given time	C
Cache-span Level	l
Location size to object size ratio	r
Avg. path length from consumer to publisher	d
Degree of graph	ω
Fraction of content store for caching objects	ρ

We assume that every cached object has same size and every router has content store of size S. The Table 2 summarizes notations and symbols used for deriving expression for location cache framework. Let the location size to object size ratio be r. i.e. r number of locations can be stored instead of one cache object in content store. Level of cache-span determines the number of neighboring content stores to be searched. As the level increases number of cache searches increases exponentially in relation with average degree of graph. For level 1 of cache-span the cache searches become $\omega - 1$, where ω corresponds to degree of graph in given network topology. We assume that fraction ρ of content store is used for caching objects and remaining for location cache.

To minimize round trip latency, we need to derive expression for fraction ρ. To precisely determine the value of ρ, we need to consider size of cache of each router expressed in S, size of network with number of nodes and value of r.

By using basic caching model described above we now present expression for expected latency (based on Eq. 12) as

$$L_p = \sum_{i=1}^{d} i P_r\{H_i|M_{i-1}\}P_r\{M_{i-1}\} \tag{13}$$

If cached objects are occupying fraction ρ content store, then $S(1-\rho)$ is occupied for storing location information. This equals to $Cr(1-\rho)$ locations of neighboring objects. If each content store at NDN router can store upto ρC number of objects, $Cr(1-\rho)$ locations can store locations from $(1-\rho)r/\rho$ number of nodes. This implies dtour length of maximum upto $log_{\delta-1}(1-\rho)\alpha/\rho$. The one-way length of detour is then given by

$$f = \sum_{x}^{F} x\frac{(\omega-1)^x \rho C}{(1.\rho)Cr} \tag{14}$$

Where x represents level of cache-span and F is $log_{\delta-1}(1-\rho)r/\rho$

In optimal partitioning, the location information of nearest routers is searched first. In this case, we observe that, a hit for required data packet occurs in cache-span caches in searching only half of caches. The expected length of detour in this case is derived from Eq. 14 as

$$f = log_{\omega-1}\frac{Cr(1-\rho)}{2\rho C} \tag{15}$$

Let $\mathcal{P}$ denote number of popular objects requested by consumer and let the number of objects checked upto $i-1$ from requesting consumer be $\mathcal{N}_{i-1}$. We assume that all NDN routers have same cache size equal to S, such that for all $S_i = S$ and this implies that $\mathcal{N}_{d-1} = (d-1)S$

From above assumptions, the one-way latency in our model is given as

$$L_p = \sum_{i=1}^{d} i P_r\{H_i \cap M_{i-1}\} \tag{16}$$

$$= \sum_{i}^{d} i\frac{S_i}{P} + d(1-(d-1)\tfrac{S}{P}) \tag{17}$$

Content store in every router are updated independently regardless of neighbor location cache information. Let μ denote set of objects in remote location cache which are unique, so the total number of objects searched at any given router are Cp local objects and $\mu Cr(1 - \rho)$ remote objects.

4   Experimentation and Results

We use ndnSIM which is NS-3 based NDN simulator [8]. For evaluating the performance of proactive caching strategy and optimal partitioning we use different set of parameters with wide range of topology choices differing in number of NDN routers, avg in-degree and out-degree of each node and varying ratio of Intra-AS and AS nodes.

4.1   Performance Metrics

Cache Hit Ratio
We define performance metrics of cache hit ratio for all caches in given AS instead of individual cache. It is measured as aggregate of requests served cumulatively by caches in an AS. Let the number of nodes in an AS be N and each consumer C generates K number of interest packets, then

$$HR_C = \frac{\sum_{i=1}^{n} H_i}{\sum_{i=1}^{k} K_i}$$

The ratio HR_C is affected by content diversity and popularity of cached objects.

Content Delivery Time
The content delivery time is time taken by consumer to receive the data packet for requested interest packet. It is measured as

$$DT = \frac{DataPktRevTime - IntPktTranTime}{TotalIntPkt}$$

Where C_h is number of hops to nearest content store with hit and C_H is number of hops to original publisher.

4.2   Analysis of Cache Hit Ratio

For simulation purpose we assume that consumers are connected only to AS routers and each routers content store can store max up to 500 objects. All routers in AS has same content store size as [11] proved that perceived gain of heterogeneous content store size is limited. The size of each cached object is fixed at 1024 bytes. Each consumer generates 100 requests per second and follow Poisson distribution.

Figure 3 shows the simulation results for cache hit ratio. Cache hit ratio is correlated with content object popularity. As popular content is uniformly distributed in cache-span, our proposed method shows high hit ratio compared with other schemes.

The interest request from consumer has hit as the cache size in routers in forwarding path grows. As can be seen from simulation results, as the number of content objects increase, cache hit ratio also increase. This also results in lower round-trip latency. As cache size grows, proactive caching strategy shows nearly 58% cache hit ratio within cache-span radius of given AS. This is equivalent to gain of 60 to 80% over default methods such as LCE and LCProb and marginal gain around 8% compared with EMC.

4.3 Analysis of Round-Trip Latency

Figure 2 represents performance of our Proactive caching strategy against LCE, LCProb and EMC in terms of reduction in round trip latency against cache size of routers in cache-span with radius = 6. The graph shows that on average, interest packet traversed maximum 4 hops to reach the required object cached in node in AS, given that the object requested is popular and cached en-route to publisher.

4.4 Impact of Zipf Parameter

Interest packets are generated with zipf distribution [9] with popularity measure of $\alpha = 0.7$. As value of α approaches to 0, all content objects become equally popular and value close to 1 indicates concentration of popularity on several content objects. As current internet trend is skewed [10] towards few popular objects, the evaluation is performed on popularity range of $0.6 \leq \alpha \leq 0.9$ with median 0.7 fixed as value of α.

Figure 4 shows how parameter α impacts proactive caching performance. It clearly shows that proactive caching outperforms other caching schemes in terms of cache hit ratio. As indicated by graph, as the value of α increases, the content popularity also increases, thus proactive caching scheme caches most popular objects in cache-span, which results in more cache hits. In terms of cache hits and other parameters only EMC has comparative performance.

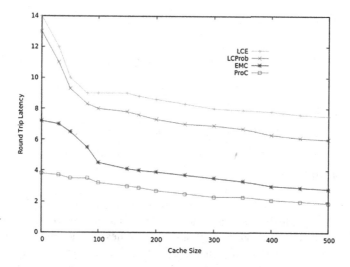

Fig. 2. Round trip latency

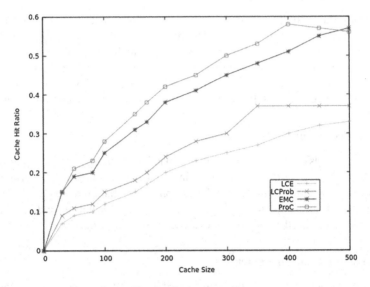

Fig. 3. Cache hit ratio

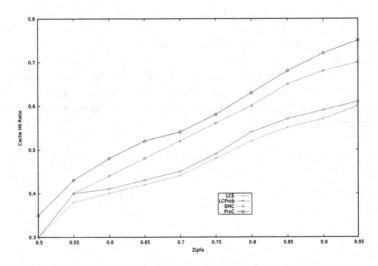

Fig. 4. Effect of Zipf parameter α on cache hit ratio

5 Conclusion

This work is concerned with effective in-network cache placement using cache-span radius with low complexity to reduce the round-trip latency for requested object. This method is compared with modern methods which relies on caching in either edge or core routers. The scheme also defines an optimal cache partitioning for storing neighbor location information which results in high hit ratio for interest request.

Popular objects are placed near to consumer in its forwarding path and within cache-span radius to serve major requests from caches. Our simulation results and analysis show that our proposed method is more beneficial across wide range of NDN topology and access patterns. It makes more efficient use of available content store space. It increases cache hit ratio upto 65% and also reduces avg hop count for user's request. In future we would like to evaluate proactive caching on more realistic testbeds.

References

1. Jacobson, V., Smetters, D.K., Thornton, J.D., Plass, M., Briggs, N., Braynard, R.: Networking named content. Commun. ACM. (2012)
2. Zhang, L.: Named Data Networking (NDN) Project, October 2010
3. Zhang, G., Li, Y., Lin, T.: Caching in information centric networking: a survey. Comput. Netw. 57, 3128–3141 (2013)
4. Choi, H., Yoo, J., Chung, T., Choi, N., Kwon, T., Choi, Y.: CoRC: coordinated routing and caching for named data networking. In: Proceedings of Tenth ACM/IEEE Symposium Architecture Network Communication System - ANCS 2014 (2014)
5. Wu, H., Li, J., Wang, Y., Liu, B.: EMC: the effective multi-path caching scheme for named data networking. In: 2013 22nd International Conference on Computer Communications and Networks (ICCCN) (2013)
6. Hu, X., Gong, J., Cheng, G., Fan, C.: Enhancing in-network caching by coupling cache placement, replacement and location. In: IEEE International Conference on Communications (2015)
7. Podlipnig, S., Böszörmenyi, L.: A survey of Web cache replacement strategies. ACM Comput. Surv. 35, 374–398 (2003)
8. Mastorakis, S., Afanasyev, A., Moiseenko, I., Zhang, L.: ndnSIM2: an updated NDN simulator for NS-3. Dept. Comput. Sci., Univ. California, Los Angeles, Los Angeles, CA, USA, Technical report. NDN-0028 (2016)
9. Breslau, L., Cao, P., Fan, L., Phillips, G., Shenker, S.: Web caching and zipf-like distributions: Evidence and implications. In: Proceedings - IEEE INFOCOM (1999)
10. CISCO: Cisco visual networking index: Forecast and methodology. Cisco (2017)
11. Rossi, D., Rossini, G.: On sizing CCN content stores by exploiting topological information. In: Proceedings - IEEE INFOCOM (2012)
12. Psaras, I., Chai, W.K., Pavlou, G.: In-network cache management and resource allocation for information-centric networks. IEEE Trans. Parallel Distrib. Syst. 25, 2920–2931 (2014)
13. Mahmood, A., Casetti, C., Chiasserini, C.F., Giaccone, P., Härri, J.: Efficient caching through stateful SDN in named data networking. Trans. Emerg. Telecommun. Technol. 29, e3271 (2018)

Hybrid Multimodal Medical Image Fusion Algorithms for Astrocytoma Disease Analysis

B. Rajalingam$^{(\boxtimes)}$, R. Priya, and R. Bhavani

Department of Computer Science and Engineering, Annamalai University,
Chidambaram, India
rajalingam35@gmail.com

Abstract. Astrocytoma is a type of cancer that can form in the brain or spinal cord. It is begins in cells called astrocytes that support nerve cells. Astrocytoma signs and symptoms depend on the location of the tumor. In the analysis of such indicative patients, these tumors of brain can be visualized using a feature based fusion of input images. Multimodality image fusion has played an important role to diagnose the diseases for clinical treatment analysis and enhancing the performance and precision of the computer assisted system. In a recent development of medical field single multimodal medical image cannot provide all the details of human body. For example, the soft tissue information can be represented by magnetic resonance imaging, computed tomography imaging represent the bones dense structure with less distortion. In this paper, proposed method to merge the discrete fractional wavelet transform (DFRWT) with dual tree complex wavelet transform (DTCWT) based hybrid fusion technique for multimodality medical images. The developed fusion algorithm is experienced on the pilot study datasets of patients affected with astrocytoma disease. The fused image conveys the superior description of the information than the source images. Experimental results are evaluated by the number of well-known performance evaluation metrics.

Keywords: Astrocytoma · CT · PET · SPECT · DTCWT · DFRWT

1 Introduction

Multimodal medical image fusion has become a crucial area of research due to its importance in delivering a high quality output for diagnostics and treatments in the medical field. It is due to the fact that some type of image representations provides unique details within a restricted domain and some other imaging modality provides information that is common. For instance, computed tomography (CT) images are best suited for studying dense structures like bones with minimal distortions, but it gives less accuracy while detecting any physiological changes. Likewise, Magnetic resonance (MR) imaging helps to observe the fine structure of soft tissues [27]. Multimodality medical image fusion is useful in almost all of the modern health care practices. It is aims to combine the data from various imaging modalities to obtain fused medical images with higher accuracy and complete description of the given object. It makes radiologists more comfortable in reporting CT, MRI, PET and SPECT

© Springer Nature Singapore Pte Ltd. 2019
A. K. Somani et al. (Eds.): ICETCE 2019, CCIS 985, pp. 336–348, 2019.
https://doi.org/10.1007/978-981-13-8300-7_28

studies quickly with better accuracy [12, 13]. Astrocytoma is a brain disease, which is a type of cancer in brain. They start off in a star-shaped glial cell in cerebrum called astrocytes. This type of tumor grows only in brain and spinal cord and it does not affect other organs. Seizures, headaches and nausea are the symptoms identified after the diseases occur in brain. Sometime it occurs in spinal cord and cause disability in the area affected. It grows quickly if it is aggressive cancer otherwise can be a slow-growing tumor [25, 26, 28].

This paper is structured as follows: Sect. 2 explains the recent literature works on image fusion. Section 3 explains the proposed work. Section 4 discusses the implantation results. Section 5 describes the conclusion of the research work.

2   Related Works

Gupta [1] proposed the medical image fusion in NSST domain using the adaptive spiking neural model. Daniel [2] proposed an image fusion system based on hybrid genetic grey wolf optimization. Shahdoosti, et al. [3] proposed the tetrolet transform for multimodality image fusion. El-Hoseny, et al. [4] examines some of medical image fusion techniques to develop the hybrid fusion algorithm for enhancing the quality of fused image. Xi, et al. [5] proposed a multimodality image fusion algorithm combined with sparse representation and PCNN for clinical treatment analysis. Chavan, et al. [6] proposed the NSxRW transform based image fusion used for the analysis and post treatment review of neurocysticercosis. Sharma, et al. [7] proposed algorithm for image fusion based on NSST with simplified model of PCNN. Sreeja, et al. [8] proposed fusion algorithm to fuse the medical image and enhance the quality of the fused image. Xua [9] proposed the DFRWT method for medical image fusion. Liu, et al. [10] proposed the structure tensor and NSST to extract geometric feature and apply the unified optimization model to perform the image fusion. Liu, et al. [11] proposed NSST based fusion algorithm exploiting moving frame based decomposition.

3   Proposed Work

3.1   Conventional Fusion Algorithms

This research paper experiment the some of the conventional and hybrid fusion algorithms for different types of input medical images [14].

1. Discrete Fractional Wavelet Transform
i. Steps of the DFRWT algorithm for image fusion

Step 1: Get the two input images.
Step 2: images resized into 512 × 512.
Step 3: Both images are to be apply 1 layer DFRWT in the same p order and get sub bands coefficients of 3 1 + 1 pieces.

Step 4: Region has K × L size (i.e., 3 × 3, 5 × 5, and 7 × 7; here, take 3 × 3), and center is p(a, b).

$$var_n^\tau(a,b) = \frac{1}{K \times L} \sum_{i=1}^{K} \sum_{j=1}^{L} \left[abs\left(f_n^\tau(a+i, b+j) - \mu \right) \right]^2 \qquad (1)$$

Where, local variance - $var_n^m(a,b)$, $f_n^\tau(a,b)$ - subband coefficient, i, j – position, μ - average coefficients, K and L - number of rows and columns, Direction - τ (τ = x, v, d, h), local variance - $var_{n.X}^\tau(a,b)$, $var_{n.Y}^\tau(a,b)$.

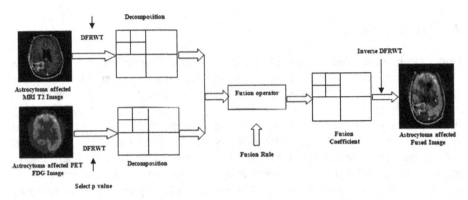

Fig. 1. DFRWT algorithm for image fusion

Step 5: Based on the regional variance, the weight of current fusion coefficient is

$$\omega_{n.X}^\tau = \frac{var_{n.X}^\tau(A,B)}{var_{n.X}^\tau(A,B) + var_{n.Y}^\tau(A,B)} \qquad (2)$$

$$\omega_{n.Y}^\tau = \frac{var_{n.Y}^\tau(A,B)}{var_{n.X}^\tau(A,B) + var_{n.Y}^\tau(A,B)} \qquad (3)$$

Where, DFRWT coefficients weights - $\omega_{n.X}^\tau$ and $\omega_{n.Y}^\tau$, and fusion coefficients - $R_{n.X}^\tau$

$$R_{n.X}^\tau = f_{n.X}^\tau(a,b) X \omega_{n.X}^\tau + f_{n.X}^\tau(a,b) X \omega_{n.X}^\tau \qquad (4)$$

Step 6: Apply the IDFRWT to reconstructed fused medical image. The Fig. 1 demonstrates the flow diagram of the DFRWT based image fusion.

2. Dual Tree Complex Wavelet Transform

It is uses two real DWT in parallel. One DWT generates real part and the other generates imaginary part of DTCWT. Dual tree complex wavelet transforms having following important properties: High Directionality, Shift Invariance, Perfect Reconstruction and Computational Efficiency. The analysis and synthesis filter banks of DTCWT are described in Figs. 2 and 3.

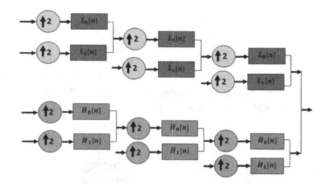

Fig. 2. Synthesis Filter bank for DTCWT

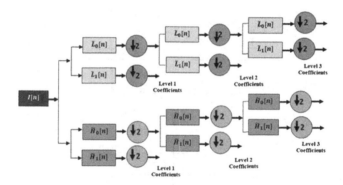

Fig. 3. Analysis Filter bank for DTCWT

DTCWT perform medical image fusion based on the following procedure. First, the disease affected input multimodal medical images are decomposed by DTCWT into coefficient sets and then, the transformed coefficients are fused using an appropriate fusion rule. A new threshold based fusion rule has been applied. The overall structure of the DTCWT based fusion method is shown in Fig. 4.

i. *The Threshold and Fusion Rule*

The threshold value depends on the statistical properties and decomposition levels of the wavelet coefficients. Generally, threshold is used for denoising but, in this DTCWT method is use the image fusion. In wavelet based thresholding, the coefficients having absolute values lower than the threshold are discarded, because the noise affects the small value wavelet coefficients substantially more than the high valued wavelet coefficients. In multimodal medical image fusion also, the wavelet coefficients having high absolute values are selected. Therefore, the wavelet coefficients whose absolute difference from the threshold is higher are selected. The threshold is defined in equation as,

$$\lambda = \frac{1}{2^{(l-1)}} \frac{\sigma}{\mu} M \tag{5}$$

Where, σ – 'standard deviation of wavelet coefficients, μ – mean, M – median of absolute DTCWT coefficients, l – level of decomposition. These statistical parameters jointly represent the variation in the intensity of wavelet coefficients which is used in the fusion algorithm to select the better coefficients.

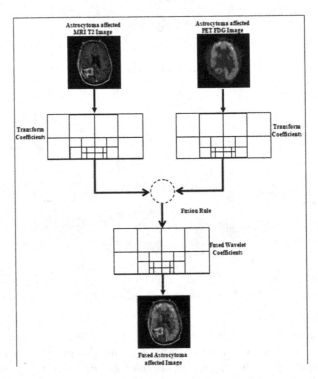

Fig. 4. Overall structure of the DTCWT

ii. *Procedural Steps of the DTCWT Fusion Algorithm*

Step 1: Let Image$_1$ and Image$_2$ be two input multimodal medical images. Medical Images are decomposed by dual tree complex wavelet transform into complex coefficient sets *Cof$_1$* and *Cof$_2$*.

$$Image\ 1 \xrightarrow{DTCWT} cof_1$$

$$Image\ 2 \xrightarrow{DTCWT} cof_2$$

Step 2: For both the coefficient sets, thresholds are calculated for each decomposition level.

Step 3: Absolute difference of all wavelet coefficients from their corresponding threshold are calculated, as below.

$$D_1 = |Cof_1| - |\lambda_1| \tag{6}$$

$$D_2 = |Cof_2| - |\lambda_2| \tag{7}$$

Step 4: Absolute differences of corresponding coefficients of both medical images are compared and coefficient having larger value of absolute difference from the threshold is selected, to form coefficient set of the fused multimodal medical image.

$$Cof(i,j) = \begin{cases} Cof_1 \text{ if } |D_1| \geq |D_2| \\ Cof_2 \text{ if } |D_1| < |D_2| \end{cases} \tag{8}$$

Step 5: Finally, inverse DTCWT is applied on the fused coefficient set to get final output image.

$$Cof \xrightarrow{Inverse\ DTCWT} Fused\ medical\ image$$

3.2 Hybrid Fusion Algorithm

Existing methods require the potential to get superior quality images. To enhance the visual quality of the output the proposed algorithm is used to combine the DFRWT with DTCWT. Before fusion process the two level conversions on source images are applied. These outcomes give best quality, superior handling of curved shapes and improved characterization of input images [15, 16].

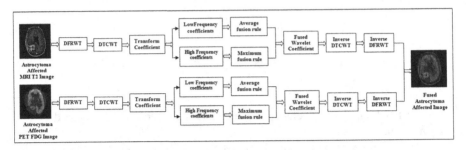

Fig. 5. proposed diagram of the hybrid algorithm (DFRWT-DTCWT)

(i) *Steps for hybrid fusion Algorithm (DFRWT-DTCWT)*

1. Get the two images.
2. Input images resized into 256×256.
3. Obtain highpass directional subband coefficients and lowpass sub and coefficients of input images at each scale and each direction by DFRWT.
4. Multimodal Medical Images are decomposed by dual tree complex wavelet transform into complex coefficient sets. For both the coefficient sets, thresholds are calculated for each decomposition level.
5. Absolute difference of all wavelet coefficients from their corresponding threshold is calculated.
6. Absolute differences of corresponding coefficients of both the input medical image are compared and the coefficient having larger value of absolute difference from the threshold is selected, to form coefficient set of the outcome.
7. Finally, inverse DFRWT and inverse DTCWT is applied on the fused coefficient set to obtain the final output. Figure 5 illustrates the proposed diagram of the hybrid algorithm.

4   Implementation Results and Evaluations

The traditional and hybrid techniques are applied into to input images and get the fused image. [17–19] Input images are collected from the harvard medical school [24] and radiopedia.org [23]. The experimental performance evaluation metrics results of the conventional and hybrid fusion techniques are shown in Table 1. For the experimental work three set of Astrocytoma disease affected multimodal medical images are taken.

4.1   Performance Evaluation Metrics

The similarities and dissimilarities, the output images are compare to source images. Some of the performance evaluation metrics are as follows. [20–22]

1. Fusion Factor

Fus Fact says the difference between the two images i.e. the fused image and input images is estimated by calculating how much information contained in them. It depends on the mutual information of original medical images.

$$FusFact = MI_{input1,Fu} + MI_{input2,Fus}$$

$$MI_{X,Y} = \sum_{k,l} P_{X,Y}(k,l) \log \frac{P_X(k,l)}{P_X(k)P_Y(l)} \tag{9}$$

Here, $MI_{X,B}$ is the related data between input images, p(.) is probability distribution function. Fus fact parameter represents higher value means the outcome is superior quality.

2. Image quality index

The output of the IQI gives the similar information between two input images. The fused output image is declare visually superior quality if IQI value is near to one.

$$IQI(X, Y) = \frac{2\sigma_{X,Y} \cdot 2\mu_X \mu_Y}{(\sigma_X^2 + \sigma_Y^2)(\mu_X^2 + \mu_Y^2)} \tag{10}$$

3. Edge quality measure

To evaluate the fused image edge quality the EQM is better tool. $EQM_{x,y}^{fus}$ is denoted as edges of input images.

$$EQM_{x,y}^{fus} = \frac{\sum_{a=0}^{K-1} \sum_{b=0}^{L-1} EQ_{x,fus}(a,b)w_x(a,b) + EQ_{y,fus}(a,b)w_y(a,b)}{\sum_{m=0}^{K-1} \sum_{n=0}^{L-1} w_x(m,n) + w_y(m,n)} \tag{11}$$

4. Mean structural similarity index measure

If mSSIM parameter gets the value nearest one means fused output image is structurally similar comparing with input images.

$$M(k,l) = \frac{(2\mu_k\mu_l + I_1)(2\sigma_{k,l} + I_2)}{(\mu_k^2 + \mu_l^2 + I_1)(\sigma_k^2 + \sigma_l^2 + I_2)} \tag{12}$$

$$mSSIM(k,l) = \frac{1}{N} \sum_{o=1}^{N} SSIM(k_o, l_o) \tag{13}$$

5. Correlation coefficient

It is representing the similarities of output with respect source images. If the CC value is near or equal to one means both input images are similar to output image.

$$CR(k,l) = \frac{2\sum_{k=0}^{M-1}\sum_{l=0}^{N-1} F_i(k,l).I_{fus}(k,l)}{\sum_{k=0}^{M-1}\sum_{l=0}^{N-1} |F_i(k,l)|^2 + \sum_{k=0}^{M-1}\sum_{l=0}^{N-1} |I_{fus}(k,l)|^2} \quad (14)$$

Table 1. Evaluation metrics achieved from various fusion methods

Methods	Algorithm	Fus Fact	IQI	EQM	MSSIM	CCR
case 1	DWT	3.1661	0.5882	0.7891	0.7181	0.7276
	CVT	3.3411	0.6165	0.8219	0.7318	0.7416
	DFRWT	3.4862	0.6371	0.8526	0.7621	0.7856
	NSST	3.6741	0.6882	0.8893	0.7982	0.7922
	DTCWT	3.5882	0.6683	0.8763	0.7768	0.8035
	Proposed	4.2991	0.8621	0.9012	0.8127	0.8856
Case 2	DWT	1.0622	0.8276	0.6218	0.6873	0.6175
	CVT	1.2871	0.8476	0.6376	0.7183	0.6375
	DFRWT	1.4562	0.8789	0.6598	0.7329	0.6593
	NSST	1.6987	0.8998	0.7176	0.7687	0.6857
	DTCWT	1.4281	0.8852	0.6854	0.7459	0.7183
	Proposed	2.2180	1.8723	0.9198	0.9159	0.8838
Case 3	DWT	4.2694	0.6261	0.7126	0.6932	0.8443
	CVT	4.3678	0.7162	0.7376	0.7027	0.8632
	DFRWT	4.4753	0.7367	0.7821	0.7328	0.8732
	NSST	4.6234	0.7872	0.8129	0.7672	0.8963
	DTCWT	4.5762	0.7528	0.7902	0.7472	0.8821
	Proposed	5.1927	0.8052	0.8782	0.8032	0.9562

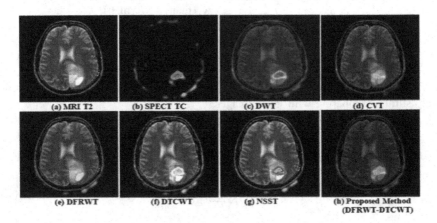

(a) MRI T2   (b) SPECT TC   (c) DWT   (d) CVT

(e) DFRWT   (f) DTCWT   (g) NSST   (h) Proposed Method (DFRWT-DTCWT)

Fig. 6. Implemented output results case 1

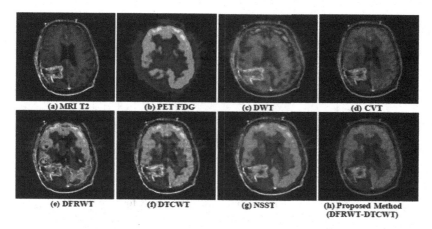

Fig. 7. Implemented output results case 2

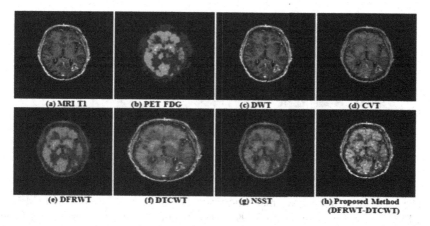

Fig. 8. Implemented output results case 3

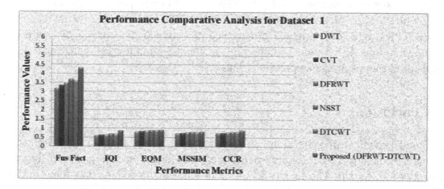

Fig. 9. Performance comparative analysis for case 1

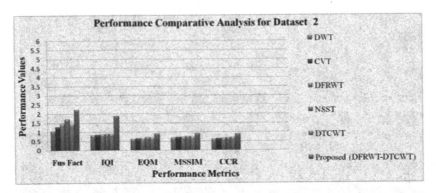

Fig. 10. Performance comparative analysis for case 2

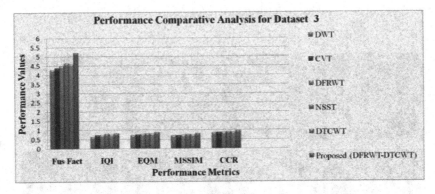

Fig. 11. Performance comparative analysis for case 3

The set 1 input images are magnetic resonance imaging - T2 weighted (MRI T2) and SPECT with perfusion agent Tc99m-HM-PAO (SPECT TC), set 2 input images are magnetic resonance imaging - T2 weighted (MRI T2) and positron emission tomography with Fluorine-18 Deoxyglucose (PET FDG) and set 3 input images are magnetic resonance imaging – T1 weighted (MRI T1) and positron emission tomography with Fluorine-18 Deoxyglucose (PET FDG). The implemented output results are shown in Figs. 6, 7 and 8. The performance comparative analyses of the two set results are shown in Figs. 9, 10 and 11. In the Figs. 6, 7 and 8 input images represented into a, b and fused output images represented into c, d, e, f, g & h.

5 Conclusion

Multimodal medical image fusion has played a vital role to diagnose the disease for clinical treatment analysis and improving the performance and precision of the computer assisted system. The image fusion is an efficient procedure functional in diagnosis, medical analysis of levels of disease and clinical review in astrocytoma disease.

The proposed method is better for visualization, accurate interpretation and precise localization of the tumor and lesions formed in the brain. Proposed a hybrid algorithm is developed for multimodality images to analyze and review the astrocytoma disease. The proposed hybrid fusion algorithm was verified through a simulation experiment on multimodality images. The proposed hybrid fusion algorithm gives better results.

References

1. Gupta, D.: Nonsubsampled shearlet domain fusion techniques for CT–MR neurological images using improved biological inspired neural model. Biocybern. Biomed. Eng. **38**, 262–274 (2017)
2. Daniel, E., Anitha, J., Kamaleshwaran, K.K., Rani, I.: Optimum spectrum mask based medical image fusion using Gray Wolf Optimization. Biomed. Signal Process. Control **34**, 36–43 (2017)
3. Shahdoosti, H.R., Mehrabi, A.: Multimodal image fusion using sparse representation classification in tetrolet domain. Digit. Signal Process. **79**, 9–22 (2018)
4. El-Hoseny, H.M., Rabaie, E.S.M.E., Elrahman, W.A., El-Samie, F.E.A.: Medical image fusion techniques based on combined discrete transform domains. In: Port Said, Egypt, Arab Academy for Science, Technology & Maritime Transport, pp. 471–480. IEEE (2017)
5. Xia, J., Chen, Y., Chen, A., Chen, Y.: Medical image fusion based on sparse representation and PCNN in NSCT domain. Comput. Math. Methods Med. (2018)
6. Chavan, S.S., Mahajan, A., Talbar, S.N., Desai, S., Thakur, M., D'cruz, A.: Nonsubsampled rotated complex wavelet transform (NSRCxWT) for medical image fusion related to clinical aspects in neurocysticercosis. Comput. Biol. Med. **81**, 64–78 (2017)
7. Ramlal, S.D., Sachdeva, J., Ahuja, C.K., Khandelwal, N.: Multimodal medical image fusion using non-subsampled shearlet transform and pulse coupled neural network incorporated with morphological gradient. Signal Image Video Process. **12**, 1479–1487 (2018)
8. Sreeja, P., Hariharan, S.: An improved feature based image fusion technique for enhancement of liver lesions. Biocybern. Biomed. Eng. **38**, 611–623 (2018)
9. Xua, X., Wang, Y., Chen, S.: Medical image fusions using discrete fractional wavelet transform. Biomed. Signal Process. Control **27**, 103–111 (2016)
10. Liu, X., Mei, W., Huiqian, D.: Multi-modality medical image fusion based on image decomposition framework and nonsubsampled shearlet transform. Biomed. Signal Process. Control **40**, 343–350 (2018)
11. Liu, X., Mei, W., Du, H.: Structure tensor and nonsubsampled shearlet transform based algorithm for CT and MRI image fusion. Neurocomputing **235**, 131–139 (2017)
12. Rajalingam, B., Priya, R.: Multimodality medical image fusion based on hybrid fusion techniques. Int. J. Eng. Manuf. Sci. **7**(1), 22–29 (2017)
13. Rajalingam, B., Priya, R.: A novel approach for multimodal medical image fusion using hybrid fusion algorithms for disease analysis. Int. J. Pure Appl. Math. **117**(15), 599–619 (2017)
14. Rajalingam, B., Priya, R.: Hybrid multimodality medical image fusion technique for feature enhancement in medical diagnosis. Int. J. Eng. Sci. Inven. **2**, 52–60 (2018)
15. Rajalingam, B., Priya, R.: Combining multi-modality medical image fusion based on hybrid intelligence for disease identification. Int. J. Adv. Res. Trends Eng. Technol. **5**(12), 862–870 (2018)

16. Rajalingam, B., Priya, R.: Hybrid multimodality medical image fusion based on guided image filter with pulse coupled neural network. Int. J. Sci. Res. Sci. Eng. Technol. **5**(3), 86–100 (2018)

17. Rajalingam, B., Priya, R.: Multimodal medical image fusion based on deep learning neural network for clinical treatment analysis. Int. J. Chem. Tech. Res. **11**(06), 160–176 (2018)

18. Rajalingam, B., Priya, R.: Review of multimodality medical image fusion using combined transform techniques for clinical application. Int. J. Sci. Res. Comput. Sci. Appl. Manag. Stud. **7**(3) (2018)

19. Rajalingam, B., Priya, R.: Multimodal medical image fusion using various hybrid fusion techniques for clinical treatment analysis. Smart Constr. Res. **2**(2), 1–20 (2018)

20. Rajalingam, B., Priya, R.: Enhancement of hybrid multimodal medical image fusion techniques for clinical disease analysis. Int. J. Comput. Vis. Image Process. **8**(3), 17–40 (2018)

21. Rajalingam, B., Priya, R., Bhavani, R.: Comparative analysis for various traditional and hybrid multimodal medical image fusion techniques for clinical treatment analysis. In: Image Segmentation: A Guide to Image Mining, ICSES Transactions on Image Processing and Pattern Recognition (ITIPPR). ICSES Publisher, Chap. 3, pp. 26–50 (2018)

22. Rajalingam, B., Priya, R., Bhavani, R.: Hybrid multimodality medical image fusion using various fusion techniques with quantitative and qualitative analysis. In: Advanced Classification Techniques for Healthcare Analysis. IGI Global Publisher, Chapt. 10, pp. 206–233 (2019)

23. https://radiopaedia.org. Accessed 2017

24. http://www.med.harvard.edu. Accessed 2017

25. https://www.mayoclinic.org/diseases-conditions/astrocytoma. Accessed 2018

26. Seth, S., Agarwal, B.: A hybrid deep learning model for detecting diabetic retinopathy. J. Stat. Manag. Syst. **21**(4), 569–574 (2018)

27. Gupta, M., Lechner, J., Agarwal, B.: Performance analysis of Kalman filter in computed tomography thorax for image denoising. Recent Pat. Comput. Sci. (2019). https://doi.org/10.2174/2213275912666190119162942

28. Mittal, M., Goyal, L.M., Kaur, S., Kaur, I., Verma, A., Hemanth, D.J.: Deep learning based enhanced tumor segmentation approach for MR brain images. Appl. Soft Comput. **78**, 346–354 (2019)

Advanced Expert System Using Particle Swarm Optimization Based Adaptive Network Based Fuzzy Inference System to Diagnose the Physical Constitution of Human Body

M. Sivaram[1]([⊠]), Amin Salih Mohammed[1], D. Yuvaraj[2], V. Porkodi[1],
V. Manikandan[1], and N. Yuvaraj[3]

[1] Department of Computer Networking, Lebanese French University, Erbil, Iraq
{Sivaram.murugan,kakshar,porkodi.sivaram,
v.manikandan}@lfu.edu.krd
[2] Department of Computer Science,
Cihan University, Duhok, Kurdistan Region, Iraq
yuvaraj.d@duhokcihan.edu.krd
[3] Department of Computer Science and Engineering,
St. Peter's Institute of Higher Education and Research, Chennai, India
yraj1989@gmail.com

Abstract. The Korean medicine has suggested 8 distinct combinations of constitutions in a human body. The eccentricity of the physicians is to diagnose the relevant constitution over a patient physical body. To assist the physicians and to improve the diagnosing quality of disease, we present an automating diagnosing method. Hence, to automate the diagnosis of 8 constitutions, an expert system is used, which predicts the constitutions based on given inputs. An automated diagnosis is carried out using rule based optimization expert system, namely Bees Swarm Optimization (BSO) based Adaptive Neuro Fuzzy Interference System (ANFIS). The BSO based ANFIS or BSO-ANFIS is recommended to automate the diagnosis process using standard datasets. The comparative results with ANFIS system and proves that BSO-ANFIS matches well with the physicians report than ANFIS system.

Keywords: Korean constitutions · BSO-ANFIS · Medical expert systems

1 Introduction

The data mining refers to the discovery of meaningful information from the several collections of data. The theory of data mining is represented in the form of data mining model that has its own framework to construct a system to acquire meaning data. The data mining framework is modelled in such a way that it suitably forms a knowledge base from the machine world to the humans. This framework provides the meaningful information to the humans. In case of medical applications, the data mining framework predicts the disease for the purpose of diagnosis [1]. It provides medical solution to

© Springer Nature Singapore Pte Ltd. 2019
A. K. Somani et al. (Eds.): ICETCE 2019, CCIS 985, pp. 349–362, 2019.
https://doi.org/10.1007/978-981-13-8300-7_29

analyse the disease in an improved way and the use of prediction framework makes the diagnostic process an easier one [2].

The Constitution medicine in medical fields play a major role in curing intractable and incurable disease [3]. The constitutional medicine divides the person into 8 different constitutions namely: Pulmotonia, Colonotonia, Renotonia, Vesicotonia, Pancreotonia, Gastrotonia, Hepatonia and Cholecystonia [4]. The 8 constitutions has different characteristics that vary in their personality, outward appearance, foods, behaviour and their lifestyle, pathophysiology of medicine, disease, food and treatments [4, 5]. The therapies in constitution medicine include life and diet adjustment method based on 8 constitution [6]. However, diagnosis of such disease is an important task before treating such ailments using constitution medicine. There exist several methods for medical diagnosis that uses optimization to carry out the prediction of disease [7–12].

Hence, the present research works on finding a suitable methods that helps in analyzing the ailments and providing solutions using the constitutional medicines to the patients. To achieve this, rules are defined with the predefined datasets using Adaptive Neuro Fuzzy Interference System (ANFIS). The training patterns of the ANFIS is improved using Bees Swarm Optimization (BSO). The proposed medical diagnosis framework uses training inputs acquired from the BSO algorithm and it trains the system to provide relevant predictions over the test data. The predefined rules in the fuzzy inference engine are improved furthermore from the input data provided by the BSO algorithm [17]. This BSO-ANFIS medical diagnosis framework is hence used as a diagnostic tool to identify the patient ailments and helps the clinicians to treat the patients with 8 constitutional medicine.

The outline of the paper is presented as follows: Sect. 2 deals with the data collected to form the rules for fuzzy inference engine. Section 3 presents with the proposed medical diagnosis model. Section 4 evaluates the proposed model and Sect. 6 concludes the entire work.

2 Data Collection

The section provides numerous parameters that is associated with each constitution. Here, each order, characteristic and symptoms' is taken as an input parameter for defining the rules. The values of all the parameters (of Tables 2 and 3) is taken from dataset that are considered to be dependent factors w.r.t the constitutions. The collected data is grouped to form a matrix (Ni xMj) and taken as an input to the ANFIS controller. Among 300 cases, 250 cases are used as training sets and the rest of the cases are used for testing purpose. A 10-fold cross validation is performed to separate the testing and training samples. The other pre-processing steps involves checking the reliability of datasets and extraction of hidden features from the datasets. The reliability is carried out with Cronbach's alpha (a) test using 95% confidence interval on the whole dataset. The extraction of hidden features from multidimensional datasets is calculated from [13]. The proposed scheme is shown in Fig. 1.

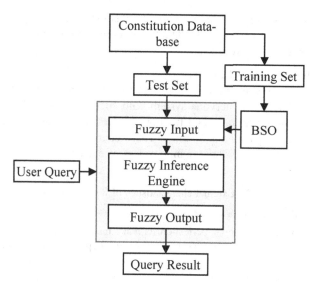

Fig. 1. Proposed model

2.1 Input Parameter: Constitutions

The Table 1 gives the characteristics of diseases based on the dynamic characterization of organs, which is placed in an order. The diagnosis of disease is ordered or ranged from the Inferior or Weakest – Considerably Weak – Normal – Relatively Strong – Superior or Strongest. The organ that is place first is treated as inferior; the organ at the end is superior and so on. For example, the Renotonia constitution has strongest K, relatively strong Lu, normal L, considerably weak H and weakest P. Similarly, the other constitutions are ordered in the same manner.

Table 1. Dynamic orders of various constitutions

Constitutions	Orders of 12 organs
Pulmotonia	L > K > H > P > Lu
Colonotonia	Gb > Si > S > B > Li
Renotonia	P > H > L > Lu > K
Vesicotonia	S > Li > Si > Gb > B
Pancreotonia	Lu > P > H > K > L
Gastrotonia	Li > B > S > Si > Gb
Hepatonia	K > Lu > L > H > P
Cholecystonia	B > Gb > Si > Li > S

*L – Liver, K- Kidney, H – Heart, P – Pancreas, Lu – Lungs, Gb - Gall bladder, Si - Small intestine, Li – Large intestine, S – Stomach and B –Bladder.

The other two organs associated with these constituents are Sympathetic (SN) and Parasympathetic nervous system (PN). The SN responds to concentration and tension, PN responds to digestion and relaxation.

Input Parameter: External Characteristics

The association of various characteristics related to the responses related to SN and PN is discussed in Table 2. The nervous system with such responses plays a major role during the formation of regimens and finding out the formulation for better cure and prevention measures of each constitution's. These characteristics is already defined in the ANFIS system along with the dynamic orders of each constitution's.

Table 2. External characteristics of 8 constitutions

Constitutions	External characteristics of a patient
Pulmotonia	Naturally creative, perfectionists make mistakes unlikely, principled, fairly consistent, professional, music listeners, hyper-sensitive, unrealistic idealists, rare keep friends, have meaningful relationships with a few
Colonotonia	Cheerful, emotional, adventurous, sensitive, careful discretionary, consistent nature different talents, expertise, music listeners, have good friendships, negative, stubborn, passive, suspicious and consumes inappropriate diets
Renotonia	Composed, cooperative, patient, perfectionist, good listeners, express anger rarely, posses social skill, work in service related fields, detail-oriented personality, suspicious personality, maintains secrecy, do not trust others
Vesicotonia	Gentle, cool, collected, realistic, reserved, patient, good listeners, patience towards work, make mistake in work, attentiveness to details and sensitive, negative, stubborn, grow greedy, cold, suspicious and close-minded easily
Pancreotonia	Restless, proactive, and sensible, naturally positive, passionate, achiever, color sense, interested in arts and fashion, socially active, straightforward, positive, posses strong sense of justice, impatient, misspeak, offended, getting upset easily, quickly overcomes the emotions
Gastrotonia	Honest, active, meticulous, positive, sensible, responsible, conscientious, devoted workers, efficient, accurate vision, good sense over aesthetics, pretend to be generous, straightforward, strict, impatient, sensitive and not social
Hepatonia	Quiet, reserved, realistic, skilled in predicting the situations, has large friends' circle, overly ambitious, works for results, unsuccessful outcomes
Cholecystonia	Simple and quick in decisions making, adaptable, charitable, sociable, large friends' circle, adapts easily to social environments, natural athleticism, developed muscles, meat diet, excellent athletes, healthy and requires more sleep

Input Parameter: Health Related Issues

The health related issues associated with each constitutions is shown in Table 3. Here, each disease or symptoms' are set as a rule based function along with the parameters taken from Tables 1 and 2.

Table 3. Internal characteristics of each constitutions

Constitutions	Internal characteristics of a patient
Pulmotonia	Weak liver and feels weak after perspiring, feeling illness due to meat consumption
Colonotonia	Chronic fatigue, kidney failure, indigestion, lupus, multiple sclerosis, Parkinson's disease and excessive perspiration leads easily to get fatigued
Renotonia	Mild chronic constipation, sweat-less, perspire, digestive problems from cold food and drinks intake
Vesicotonia	Weak stomach due to cold foods intake, illnesses due beverages, warm foods, eats less and infrequent perspiration with good health
Pancreotonia	Hearty appetite, consuming spicy foods, apples, chicken and herbal medicine heats the strong pancreas, diabetes and grows impatient
Gastrotonia	Indigestion, chronic headaches, body pain, heavy side effects over antibiotics and herbal medicine, anxiety disorder and nervous system malfunction
Hepatonia	Sweats frequently, fall asleep, high blood pressure, lower abdominal indigestion, skin diseases, arthritis, high cholesterol, depression
Cholecystonia	Discomfort in lower abdomen, frequent bowel movement, weak large intestine, excited liver and gall bladder, arthritis, high cholesterol, skin problems, insomnia and obesity

The input parameter is set as a rule based function for the user to determine the accurate constitutions, when the inputs are given from the user end. So that the system requires no assistance of doctors for diagnosing the constitution.

3 ANFIS

ANFIS algorithm combines the inference feature of fuzzy logic and the learning ability of neural networks. The structure of ANFIS possesses input-output (IO) data pairs, which gets trained by available/existing IO pairs. The rule for ANFIS is created using IF-THEN condition and expert solutions are added further in the ANFIS inference engine [14]. The structure of the ANFIS has antecedent and conclusion part, and rule part acts as an interconnection. The ANFIS is often updated using these parameters that are explained below:

Initially, the ANFIS system requires the use of rule selection strategy based on the available parameters that decides the diagnosis of constituents. To attain effective diagnosis, a rule based selection or strategy is set with the available input parameters. This includes the constitutions associated with various organs, external characteristics of each constitution and the disease or symptoms related to each constitution.

3.1 Fuzzification

The parameters chosen are fuzzified to convert the crisp values to a fuzzy membership values. The input variable consist of one triangular and two trapezoidal membership functions and the output variable consist of one trapezoidal and two triangular membership function. The range of fuzzy sets are defined based on the values obtained from the output membership function, which is defined in Table 4.

Table 4. Output values from the membership function

Constitutions	Membership functions		
	Low	Medium	High
Pulmotonia	0–40	30–70	60–110
Colonotonia	0–40	30–70	60–110
Renotonia	0–40	30–70	60–110
Vesicotonia	0–40	30–70	60–110
Pancreotonia	0–40	30–70	60–110
Gastrotonia	0–40	30–70	60–110
Hepatonia	0–40	30–70	60–110
Cholecystonia	0–40	30–70	60–110

3.2 Rule Selection for ANFIS System

The rule selection or knowledge base for ANFIS is based on IF-THEN condition which is represented as:

IF (A AND B AND C) THEN (show the constitution)
Where, A is the constitution (Table 1), B is the external characteristics (Table 2) and C is the health related issue (Table 3).

3.3 Inference Process

This system uses Mamdani Interference Process that is shown as:

$$f(z) = \frac{\max[\min[f(A), f(B), f(C)]]}{k} \tag{1}$$

where, $k = 1, 2, \ldots, n$.

3.4 Fuzzy Operators

The membership function $\mu(x)$ is defined as:

$$\mu(x) = \frac{1}{1 + \left[\left(\frac{x - \beta_i}{\alpha_i}\right)^2\right]^{\gamma_i}} \tag{2}$$

Where, α_i, β_i and γ_i are the parameters of membership function used to attain reasonable results. The AND, OR operators are applied over membership function to combine the three linguistic variables. The fuzzy union (OR) and intersection

(AND) operators are applied on M and N (two fuzzy sets)that aggregates two membership functions as described in following Equations.

$$\mu_{AND}(x) = \min(\mu_M(x), \mu_N(x)) \tag{3}$$

$$\mu_{OR}(x) = \max(\mu_M(x), \mu_N(x)) \tag{4}$$

3.5 Firing Strength

The firing strength (w) for a specific rule is generated with the membership values and the computations are carried out over Fuzzification layer. The firing values are calculated using:

$$w = \mu_{AND}(x) \cdot \mu_{OR}(x) \tag{5}$$

3.6 Normalization Layer

The normalization of reach firing strengths for each rule is defined as the ratio of ith firing strength to the overall firing strength.

$$\bar{w}_i = \frac{w_i}{w_i + w_{i+1}} \tag{6}$$

The output is then expressed as:

$$\bar{w}_i f_i = \bar{w}_i(p_i m + q_i n + r_i) \tag{7}$$

where p_i, q_i and r_i are parameter sets that is identified by least square algorithm [15].

3.7 Defuzzification Layer

The aggregation of the fuzzy sets are defuzzified to obtain a crisp values using centroid method in the following equation.

$$Z = \frac{\int \mu_{\bar{w}}(\bar{w})\bar{w}d\bar{w}}{\int \mu_{\bar{w}}(\bar{w})\bar{w}d\bar{w}} \tag{8}$$

where the integral function is the algebraic integration function.

4 BSO Algorithm

The BSO algorithm is used to reduce the errors in the training phase and provides the required input to the Fuzzy inference engine to provide accurate results during testing. This supervised learning provides high accurate diagnosis of disease.

Initialize:
Find number of bees, percentage of experienced foragers, scouts, onlookers, dimension, radius and end condition
For all the bees
       Initialize the bees randomly inside the search space
Do
       Compute fitness of bees
       Sort bees using its fitness value
       Partition the swarm into the experienced forager, onlooker and scout
End
For all the experienced forager bees
       For D-dimensional search space
              Update the previous best position
              Select elite bee for all the experienced forager bees
              For all experienced forager bee
                     Update the position of an experienced forager bee
              End
       End
End
For each onlooker bee
       Select an elite bee from experienced forager bee for onlooker
       For D-dimensional search space
               Update the position of an onlooker bee
       End
End
For each scout bee
       For D-dimensional search space
               Walk randomly around the search space
       End
End
Adjust the radius and step size of the search space for scout bees
Enduntil the termination criterion is met.

4.1 BSO-ANFIS Process

The steps for implementing the BSO [16] ANFIS algorithm is shown in the following steps:

Step 1: The input and output data are trained based on ANFIS model. This model uses the combination of least square method and back-propagation gradient.
Step 2: An N-dimensional vector is created and that represents the total number of membership function.
Step 3: The N-dimensional vector holds the parameters related to the membership function.

Step 4: The ANFIS parameters are optimized with the help of BSO.

Step 5: The fitness function is defined using mean square error.

Step 6: The parameters are then defined using BSO algorithm (Table 5).

Step 7: Initialize randomly the BSO parameters and update it at every iteration.

Step 8: During each iteration, a single set of parameter is updated (say α_i, then β_i and so on.)

5   Results and Discussions

The proposed medical diagnosis system is tested against different meta-heurisitc algorithms, namely Gray Wolf Optimizaiton (GWO), Flower Pollination Algorithm (FPA), Particle Swarm Optimization (PSO), Ant Colony Optimization (ACO) and Genetic Algorithm (GA). The proposed system and other system is trained using 250 collected data and it is tested using 50 data. The datasets are collected from several repositories and it is also collected directly from several medical institutions. The data for testing is not considered from the trained data. The results are tested in terms of accuracy, recall, precision, f-measure against other methods. Further, the proposed method is evaluated to check the errors during training and testing.

Table 5. Parameters for training BSO-ANFIS

Parameter	Value
Total number of inputs	8 constituents with its internal and external characteristics
Total number of output	1
Total training data	250
Total testing data	50
Total number of Bee population	250
Total number of iterations	1000

The Fig. 2 shows the results of Accuracy Plot between the Proposed BSO-ANFIS and other ANFIS optimization techniques. The Fig. 3 shows the results of F-measure between the Proposed BSO-ANFIS and other ANFIS optimization techniques. The Fig. 4 shows the results of Precision Plot between the Proposed BSO-ANFIS and other ANFIS optimization techniques. The Fig. 5 shows the results of Sensitivity Plot between the Proposed BSO-ANFIS and other ANFIS optimization techniques. The Fig. 6 shows the results of Specificity Plot between the Proposed BSO-ANFIS and other ANFIS optimization techniques. The result shows that the proposed method has higher accuracy, precision, recall and F-measure than other methods.

Likewise, the Fig. 7 shows the results of MSE for the proposed medical diagnosis system and Fig. 8 shows the results of Regression Plot for the proposed medical diagnosis system. The result shows that the errors occurred during testing is very much reduced and the error occurred is very nearer to zero. This indicates that the proposed method attains reduced error rate in terms of testing. Hence, the proposed method is tested in terms of percentage error (Fig. 9) between the Proposed BSO-ANFIS and other ANFIS optimization techniques. The result shows that the proposed method has reduced percentage error than other methods.

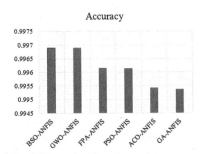

Fig. 2. Results of accuracy plot between the proposed BSO-ANFIS and other ANFIS optimization techniques.

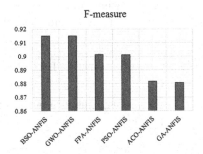

Fig. 3. Results of F-measure between the proposed BSO-ANFIS and other ANFIS optimization techniques.

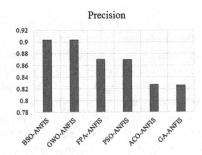

Fig. 4. Results of precision plot between the proposed BSO-ANFIS and other ANFIS optimization techniques.

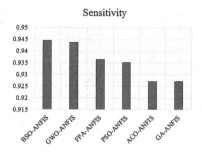

Fig. 5. Results of sensitivity plot between the proposed BSO-ANFIS and other ANFIS optimization techniques.

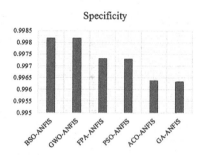

Fig. 6. Results of specificity plot between the proposed BSO-ANFIS and other ANFIS optimization techniques.

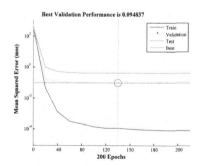

Fig. 7. Results of MSE for the proposed medical diagnosis system

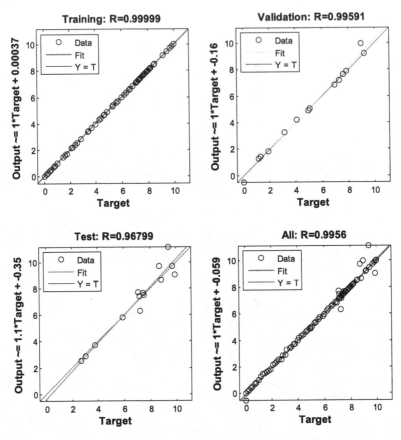

Fig. 8. Results of regression plot for the proposed medical diagnosis system

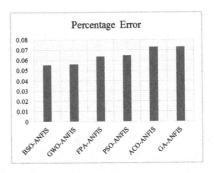

Fig. 9. Results of percentage error plot between the proposed BSO-ANFIS and other ANFIS optimization techniques.

6 Conclusions

In this paper, a medical diagnosis framework is proposed using BSO-ANFIS for the clinicians practicing Korean medicine. The proposed method uses BSO for improving the training pattern with reduced error rate. The reduced errors with trained data is sent as input to ANFIS system that uses fuzzy rule engine with rules defined from internal and external characteristics to predict the required outcomes. This machine learning system is tested in terms of various performance metrics like precision, recall, sensitivity, F-measure, training errors and testing errors. The results shows that the proposed method is effective in predicting the conditions and providing input to the clinicians than the other combined methods.

References

1. Tang, P.N., Steinbach, M., Kumar, V.: Introduction to Data Mining. Pearson Addison: Wesley, Bostan (2005)
2. Bellazzi, R., Zupan, B.: Predictive data mining in clinical medicine: current issues and guidelines. Int. J. Med. Inf. **77**(2), 81–97 (2008)
3. http://ecmed.org/board/content.asp?bsNo=18&lng=en
4. Kuon, D.W.: Eight-Constitution Medicine: An Overview. IMKS Occasional Papers No. 2. edited by Institute for Modern Korean studies, pp. 7–41. Yonsei University Press, Seoul (2003)
5. Kuon, D.W.: Studies on Constitution-Acupuncture Therapy. J. Myong-Ji Univ. **7**, 582–606 (1974)
6. Kim, Y.W., Lee, K.M., Kim, S.W.: The effect of 8 constitution acupuncture on neck pain by pain disability index and visual analogue scale. J. Korean Acupunct. Moxibustion Soc. **20**, 202–208 (2003)
7. Ranjit, K., Vishu, M., Prateek, A., Kumar, S.S., Amandeep, K.: Fuzzy expert system for identifying the physical constituents of a human body. Indian J. Sci. Technol. **9**(28) (2016)
8. Begic, F.L., Avdagic, K., Omanovic, S.: GA-ANFIS expert system prototype for prediction of dermatological diseases. Stud. Health Technol. Inf. **210**, 622–626 (2014)
9. Farooque, M.M., Aref, M., Khan, M.I., Mohammed, S.: Data mining application in classification scheme of human subjects according to Ayurvedic Prakruti–Temperament. Indian J. Sci. Technol. **9**(13) (2016)
10. Yim, J., Joo, J.: Implementation of an inference system to classify persons into eight constitutions. Indian J. Sci. Technol. **8**(26) (2015)
11. Chattopadhyay, S.: A neuro-fuzzy approach for the diagnosis of depression. Appl. Comput. Inf. **13**(1), 10–18 (2014)
12. Appiah, R., Panford, J.K., Riverson, K.: Implementation of adaptive neuro fuzzy inference system for malaria diagnosis (case study: Kwesimintsim polyclinic). Int. J. Comput. Appl. **115**(7), 33–37 (2015)
13. Pearson, K.: On Lines and Planes of Closest Fit to Systems of Points in Space, Philosophical Magazine, vol. 6, pp. 559–572 (1901). http://stat.smmu.edu.cn/history/pearson1901.PDf
14. Jang, J.S.: ANFIS: adaptive-network-based fuzzy inference system. IEEE Trans. Syst. Man Cybern. **23**(3), 665–685 (1993)

15. Miller, S.J.: The method of least squares. Mathematics Department Brown University, pp. 1–7 (2006)
16. Biswas, D.K., Panja, S.C., Guha, S.: Multi objective optimization method by BSO. Procedia Mater. Sci. **31**(6), 1815–1822 (2014)
17. Kaur, S., et al.: Mixed pixel decomposition based on extended fuzzy clustering for single spectral value remote sensing images. J. Indian Soc. Remote Sens. (2019). https://doi.org/10.1007/s12524-019-00946-2

Author Index

Printed in the United States
By Bookmasters